Tharwat F. Tadros
Polymeric Surfactants
De Gruyter Graduate

Also of Interest

Suspension Concentrates.
Preparation, Stability and Industrial Applications
Tadros, 2017
ISBN 978-3-11-048678-0, e-ISBN 978-3-11-048687-2

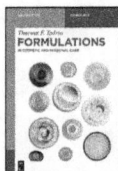

Formulations.
In Cosmetic and Personal Care
Tadros, 2016
ISBN 978-3-11-045236-5, e-ISBN 978-3-11-045238-9

Emulsions.
Formation, Stability, Industrial Applications
Tadros, 2016
ISBN 978-3-11-045217-4, e-ISBN 978-3-11-045224-2

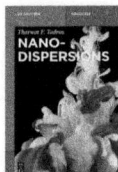

Nanodispersions.
Tadros, 2015
ISBN 978-3-11-029033-2, e-ISBN 978-3-11-029034-9

An Introduction to Surfactants.
Tadros, 2014
ISBN 978-3-11-031212-6, e-ISBN 978-3-11-031213-3

Tharwat F. Tadros

Polymeric Surfactants

Dispersion Stability and Industrial Applications

DE GRUYTER

Author
Prof. Tharwat F. Tadros
89 Nash Grove Lane
Workingham RG40 4HE
Berkshire, UK
tharwat@tadros.fsnet.co.uk

ISBN 978-3-11-048722-0
e-ISBN (PDF) 978-3-11-048728-2
e-ISBN (EPUB) 978-3-11-048752-7

Library of Congress Cataloging-in-Publication Data
A CIP catalog record for this book has been applied for at the Library of Congress.

Bibliographic information published by the Deutsche Nationalbibliothek
The Deutsche Nationalbibliothek lists this publication in the Deutsche Nationalbibliografie;
detailed bibliographic data are available on the Internet at http://dnb.dnb.de.

© 2017 Walter de Gruyter GmbH, Berlin/Boston
Cover image: Daniel_Kay/iStock/Thinkstock
Typesetting: PTP-Berlin, Protago-TEX-Production GmbH, Berlin
Printing and binding: CPI books GmbH, Leck
♾ Printed on acid-free paper
Printed in Germany

www.degruyter.com

Preface

Polymeric surfactants are essential materials for the preparation of many industrial disperse systems and their stabilization, of which we mention dyestuffs, paper coatings, inks, agrochemicals, pharmaceuticals, personal care products, ceramics and detergents. The most effective polymeric surfactants for stabilization of disperse systems are those of the A–B, A–B–A block and BA_n (or AB_n) graft types. The chain B (referred to as the "anchor" chain) is chosen to be highly insoluble in the medium and has a strong affinity (strong adsorption) to the surface of the particle or droplet. The chain A (referred to as the "stabilizing chain") is chosen to be highly soluble in the medium and strongly solvated by its molecules. These block and graft polymers provide effective steric stabilization thus preventing any flocculation and/or coalescence. One of the most important applications of polymeric surfactants is in the preparation of oil-in-water (O/W) and water-in-oil (W/O) emulsions as well as solid/liquid (S/L) dispersions. In this case, the hydrophobic portion of the surfactant molecule should adsorb "strongly" at the O/W or the S/L interface, leaving the hydrophilic components in the aqueous medium where they become strongly solvated by the water molecules, thus providing effective steric stabilization. There are generally two methods for preparation of suspensions, referred to as the condensation and dispersions methods. In the first case, one starts with molecular units and builds up the particles by a process of nucleation and growth, e.g. the preparation of polymer lattices. In the dispersion methods, preformed particles (usually powders) are dispersed in an aqueous solution containing a surfactant. The latter is essential for adequate wetting of the powder (both external and internal surfaces of the powder aggregates and agglomerates must be wetted). This is followed by dispersion of the powder using high speed stirrers and finally the dispersion is "milled" to reduce the particle size to the appropriate range. Polymeric surfactants are also effective here for the long-term stability of the resulting suspension.

This book starts with a general introduction (Chapter 1) highlighting the importance of polymeric surfactants for the preparation and stabilization of various disperse systems. Chapter 2 gives a general classification of polymeric surfactants. It starts with a section on homopolymers, which are made from the same repeating units, e.g. poly(ethylene oxide), poly(vinylpyrrolidone) and poly(acrylic acid). The next section deals with block and graft copolymers which are essential for stabilizing disperse systems. Examples of several block and graft copolymers are given to illustrate their use as stabilizers. A section is devoted to polymeric surfactants based on natural materials such as hydroxyethyl cellulose and polysaccharides. The final section in this chapter describes block and graft copolymers that are used in nonaqueous disperse systems. Chapter 3 deals with the solution properties of polymeric surfactants. It starts with the thermodynamic treatment of the free energy of mixing a homopolymer with a solvent on the basis of the Flory–Huggins treatment of the free energy. The modification of this

DOI 10.1515/9783110487282-001

treatment when using block and graft copolymers is described. Chapter 4 deals with the adsorption and conformation of polymeric surfactants at interfaces with particular reference to block and graft copolymers. Understanding the adsorption and conformation of polymeric surfactants at interfaces is key to describing how these molecules act as stabilizers. Theories of polymeric surfactant adsorption are described with particular reference to the adsorption parameters and their experimental determination. Several examples of polymeric surfactant adsorption are given in this chapter. Chapter 5 deals with stabilization of disperse systems using polymeric surfactants. The interaction between particles or droplets containing adsorbed or grafted polymeric surfactants is described in terms of the unfavourable mixing of the chains (when these are in good solvent conditions) and the loss of configurational entropy (elastic interaction) on considerable overlap of the chains. Combining steric repulsion with van der Waals attraction results in energy-distance curves that show very steep repulsion when the surface-to-surface distance between the particles or droplets become smaller than twice the adsorbed layer thickness. Chapter 6 describes the flocculation of disperse systems containing polymeric surfactants. Four different types of flocculation mechanisms are described: (i) weak and reversible flocculation obtained when using "thin" adsorbed layers (< 5 nm) that results in an attractive minimum at separation distances compared to twice the adsorbed layer thickness; (ii) incipient (strong) flocculation obtained when the solvency of the medium for the chains becomes worse than a theta solvent; (iii) depletion flocculation induced by the addition of "free" nonabsorbing polymer chains in the continuous medium; (iv) bridging flocculation obtained by simultaneous adsorption of the polymer chain on two or more particles or droplets. Chapter 7 deals with applications of polymeric surfactants in emulsion and suspension polymerization for use in the preparation of polymer colloids (latexes). The theories of emulsion and suspension polymerization are described at a fundamental level with particular emphasis on the use of polymeric surfactants for stabilization of the resulting particles against aggregation. Chapter 8 describes the role of polymeric surfactants in stabilizing preformed particles. The importance of strong adsorption that results in a high affinity adsorption isotherm is described at a fundamental level. Chapter 9 deals with the use of polymeric surfactants for stabilizing emulsions and nanoemulsions. The origin of stability against coalescence is discussed in terms of the disjoining pressure of the thin film produced on approach of two emulsion droplets. Another important use of polymeric surfactants is to reduce Ostwald ripening, in particular with nanoemulsions. Chapter 10 gives a summary of applications of polymeric surfactants in pharmacy. Several polymeric surfactants that have been approved by the Food and Drug Administration (FDA) are described. These polymeric surfactants are used for formulating various pharmaceutical systems, such as suspensions, emulsions and multiple emulsions. A section is devoted to applications involving polymeric surfactants in the preparation of biodegradable nanoparticles for targeted delivery of drugs. Chapter 11 deals with applications of polymeric surfactants in cosmetics and their advantages in formulating emulsions in hand creams.

Polymeric surfactants are also used to stabilize liposomes and vesicles that are incorporated in many cosmetic formulations to efficiently deliver active ingredients such as anti-wrinkle agents. Chapter 12 gives examples of applications of polymeric surfactants in paints and coatings. Particular attention is paid to the steric stabilization obtained when using polymeric surfactants in nonaqueous paint systems, where electrostatic stabilization is not possible. Chapter 13 deals with applications of polymeric surfactants in agrochemical formulations, in particular their use in formulating concentrated suspensions and emulsions. Chapter 14 gives examples of applications of polymeric surfactants such as gums and proteins in the food industry. The interaction between proteins and polysaccharides is also described.

The present text gives a comprehensive description of polymeric surfactants, their properties, adsorption at interfaces and their applications for the stabilization of various industrial disperse systems. The book is valuable for formulation scientists and chemical engineers who are engaged in formulating various disperse systems. It is also extremely valuable for academic researchers who are studying the role of polymeric surfactants in the preparation and stabilization of model disperse systems.

April, 2017
Tharwat Tadros

Contents

1 Polymeric surfactants and colloid stability – general introduction

1.1 Polymeric surfactants in disperse systems

Polymers, sometimes referred to as polymeric surfactants, are essential materials for preparing and stabilizing many disperse systems, of which we mention dyestuffs, paper coatings, inks, agrochemicals, pharmaceuticals, personal care products, ceramics and detergents [1]. The most effective polymeric surfactants for stabilizing disperse systems are those of the A–B, A–B–A block and BA_n (or AB_n) graft types. The chain B (referred to as the "anchor" chain) is chosen to be highly insoluble in the medium and has a strong affinity (strong adsorption) to the surface of the particle or droplet. The chain A (referred to as the "stabilizing chain") is chosen to be highly soluble in the medium and strongly solvated by its molecules. These block and graft polymers provide effective steric stabilization, thus preventing any flocculation and/or coalescence. One of the most important applications of polymeric surfactants is in the preparation of oil-in-water (O/W) and water-in-oil (W/O) emulsions as well as solid/liquid dispersions [2–4]. In this case, the hydrophobic portion of the surfactant molecule should adsorb "strongly" at the O/W interface or become dissolved in the oil phase, leaving the hydrophilic components in the aqueous medium, where they become strongly solvated by the water molecules.

The other major application of surfactants is for the preparation of solid/liquid dispersions (usually referred to as suspensions). There are generally two methods for preparation of suspensions, referred to as the condensation and dispersions methods. In the first case, one starts with molecular units and builds up the particles by a process of nucleation and growth [1]. A typical example is the preparation of polymer lattices. In this case, the monomer (such as styrene or methylmethacrylate) is emulsified in water using an anionic or nonionic surfactant (such as sodium dodecyl sulphate or alcohol ethoxylate). An initiator such as potassium persulphate is added and when the temperature of the system is increased, initiation occurs resulting in the formation of the latex (polystyrene or polymethylmethacrylate). Polymeric surfactants can be used either alone or in combination with low molecular weight surfactants for preparing and stabilizing polymer lattices. In the dispersion methods, preformed particles (usually powders) are dispersed in an aqueous solution containing a surfactant. The latter is essential for adequate wetting of the powder (both external and internal surfaces of the powder aggregates and agglomerates must be wetted) [1]. This is followed by dispersion of the powder using high speed stirrers and finally the dispersion is "milled" to reduce the particle size to the appropriate range. Polymeric surfactants are also effective here for the long-term stability of the resulting suspension.

DOI 10.1515/9783110487282-002

For the stabilization of emulsions and suspensions against flocculation, coalescence and Ostwald ripening the following criteria must be satisfied:

(i)   Complete coverage of the droplets or particles by the surfactant. Any bare patches may result in flocculation as a result of van der Waals attraction or bridging.

(ii) Strong adsorption (or "anchoring") of the surfactant molecule to the surface of the droplet or particle.

(iii) Strong solvation (hydration) of the stabilizing chain to provide effective steric stabilization.

(iv) Reasonably thick adsorbed layer to prevent weak flocculation [1].

As mentioned above, most of the above criteria for stability are best served by using a polymeric surfactant of the type A–B, A–B–A blocks and BA_n (or AB_n) grafts, which are the most efficient ones for stabilizing emulsions and suspensions. These block and graft copolymers are ideal for preparing concentrated emulsions and suspensions, which are needed in many industrial applications.

1.2 Polymers (macromolecules)

Polymers, also referred to as macromolecules, are built up of a large number of molecular units that are linked together by covalent bonds. For example, polyethylene consists of repeat units of CH_2-(methylene) groups and their number determines the degree of polymerization N. With polystyrene the phenyl groups are attached as side groups to the C–C backbone chain as illustrated below:

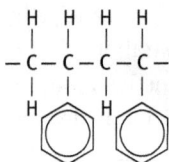

The above macromolecules are uncharged and are referred to as nonionic. If the chains are built up of monomers that contain an ionizable group (that can dissociate to a chain-fixed cation or anion with a mobile counterion) the molecule is referred to as polyelectrolyte.

In most synthetic methods for the preparation of macromolecules, the result is a mixture with various molar masses. Thus, one needs to characterize the molar mass distribution (that can be determined for example by gel permeation chromatography). If the molar mass is denoted by M, then the number density distribution function p(M) and hence the fraction of polymer with molar mass p(M) dM in the range M to M + dM can be normalized to by

$$\int_0^\infty p(M)\,dM = 1. \tag{1.1}$$

The number average molar mass is given by

$$\overline{M}_n = \int_0^\infty p(M)M\,dM. \tag{1.2}$$

The weight average molar mass is given by

$$\overline{M}_w = \frac{\int_0^\infty p(M)M \cdot M\,dM}{\int_0^\infty p(M)M\,dM}. \tag{1.3}$$

The polydispersity coefficient U (which is a measure of polydispersity) is given by

$$U = \frac{\overline{M}_w}{\overline{M}_n} - 1. \tag{1.4}$$

The molar mass distribution varies greatly between different polymeric compounds. In step polymerization, where monomers react in such a way that groups linked together can be coupled with other groups, the molar mass distribution is broad. Chain polymerization, where reactive centers are created that react at the beginning and become shifted after the reaction to the new end of the chain, thus growing, produces a much narrower distribution.

Large variations in the chemical structure can be obtained by a combination of different monomers. This procedure, referred to as copolymerization, produces statistical and block copolymers. In the first case, coupling is statistical and is determined by the probabilities of attachment of two monomers in a growing chain. The resulting polymer is sometimes referred to as a random copolymer. In the second case, the molecule is produced by coupling long macromolecular sequences of uniform composition. Depending on the number of sequences, di-, tri- or multiblock copolymers may be prepared.

The above description refers to linear chains. However, several other structures that contain short or long chain branches are produced. These side chains can be short or long, thus producing grafted chain polymers. A special type is that of star polymers where several polymer chains emanate from one common multifunctional center.

1.3 Single chain conformations

As each macromolecule possesses a large number of internal degrees of freedom, the analysis of the properties of the individual polymer becomes an important point of concern [5]. It is obvious that understanding the behaviour of single chains is a necessary prerequisite for the treatment of aggregate properties. Phenomena that are dominated by intermolecular forces, such as the phase behaviour of binary polymer mixtures and the structure of polyelectrolyte solutions, are important for the treatment of polymer solutions. Other important phenomena such as the viscoelasticity of most polymer solutions also deserve particular attention.

The conformational state of a single polymer chain can be considered in terms of its full steric structure. For example, a polymer chain like polyethylene possesses a great internal flexibility and is able to change its conformation totally. Basically, the number of degrees of freedom of the chain is given by three times the number of atoms and it is convenient to split them into two different groups [5]. A first group concerns changes in valence angles and bond lengths, because they occur during molecular vibrations with frequencies in the infrared range. These movements are limited and do not affect the overall form of the chain. The second group of motions is of a different character, in that they have the potential to alter the form. These are the rotations about the C–C bond, which can convert the stretched chain into a coil and accomplish the transitions between all conformational states. In dealing with the conformational properties of a given polymer, only the latter group of degrees of freedom has to be considered

The huge number of rotational isomeric states that a polymer chain may adopt becomes effective in the fluid phase. Polymers in solution or the melt change between the different states and these are populated according to the law of Boltzmann statistics [5]. Because the large majority of conformations are coil-like, it is said that the polymers in fluid state represent random coils. Figure 1.1 shows a polymer coil as it might look at limited resolution; a bent chain with a continuous appearance.

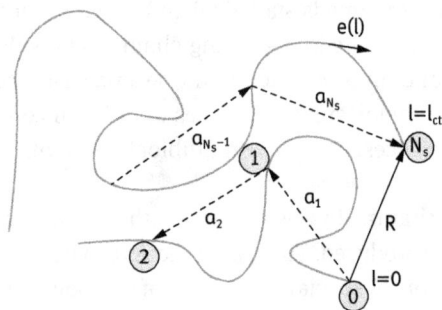

Fig. 1.1: Polymer chain in low resolution (contour length l_{cr}, local chain direction $e(l)$ together with an associated chain of N_s freely joined segments, connecting the junction points 0 to N_s) [5].

For the representation of the polymer coil, one can choose a curvilinear coordinate l from l = 0 to l = l_{cr} at the other end and describe the varying local chain direction by unit vectors e(l). The chains may be stiff, i.e. opposing strong bending, or highly flexible, thus facilitating coiling. The polymer chains possess, on length scale in the order of some nanometres, properties that are independent of the chemical structure. They may be grouped into two main classes, namely Gaussian or ideal chains (for vanishing excluded volume interactions) and expanded chains. This classification holds only for chains with large molar masses. Real chains with moderate molar masses often exhibit intermediate structures depending on the selected length scale. For example, chains can be expanded over their whole length but follow Gaussian statistics with parts, or chains can resemble straight rods for low molar masses and turn into ideal coils for high molar masses.

1.4 Polymer properties

Polymer properties are governed by the choice of monomers. In the polymerization process, monomers constitute the repeat units, for example, acrylic acid is polymerized into poly(acrylic acid). A polymer can be linear, branched or cross-linked as illustrated in Fig. 1.2 [6].

If the polymer is synthesized with more than one kind of polymer, it is called a copolymer (Fig. 1.2 (b)). The monomer units in a copolymer can be either (i) randomly distributed, (ii) distributed in blocks, or (iii) distributed such that one of the monomers is grafted in chains onto the backbone of the other monomer chain (graft copolymer) as illustrated in Fig. 1.2 (b). The polarity of the monomer units is used to categorize the polymer:
(i) nonpolar polymers such as polystyrene and polyethylene;
(ii) polar, but water insoluble polymers such as poly(methyl methacrylate);
(iii) water-soluble polymers such as polyethylene oxide and polyvinyl alcohol;
(iv) ionizable polymers, or polyelectrolytes, such as poly(acrylic acid).

The configuration of a polymer in solution depends on the balance between the interaction of the segments with the solvent and the interaction of the polymer segments with each other. As discussed above, a polymer can form a random coil or an extended configuration.

Dissolution of a polymer in a given solvent can be a problem, in particular at an industrial scale. Firstly, when the polymer is added to a solvent, the latter has to penetrate the polymer coil and the coil swells with the solvent forming a highly viscous and sticky mass (gel). The next step requires disentanglement of the chains from the gel resulting in the diffusion of the polymer chains into the solvent. This can be a slow process and it may take several hours or even days depending on the molar mass of the polymer and its chemical structure.

Linear

Branched

Network

Randomcopolymer

Block copolymer

Grafted copolymer

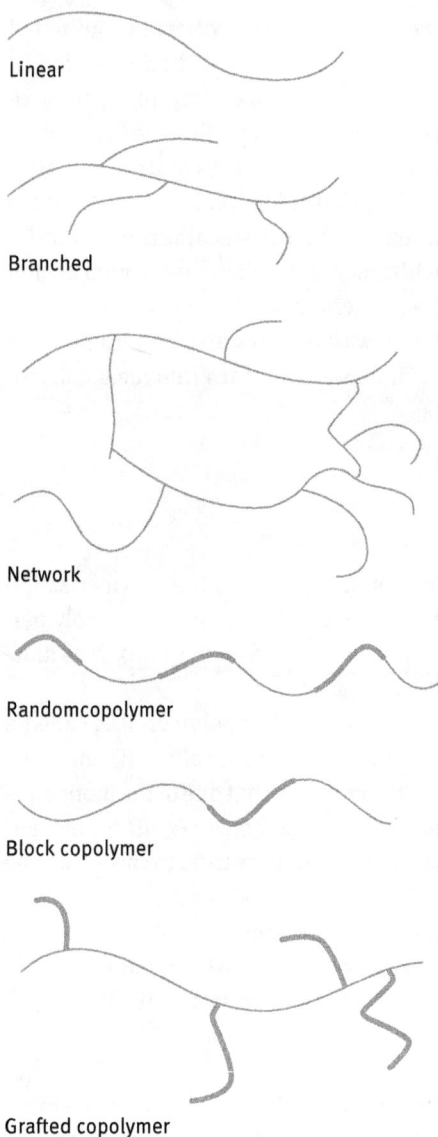

Fig. 1.2: Schematic representation of polymer structures.

1.5 Outline of the book

Chapter 2 gives a general classification of polymeric surfactants. It starts with a section on homopolymers which are made from the same repeating units. Water soluble homopolymers such as poly(ethylene oxide), poly(vinylpyrrolidone) and poly(acrylic acid) are described. These homopolymers do not show much surface activity at the oil/water interface and hence they are seldom used for preparing and stabilizing emulsions. However, these homopolymers may adsorb at a solid/liquid interface such as silica or alumina surfaces. The next section deals with block and graft copolymers which are essential for the stabilization of disperse systems. Examples of several block and graft copolymers are given to illustrate their use as stabilizers. A section is devoted to polymeric surfactants based on natural materials such as hydroxyethyl cellulose and polysaccharides. The final section in this chapter describes block and graft copolymers used in nonaqueous disperse systems.

Chapter 3 deals with the solution properties of polymeric surfactants. It starts with the thermodynamic treatment of the free energy of mixing a homopolymer with a solvent on the basis of the Flory–Huggins treatment of the free energy. The modification of this treatment when using block and graft copolymers is described. Particular attention is paid to the association of polymeric surfactants in solution and the effect of solvency of the medium for the chain.

Chapter 4 deals with the adsorption and conformation of polymeric surfactants at interfaces with particular reference to block and graft copolymers. Understanding the adsorption and conformation of polymeric surfactants at interfaces is key to knowing how these molecules act as stabilizers. Most basic ideas on adsorption and conformation of polymers have been developed for the solid/liquid interface. The process of polymer adsorption is fairly complicated. In addition to the usual adsorption considerations such as polymer/surface, polymer/solvent and surface/solvent interactions, one of the principal problems to be resolved is the configuration (conformation) of the polymer at the solid/liquid interface. A model in which each polymer molecule is attached in sequences separated by bridges that extend into solution is described. The segments in direct contact with the surface are termed "trains"; those in between and extended into solution are termed "loops"; the free ends of the macromolecule also extending in solution are termed "tails". Theories of polymeric surfactant adsorption are described with particular reference to the adsorption parameters and their experimental determination. Several examples of polymeric surfactant adsorption are given in this chapter.

Chapter 5 deals with the stabilization of disperse systems using polymeric surfactants. It emphasizes the importance of polymeric surfactants in the preparation of concentrated aqueous and nonaqueous disperse systems. The interaction between particles or droplets containing adsorbed or grafted polymeric surfactants is described at a fundamental level. The strong steric repulsion obtained when the adsorbed layers of adsorbed or grafted polymers begin to overlap is described in terms of the unfavourable mixing of the chains (when these are in good solvent conditions) and the loss of configurational entropy (elastic interaction) on considerable overlap of the chains. Combining the steric repulsion with the van der Waals attraction results in energy-distance curves that show very steep repulsion when the surface-to-surface distance between the particles or droplets become smaller than twice the adsorbed layer thickness. The criteria of effective steric stabilization are described in terms of full coverage of the particles or droplets by the absorbed polymer, strong adsorption ("anchoring") of the chains, good solvency of the stabilizing chains by the molecules of the medium and optimum adsorbed layer thickness.

Chapter 6 describes the flocculation of disperse systems containing polymeric surfactants. Four different types of flocculation mechanisms are described:

(i)   weak and reversible flocculation obtained when using "thin" adsorbed layers (< 5 nm) that results in an attractive minimum at separation distances compared to twice the adsorbed layer thickness;

(ii) incipient (strong) flocculation obtained when the solvency of the medium for the chains becomes worse than a theta solvent;

(iii) depletion flocculation induced by the addition of "free" nonabsorbing polymer chains in the continuous medium;

(iv) bridging flocculation obtained by simultaneous adsorption of the polymer chain on two or more particles or droplets. The mechanism of each type of flocculation is described at a fundamental level.

Chapter 7 deals with applications of polymeric surfactants in emulsion and suspension polymerization that are used for the preparation of polymer colloids (latexes). The theories of emulsion and suspension polymerization are described at a fundamental level with particular emphasis on the use of polymeric surfactants for stabilizing the resulting particles against aggregation. This is particularly the case with nonaqueous latex dispersions. Chapter 8 describes the role of polymeric surfactants in stabilizing preformed particles. The importance of strong adsorption that results in a high affinity adsorption isotherm is described at a fundamental level. A section is devoted to the kinetics of polymeric surfactant adsorption with special reference to the effect of the molar mass of the polymer chain, the solvency of the medium and temperature. The assessment of the stability of the resulting dispersion is described using rheological techniques.

Chapter 9 deals with the use of polymeric surfactants for the stabilization of emulsions and nanoemulsions. The origin of stability against coalescence is discussed in terms of the disjoining pressure of the thin film produced on the approach of two emulsion droplets. This disjoining pressure consists of three contributions, namely a positive electrostatic repulsion, a positive steric repulsion and a negative van der Waals attraction. To produce a stable film that resists coalescence, the positive contributions must exceed the negative one. The high net positive disjoining pressure is obtained using polymeric surfactants for the preparation of emulsions. Another important use of polymeric surfactants is to reduce Ostwald ripening, in particular with nanoemulsions. This reduction is obtained by strong adsorption of the polymeric surfactant at the oil/water interface and increasing the dilatational elasticity of the film.

Chapter 10 gives a summary of applications of polymeric surfactants in pharmacy. Several polymeric surfactants have been approved by the Food and Drug Administration (FDA), such as the poloxamers consisting of A–B–A block copolymers of polyethylene oxide (A) and polypropylene oxide (PPO). These polymeric surfactants are used for the formulation of various pharmaceutical systems, such as suspensions, emulsions and multiple emulsions. Chapter 11 deals with applications of polymeric surfactants in cosmetics. The advantages of using polymeric surfactants in formulating emulsions in hand creams are discussed in terms of lack of skin irritation and high stability of the resulting system. Polymeric surfactants are also used for the stabilization of liposomes and vesicles that are incorporated in many cosmetic formulations for efficient delivery of active ingredients such as anti-wrinkle agents. Chapter 12 gives examples of applications of polymeric surfactants in paints and coatings. Particular attention is paid to the steric stabilization obtained when using polymeric surfactants in nonaqueous paint systems, where electrostatic stabilization is not possible. Chapter 13 deals with applications of polymeric surfactants in agrochemical formulations, in particular their use in the formulation of concentrated suspensions and emulsions. The interaction of polymeric surfactants with low molecular surfactants (used as wetters) is discussed to illustrate the synergy between the two molecules. Chapter 14 gives examples of applications of polymeric surfactants such as gums and proteins in the food industry. The interaction between proteins and polysaccharides is also described.

References

[1] Tadros, Th. F., "Applied Surfactants", Wiley-VCH, Germany (2005).
[2] Tadros, Th. F., in "Principles of Polymer Science and Technology in Cosmetics and Personal Care", E. D. Goddard and J. V. Gruber (eds.), Marcel Dekker, NY (1999).
[3] Tadros, Th. F., in "Novel Surfactants", K. Holmberg (ed.), Marcel Dekker, NY (2003).
[4] Tadros, Th. F., "Interfacial Phenomena and Colloid Stability, Basic Principles", De Gruyter, Germany (2015).
[5] Strobel, G., "The Physics of Polymers", 3rd ed., Springer, Germany (2007).
[6] Holmberg, K., Jonsson, B., Kronberg, B. and Lindman, B., "Surfactants and Polymers in Aqueous Solution"', 2nd ed., John Wiley & Sons, USA (2003).

2 General classification of polymeric surfactants

2.1 Homopolymers

Perhaps the simplest type of polymeric surfactant is a homopolymer [1–4] that is formed from the same repeating units, such as poly(ethylene oxide) and is then referred to as polyethylene glycol which has the following chemical structure with repeating units of ethylene oxide,

$$H-(O-CH_2-CH_2)_n-OH.$$

Another homopolymer that is used in many pharmaceutical formulations is poly (vinylpyrrolidone) that is formed from repeating units of vinylpyrrolidone.

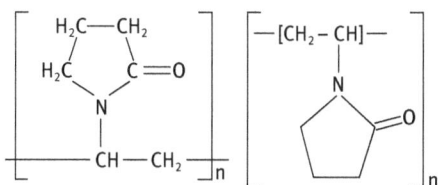

These homopolymers have little surface activity at the O/W interface, since the homopolymer segments (ethylene oxide or vinylpyrrolidone) are highly water soluble and have little affinity to the interface. However, such homopolymers may adsorb significantly at the S/L interface, e.g. on silica. Even if the adsorption energy per monomer segment to the surface is small (fraction of kT, where k is the Boltzmann constant and T is the absolute temperature), the total adsorption energy per molecule may be sufficient to overcome the unfavourable entropy loss of the molecule at the S/L interface [1–4].

2.2 Random copolymers

As mentioned above, homopolymers are not the most suitable emulsifiers or dispersants. A small variant is to use polymers that contain specific groups with high affinity to the surface that are randomly attached to the polymer chain [1–4]. This is exemplified by partially hydrolysed poly(vinyl acetate) (PVAc), technically referred to as poly(vinyl alcohol) (PVA). The polymer is prepared by partial hydrolysis of PVAc, leaving some residual vinyl acetate groups. Most commercially available PVA molecules contain 4–20 mol % acetate groups. These acetate groups, which are hydrophobic, give the molecule its amphipathic character. This blocky distribution of acetate groups

DOI 10.1515/9783110487282-003

on the poly(vinyl alcohol) backbone provides more effective anchoring of the polymeric surfactant chain on the particles or emulsion droplets. On a hydrophobic surface such as polystyrene or a hydrocarbon oil, the polymer adsorbs with preferential attachment of the acetate groups on the surface, leaving the more hydrophilic vinyl alcohol segments dangling in the aqueous medium. These partially hydrolysed PVA molecules also exhibit surface activity at the O/W interface as indicated by the reduction of the interfacial tension with increasing polymer concentration.

2.3 Block and graft copolymers

The most convenient polymeric surfactants are those of the block and graft copolymer type. A block copolymer is a linear arrangement of blocks of variable monomer composition. The nomenclature for a diblock is poly-A-block-poly-B and for a triblock is poly-A-block-poly-B-poly-A. One of the most widely used triblock polymeric surfactants are the "Pluronics" or "Poloxamers" (BASF, Germany), which consists of two poly-A blocks of poly(ethylene oxide) (PEO) and one block of poly(propylene oxide) (PPO). Several chain lengths of PEO and PPO are available as indicated in the chemical structure below.

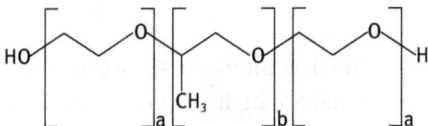

General structure
with a = 2–130 and b = 15–67

For the Pluronics the trade name is coded with a letter L (liquid), P (paste) and F (flake) that defines its physical form at room temperature. This is followed by two or three digits; the first one or two digits multiplied by 300 indicate the approximate molar mass of PPO and the last digit multiplied by 10 indicates the percentage of PEO. For example, Pluronic L61 indicates a liquid with PPO molar mass of 1800 and 10 % PEO. Pluronic F127 indicates a flake with PPO molar mass 3600 and 70 % PEO. The poloxamers (which are FDA approved) are commonly named with the letter P followed by three digits; the first two digits multiplied by 10 give the approximate molar mas of PPO and the last digit multiplied by 10 gives the percentage of PEO. For example, Poloxamer P407 has a molar mass of PPO of 400 and 70 % PEO. Later, triblocks of PPO–PEO–PPO (inverse Pluronics) became available for some specific applications. These polymeric triblocks can be applied as emulsifiers or dispersants, whereby the assumption is made that the hydrophobic PPO chain resides at the hydrophobic surface, leaving the two PEO chains dangling in aqueous solution and hence providing steric repulsion. Although these triblock polymeric surfactants have been widely used in various applications in emulsions and suspensions, some doubt has arisen on how effective these can be. It is generally accepted that the PPO chain is not sufficiently hydrophobic to provide a strong "anchor" to a hydrophobic surface or to an oil droplet. Indeed, the reason for the surface activity of the PEO–PPO–PEO triblock copolymers at the O/W interface may stem from a process of "rejection" anchoring of the PPO chain since it is not soluble in both oil and water [1–4].

Several other di- and triblock copolymers have been synthesized, although these are of limited commercial availability. Typical examples are diblocks of polystyrene-block-polyvinyl alcohol, triblocks of poly(methyl methacrylate)-block poly(ethylene oxide)-block poly(methyl methacrylate), diblocks of polystyrene block-polyethylene oxide and triblocks of polyethylene oxide-block polystyrene-polyethylene oxide [1–4]. An alternative (and perhaps more efficient) polymeric surfactant is the amphipathic graft copolymer consisting of a polymeric backbone B (polystyrene or polymethyl methacrylate) and several A chains ("teeth") such as polyethylene oxide. This graft copolymer is sometimes referred to as a "comb" stabilizer. This copolymer is usually prepared by grafting a macromonomer such methoxy polyethylene oxide methacrylate with polymethyl methacrylate. The "grafting onto" technique has also been used to synthesize polystyrene-polyethylene oxide graft copolymers.

2.4 Polymeric surfactants based on polysaccharides

Several surface active graft copolymers are available based on a hydrophilic backbone of polysaccharide to which several hydrophobic alkyl chains are attached. A good example is Emulsan, which is produced by micro-organisms (bacteria). It consists of a backbone of hetero-polysaccharide (with repeating trisaccharide carrying a negative charge). Fatty acid chains are covalently linked to the polysaccharide through ester linkages [5] as is illustrated in Fig. 2.1.

Emulsan is moderately surface active showing a small reduction of the O/W interfacial tension (for example, it reduces the interfacial tension of hexane/water from 47 to around $55\,mN\,m^{-1}$). However, it has a strong tendency to adsorb at the O/W interface and can be very effective in stabilizing emulsions of specific oils in water.

Natural polysaccharides can be chemically modified into the equivalent of lipo-polysaccharides by attachment of long alkyl or alkyl acryl chains [5]. For example, cellulose can be modified with ethylene oxide and alkyl chloride. The cellulose is swollen in strong alkali and the semi-soluble material is reacted with ethylene oxide and alkyl chloride resulting in the formation of polymeric surfactant with the structure shown in Fig. 2.2. If the alkyl group is short, e.g. ethyl, the molecule is moderately surface active. If some of the ethyl groups are replaced, long chain alkyls, a polymer with higher surface activity, is obtained. Such graft copolymers are commercially available and they are referred to as "associative thickeners". The latter are used for rheology control of many aqueous formulations, e.g. water-borne paints.

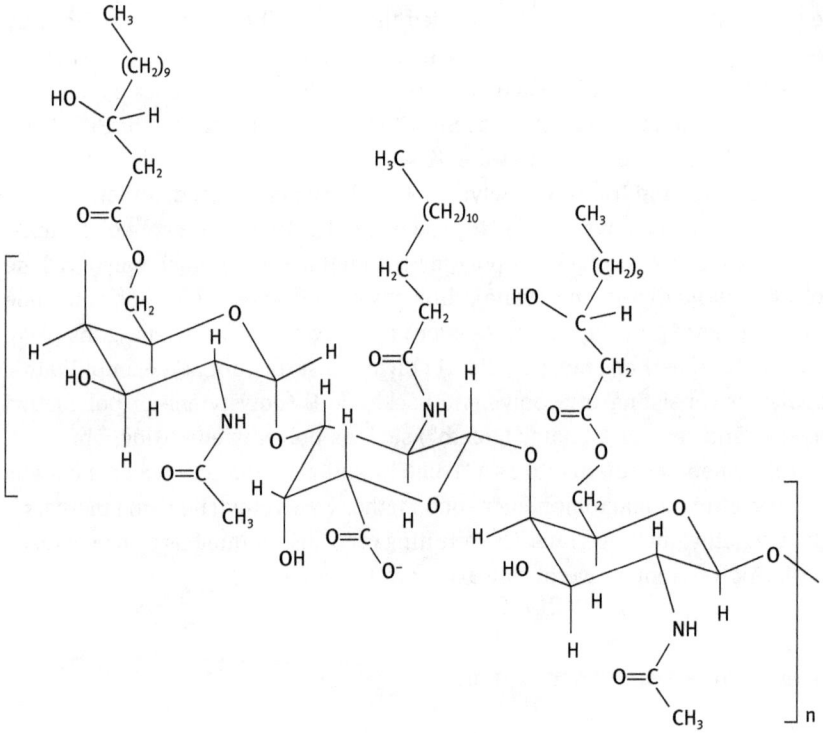

Fig. 2.1: Structure of Emulsan.

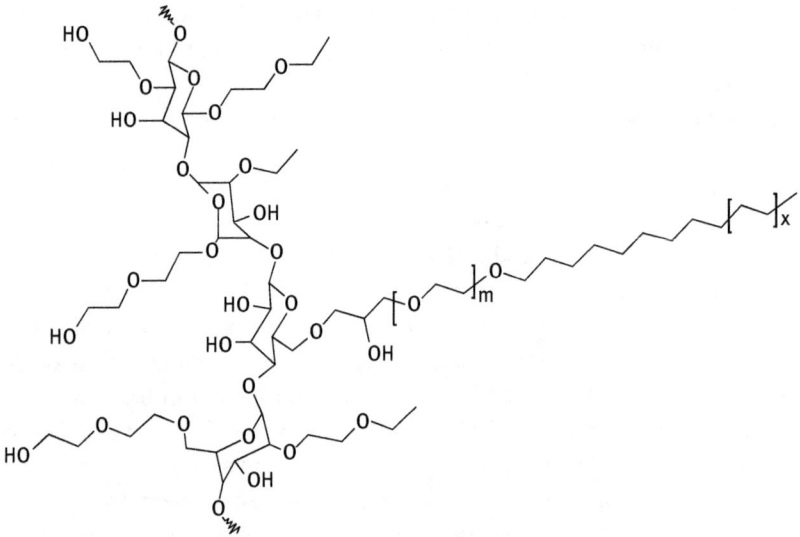

Fig. 2.2: Structure of cellulose modified with ethylene oxide and alkyl chloride.

Another example of hydrophobically modified nonionic cellulose ether (HM-EHEC) is shown in Fig. 2.3. The cellulose can be modified by a relatively random substitution of hydroxyethyl and ethyl groups to give ethyl(hydroxyethyl) cellulose (EHEC). A low fraction of hydrophobic alkyl groups is inserted to give HM-EHEC.

Another sugar-based surfactant is alkyl polyglucoside which is prepared by reacting starch or glucose with butanol using acid catalyst [6]. The resulting butyl oligoglycoside intermediate is reacted with dodecanol using acid catalyst to produce dodecyl polyglucosides as is illustrated in Fig. 2.4. The reaction yields a mixture in which on average more than one glucose unit is attached to an alcohol molecule. The average number of glucose units linked to an alcohol group is described as the (average) degree of polymerization (DP). Alkyl monoglycosides are the main group of components with a content of more than 50 % followed by the diglycosides and higher oligomers up to heptaglycosides. Due to the presence of molecules with DP > 1, the surfactant may be considered as polymeric. The alkyl polyglucosides show high surface activity at the air/water and oil/water interfaces. They also have low critical micelle concentrations comparable to those of nonionic surfactants.

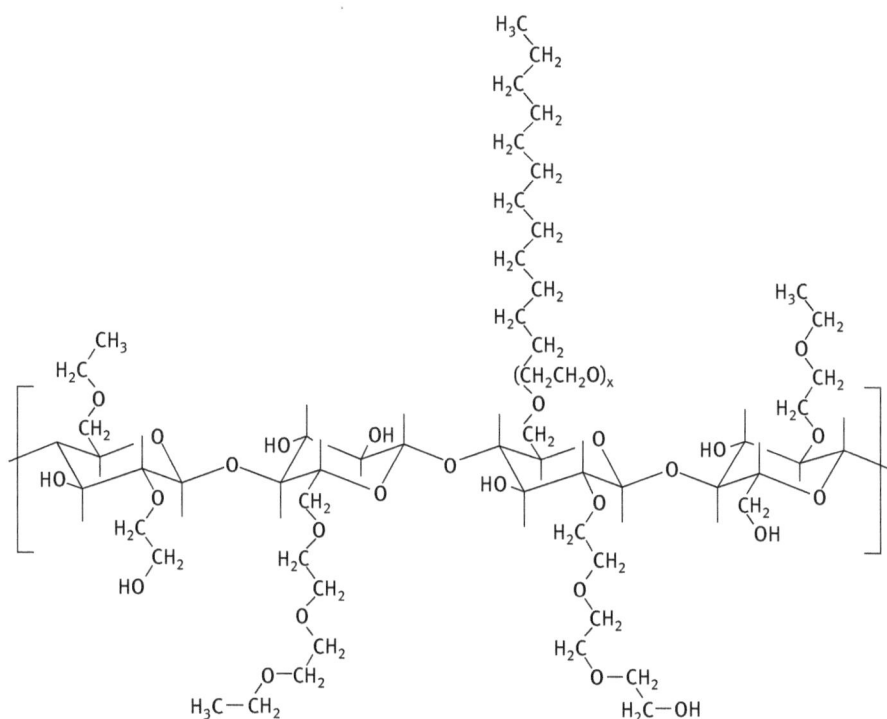

Fig. 2.3: Structure of hydrophobically modified ethylhydroxyethyl cellulose (HM-EHEC).

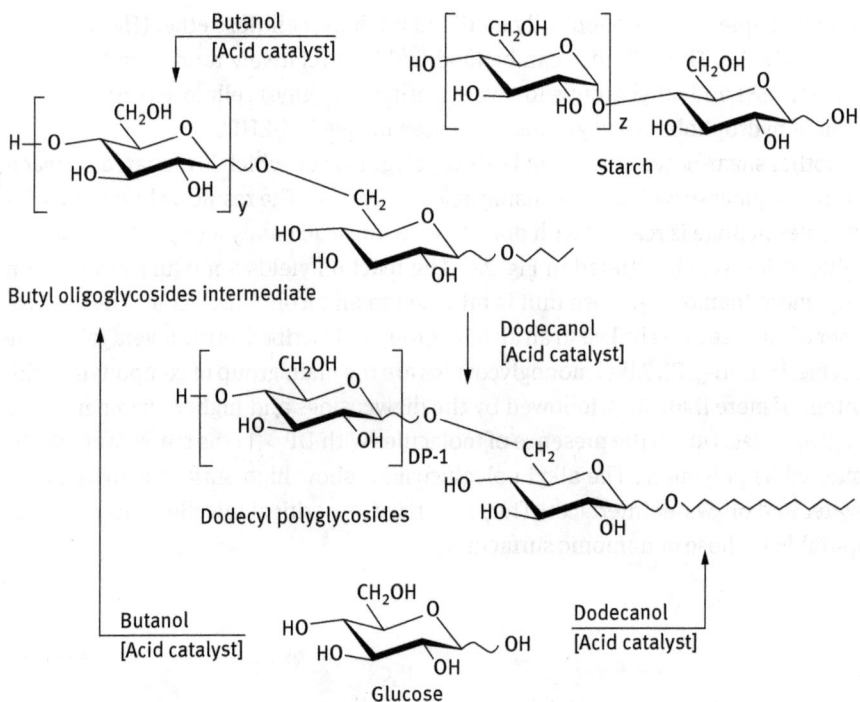

Fig. 2.4: Pathway for alkyl polyglucoside synthesis.

More recently, graft copolymers based on polysaccharides have been developed for the stabilization of disperse systems. One of the most useful graft copolymers is that based on inulin obtained from chicory roots [7–9]. It is a linear polyfructose chain with a glucose end. When extracted from chicory roots, inulin has a wide range of chain lengths ranging from 2–65 fructose units. It is fractionated to obtain a molecule with narrow molecular weight distribution with a degree of polymerization > 23 and this is commercially available as INUTEC® N25. The latter molecule is used to prepare a series of graft copolymers by random grafting of alkyl chains (using alkyl isocyanate) onto the inulin backbone. The first molecule of this series is INUTEC® SP1 (Beneo-Remy, Belgium) that is obtained by random grafting of C_{12} alkyl chains. It has an average molecular weight of ≈ 5000 Da and its structure is given in Fig. 2.5. The molecule is schematically illustrated in Fig. 2.6 which shows the hydrophilic polyfructose chain (backbone) and the randomly attached alkyl chains.

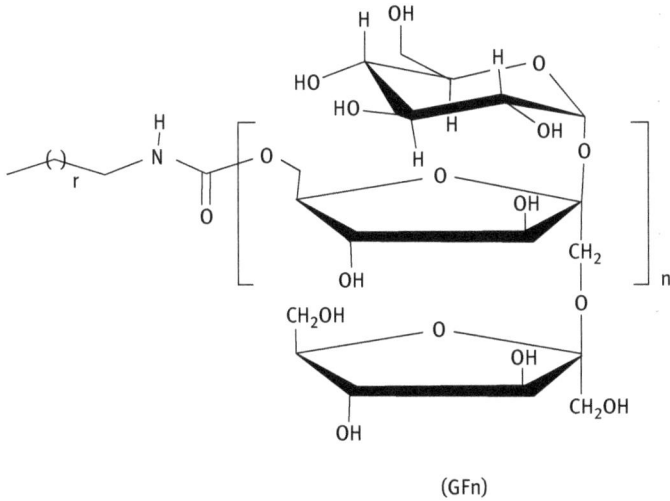

(GFn)

Fig. 2.5: Structure of INUTEC® SP1.

The main advantages of INUTEC® SP1 as a stabilizer for disperse systems are:
(i) Strong adsorption to the particle or droplet by multipoint attachment with several alkyl chains. This ensures lack of desorption and displacement of the molecule from the interface.
(ii) Strong hydration of the linear polyfructose chains both in water and in the presence of high electrolyte concentrations and high temperatures. This ensures effective steric stabilization [10].

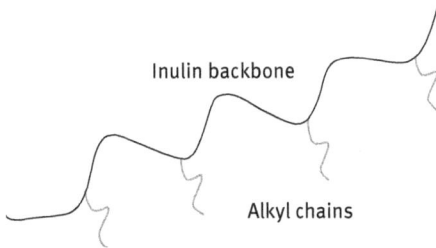

Fig. 2.6: Schematic representation of INUTEC® SP1 polymeric surfactant.

2.5 Natural polymeric biosurfactants

Most food emulsions require the use of naturally occurring polymeric surfactants that must be approved by the Food and Drug Administration (FDA). One of the most commonly used natural polysaccharide emulsifiers is gum arabic [11] which is amphiphilic. It has a nonpolar polypeptide backbone with a number of polar polysaccharide chains attached. On adsorption to oil droplet surfaces, the polypeptide chain protrudes into the oil droplet surfaces, whereas the polysaccharide chains dangle into the water. This leads to the formation of a relatively thick hydrophilic coating around the oil droplets that prevents any aggregation and coalescence of the oil droplets.

A number of other naturally occurring amphiphilic polysaccharides that are suitable for use as emulsifiers have been identified, e.g. pectin fractions isolated from beet, citrus, apple, etc. These polymeric surfactants show surface activity at the oil/water (O/W) interface and are able to stabilize the emulsion against floccu- lation and coalescence. Chitosan, a cationic polysaccharide, typically isolated from crustacean shells, is also capable of emulsion formation and stability.

One of the most plentiful naturally occurring polymeric surfactants that are com- monly used in food emulsions are the proteins. These are biopolymers consisting of strings of amino acid units covalently linked by peptide bonds [11]. The type, num- ber and position of the amino acids in the polypeptide chain determine the molar mass and functional properties of food proteins. Most proteins contain a mixture of polar and nonpolar amino acids and, therefore, the amphiphilic molecules can ad- sorb at the O/W interface thus stabilizing the emulsion. The relative balance of polar and nonpolar groups exposed on their surfaces governs the surface activity of pro- teins. If the surface hydrophobicity is too low, the driving force for protein adsorption is not strong enough to overcome the entropy loss associated with adsorption. Con- versely, if the surface hydrophobicity is too high, then the proteins tend to aggregate, become water-insoluble, and lose their surface activity. Consequently, an optimum level of surface hydrophobicity is required for a protein to be a good emulsifier. Most proteins also have a mixture of anionic, nonionic and cationic amino acids along the polypeptide chains, which determines the charge of the protein molecule under dif- ferent pH conditions. At a certain pH, referred to as isoelectric point (IEP), the protein molecule has a net zero charge. When the pH > IEP, the protein molecule is negatively charged, whereas when pH < IEP the protein molecule becomes positively charged. The charged chains play a major role in electrostatic stabilization.

Proteins may adopt various conformations in aqueous solution, depending on the balance of van der Waals forces, hydrophobic interactions, electrostatic interactions, hydrogen bonds, steric effects and entropy effects [11]. The balance is determined by the solution and environmental conditions such as pH, ionic strength and tempera- ture. Consequently, the conformation of a protein at an interface may change when these conditions are altered. The two mostly common conformations of surface ac- tive proteins used as emulsifiers are globular and random coil. Globular proteins have fairly compact spheroid structures where the majority of the nonpolar groups are lo- cated within the interior, and the majority of the polar groups are present at the ex- terior. These globular proteins have surface activity because some of the nonpolar groups remain exposed at their surfaces, which gives the driving force for adsorption at the O/W interface. Random coils have a more flexible structure, although there may still be some regions that have local order such as helical and sheet structures. The most common random coil proteins used as emulsifiers in the food industry are casein and gelatin. The structure of the protein often changes after they adsorb at the O/W interface. For example, globular proteins may unfold after they adsorb to droplet surfaces and expose groups normally located in their interiors, such as nonpolar and sulfhydryl groups. After adsorption to the oil droplet surfaces, the protein molecule may adopt a configuration where many of the hydrophilic groups protrude into the water phase, whereas most of the hydrophobic groups protrude to the oil phase. This results in high surface activity at the O/W interface and effective stabilization of the emulsion.

2.6 Silicone surfactants

Silicone surfactants are graft copolymers ("comb" type) with a backbone based on polydimethylsiloxane, which is highly hydrophobic and insoluble in water [5, 12]. The side chains ("teeth") are water soluble, charged or uncharged, and the molecule becomes surface active in aqueous solution. Poly(ethylene glycol) (EO) or poly(ethylene glycol)–poly(propylene glycol) (PEO–PPO) are by far the most common constituents of the side chains. The side chain may contain a weakly polar group such as an ester or amine or it may be an ionic group. The general structure of the silicone surfactant is shown in Fig. 2.7 with X being an ionic or nonionic polar group such as PEO or PEO–PPO. The linkage between Si and the polyether chain may be either Si–O–C or Si–C. The Si–O–C link is made by esterification of chloropolysiloxanes with hydroxylfunctional organic compounds such as PEO–PPO copolymer. This makes the molecule unstable, undergoing hydrolysis in acid or alkaline conditions. However, the Si–C linkage, where a carbon of the PEO–PPO copolymer is directly linked to the Si atom, is stable. Such a linkage is usually made by a Pt-catalysed hydrosilating addition of an Si-H function in the polysiloxane to a terminal olefinic bond in the substituted polymer [5].

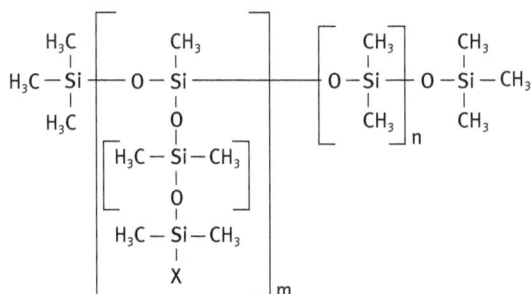

Fig. 2.7: Structure of silicone surfactants with X being an ionic or nonionic polar group such as PEO or PEO–PPO.

Silicone surfactants have unique properties when compared with hydrocarbon surfactants. They are very effective in lowering the surface tension of water to values around $20 \, \text{mN m}^{-1}$ (when compared with the value of around $30 \, \text{mN m}^{-1}$ obtained with most hydrocarbon surfactants). They also have excellent wetting on low energy surfaces such as polytetrafluroethane (PTFE). They are also powerful antifoamers.

2.7 Polymeric surfactants for nonaqueous dispersions

Block and graft copolymers based on poly(12-hydroxystearic acid) are used for nonaqueous systems such as water-in-oil (W/O) emulsions and nonaqueous dispersions, e.g. paints [13]. The poly(12-hydroxystearic acid) chains (the A chains of an A–B–A block copolymer or the A side chains of a graft copolymer), which are of low molecular weight ($\approx 1000 \, \text{Da}$), provide steric stabilization analogous to how PEO behaves in aqueous solution [5]. The B anchor chains are chosen to be highly insoluble in the nonaqueous medium and have some specific interaction with the surface of the droplet or particle. For water-in-oil (W/O) emulsions an A–B–A block copolymer of poly (12-hydroxystearic acid) (PHS) (the A chains) and poly (ethylene oxide) (PEO) (the B chain): PHS–PEO–PHS is commercially available (Arlacel P135). The PEO chain (that is soluble in the water droplets) forms the anchor chain, whereas the PHS chains form the stabilizing chains. PHS is highly soluble in most hydrocarbon solvents and is strongly solvated by its molecules. The structure of the PHS–PEO–PHS block copolymer is schematically shown in Fig. 2.8.

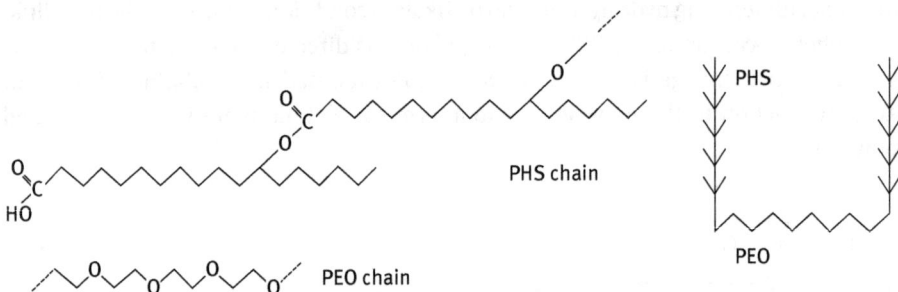

Fig. 2.8: Schematic representation of the structure of PHS–PEO–PHS block copolymer.

In the nonaqueous dispersion process (referred to as NAD) the monomer, normally an acrylic, is dissolved in a nonaqueous solvent, normally an aliphatic hydrocarbon and an oil soluble initiator and a stabilizer (to protect the resulting particles from flocculation, sometimes referred to as "protective colloid") is added to the reaction mixture. The most successful stabilizers used in NAD are block and graft copolymers. These block and graft copolymers are assembled in a variety of ways to provide the molecule with an "anchor chain" and a stabilizing chain. As mentioned above, the anchor chain should be sufficiently insoluble in the medium and have a strong affinity to the polymer particles produced. In contrast, the stabilizing chain should be soluble in the medium and strongly solvated by its molecules to provide effective steric stabilization. The length of the anchor and stabilizing chains has to be carefully adjusted to ensure strong adsorption (by multipoint attachment of the anchor chain to the particle surface) and a sufficiently "thick" layer of the stabilizing chain that prevents close approach of the particles to a distance where van der Waals attraction becomes strong. Several configurations of block and graft copolymers are possible, as is illustrated in Fig. 2.9.

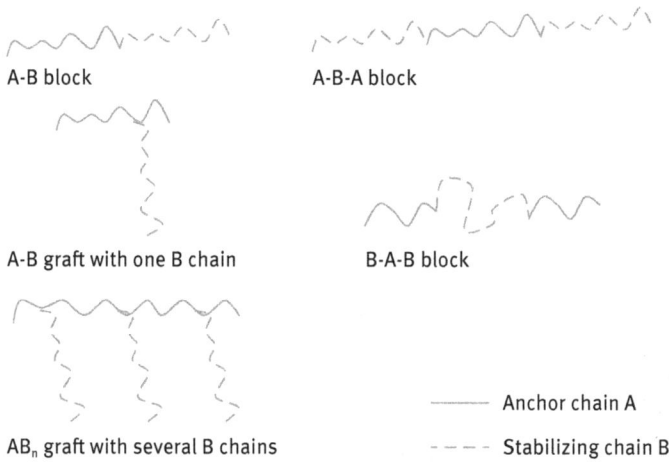

A-B block

A-B-A block

A-B graft with one B chain

B-A-B block

AB$_n$ graft with several B chains

———— Anchor chain A

– – – – Stabilizing chain B

Fig. 2.9: Configurations of block and graft copolymers.

Typical preformed graft stabilizers based on poly(12-hydroxy stearic acid) (PHS) are simple to prepare and effective in NAD polymerization. Commercial 12-hydroxystearic acid contains 8–15 % palmitic and stearic acids which limits the molecular weight during polymerization to an average of 1500–2000. This oligomer may be converted to a "macromonomer" by reacting the carboxylic group with glycidyl methacrylate. The macromonomer is then copolymerized with an equal weight of methyl methacrylate (MMA) or similar monomer to give a "comb" graft copolymer with an average molecular weight of 10 000–20 000. The graft copolymer contains on average 5–10 PHS chains pendent from a polymeric anchor backbone of PMMA. This graft copolymer can stabilize latex particles of various monomers. The major limitation of the monomer composition is that the polymer produced should be insoluble in the medium used.

Several other examples of block and graft copolymers that are used in dispersion polymerization are given in Tab. 2.1 which also shows the continuous phase and disperse polymer that can be used with these polymers.

Tab. 2.1: Block and graft copolymers used in emulsion polymerization.

Polymeric Surfactant	Continuous phase	Disperse polymer
Polystyrene-block-poly(dimethyl siloxane)	Hexane	Polystyrene
Polystyrene-block-poly(methacrylic acid)	Ethanol	Polystyrene
Polybutadiene-graft-poly(methacrylic acid)	Ethanol	Polystyrene
Poly(2-ethylhexyl acrylate)-graft-poly(vinyl acetate)	Aliphatic hydrocarbon	Poly(methyl methacrylate)
Polystyrene-block-poly(t-butylstyrene)	Aliphatic hydrocarbon	Polystyrene

2.8 Polymerizable surfactants

Polymerizable surfactants are amphipathic molecules containing somewhere in their structure a polymerizable group such as styrenic, acrylic or methacrylic, vinylic, maleic, crotonic or allylic [14]. These groups may be located in different places such as at the end or head of the hydrophilic sequence, the end of the hydrophobic moiety, between the two, or finally along the surfactant structure or pendent (side) groups. They are generally referred to as surfmers [14]. One of the main applications of polymerizable surfactants is for the stabilization of vesicles. The second major use of polymerizable surfactants is their application as stabilizers in polymerization in dispersed media, both in emulsion and dispersion polymerization. The surfactant is covalently linked to the surface of the particles, thus improving latex stability. A second kind of benefit is expected in the case of film-forming latexes. In conventional emulsion, the surfactants are not firmly attached to the particles and thus are able to migrate towards the surface of the film. This may result in defects of adhesion if the film is expected to protect the surface of a substrate. In addition, phase separation may occur during coalescence. These problems are overcome when using surfactants that are covalently linked to the surface of the latex. A third kind of benefit takes place if it is intended that the latex will release material upon flocculation. If the surfactant is covalently linked to the surface of the polymer particles, a smaller amount of it will be rejected to the water phase [13].

References

[1]   Tadros, Th. F., "Applied Surfactants", Wiley-VCH, Germany (2005).
[2]   Tadros, Th. F., in "Principles of Polymer Science and Technology in Cosmetics and Personal Care", E. D. Goddard and J. V. Gruber (eds.), Marcel Dekker, NY (1999).
[3]   Tadros, Th. F., in "Novel Surfactants", K. Holmberg (ed.), Marcel Dekker, NY (2003).
[4]   Piirma, I., "Polymeric Surfactants", Surfactant Science Series, No. 42, Marcel Dekker, NY (1992).
[5]   Holmberg, K., Jonsson, B., Kronberg, B. and Lindman, B., "Surfactants and Polymers in Aqueous Solution", 2nd edition, John Wiley & Sons, USA (2003).

[6]   von Rybinski, W. and Hill, K., "Alkyl Polyglucosides" in "Novel Surfactants", K. Holmberg (ed.), Marcel Dekker, NY (2003).

[7]   Stevens, C. V., Meriggi, A., Peristerpoulou, M., Christov, P.P., Booten, K., Levecke, B., Vandamme, A., Pittevils, N. and Tadros, Th. F., Biomacromolecules, 2, 1256 (2001).

[8]   Hirst, E. L., McGilvary, D. I. and Percival, E. G., J. Chem. Soc., 1297 (1950).

[9]   Suzuki, M., in "Science and Technology of Fructans", M. Suzuki, and Chatterton, N. J. (eds.), CRC Press, Boca Raton FL (1993), p. 21.

[10] Tadros, Th. F., Vandamme, A, Levecke, B., Booten, K., and Stevens, C. V., Advances Colloid Interface Sci., 108–109, 207 (2004).

[11] McClements, D. J. and Gumus, C. E., Advances Colloid and Interface Sci., 234, 3 (2016).

[12] Fleute-Schlachter, I. and Feldman-Krane, G., "Silicone Surfactants", in "Novel Surfactants", K. Holmberg (ed.), Marcel Dekker, NY (2003).

[13] Barrett, K. E. J. and Thomas, H. R., J. Polym. Sci., Part A1, 7, 2627 (1969).

[14] Guyot, A., "Polymerizable Surfactants", in "Novel Surfactants", K. Holmberg (ed.), Marcel Dekker, NY (2003).

3 Solution properties of polymeric surfactants

3.1 Polymer conformation and structure

Long flexible macromolecules have a large number of internal degrees of freedom. A typical primary structure of such molecules is a linear chain of units connected by covalent bonds referred to as the backbone. By rotation about the bonds in the backbone the molecule changes its shape resulting in a wide spectrum of conformations. The rotation is hindered by the side groups, so that some of these conformations may be rather unfavourable. In some macromolecules such as proteins sequences of preferred orientations show up as helical or folded sections.

For flexible linear polymers the energy barriers associated with rotation around the bonds are small with respect to thermal motion. Such molecules have a randomly fluctuating three-dimensional tertiary structure, referred to as random coil, as is illustrated in Fig. 3.1. The chain conformation is described as a random flight chain of N bonds of length ℓ. The fluctuating distance between the end points is r. The quantity $\langle s^2 \rangle^{1/2}$, referred to as the mean end-to-end distance, is a measure of the size of the chain, i.e. its mean coil diameter [1]:

$$\langle r^2 \rangle^{1/2} = N^{1/2} \ell. \tag{3.1}$$

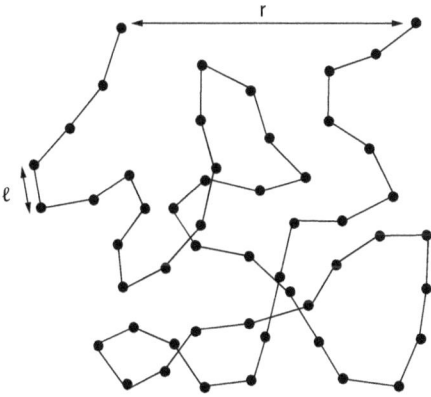

Fig. 3.1: Schematic representation of the chain conformation for a random coil.

Another useful parameter is the radius of gyration, $\langle s^2 \rangle^{1/2}$, which is a measure of the effective size of a polymer molecule (it is the root mean square distance of the elements of the chain from its center of gravity).

For linear polymers,

$$\langle s^2 \rangle^{1/2} = \frac{\langle r^2 \rangle^{1/2}}{6^{1/2}}. \tag{3.2}$$

DOI 10.1515/9783110487282-004

In real polymers the bonds cannot assume arbitrary directions but there are fixed angles between them. In addition, rotation about bonds is not entirely free, because the potential energy shows maxima and minima as a function of the rotation angle. To account for these effects the above equations are modified by introducing a rigidity parameter p (stiffness "persistence") which depends on the architecture of the chain,

$$\langle r^2 \rangle^{1/2} = 6^{1/2} p^{1/2} N^{1/2} \ell, \tag{3.3}$$

$$\langle s^2 \rangle^{1/2} = p^{1/2} N^{1/2} \ell. \tag{3.4}$$

$p = 1/6$ for a (hypothetically) fully flexible chain and p increases as the chain becomes less flexible, for example when the side groups are bulky. Typical p values for real chains are in the range 0.5–4.

A useful parameter, called the characteristic ratio, was introduced by Flory [1] and is defined as,

$$C_\infty = \frac{\langle r^2 \rangle}{N \ell_b^2}. \tag{3.5}$$

ℓ_b^2 stands for the sum of the squares of the lengths of the backbone bonds of one monomer unit,

$$\ell_b^2 = \sum_i a_i^2. \tag{3.6}$$

The main consequence of the above equations is that for ideal chains the dimensions (root mean square end-to-end distance and radius of gyration) are proportional to $N^{1/2}$. This is only valid for ideal chains where the volume of the segments and solvency effects are entirely ignored. In other words, a walk may return to its origin without any hindrance. This is unrealistic for segments which occupy a volume. In good solvents, where the chains swell, the excluded volume becomes important. The segments cannot overlap and there is an exclusion volume that automatically leads to coil expansion. In very good solvents, where the segments repel each other, the excluded volume is larger than the exclusion volume. In contrast, in a poor solvent the segments experience net attraction, the effective excluded volume is small and the ideal chain model gives a reasonable description.

3.2 Free energy of mixing of polymer with solvent – the Flory–Huggins theory

The effect of solvency for the polymer chain was considered in the thermodynamic treatment by Flory and Huggins [2], usually referred to as the Flory–Huggins theory. This theory considers the free energy of mixing of pure polymer with pure solvent, ΔG_{mix}, in terms of two contributions, namely the enthalpy of mixing, ΔH_{mix}, and the entropy of mixing, ΔS_{mix}, i.e. using the second law of thermodynamics,

$$\Delta G_{mix} = \Delta H_{mix} - T \Delta S_{mix}. \tag{3.7}$$

Assuming that the polymer chain adopts a configuration on a lattice (provided by solvent molecules) and considering that the mixing is "random", then the entropy of mixing, ΔS_{mix}, is given by the following expression:

$$\Delta S_{mix} = -k[n_1 \ln \phi_1 + n_2 \ln \phi_2], \tag{3.8}$$

where k is the Boltzmann constant, n_1 is the number of solvent molecules with a volume fraction ϕ_1, and n_2 is the number of polymer molecules with a volume fraction ϕ_2.

The enthalpy of mixing, ΔH_{mix}, is given by the following expression:

$$\Delta H_{mix} = n_1 \phi_2 \chi kT, \tag{3.9}$$

where χ is a dimensionless interaction parameter and χkT expresses the difference in energy of a solvent molecule in pure solvent compared to its immersion in pure polymer. χ is usually referred to as the Flory–Huggins interaction parameter.

Combining equations (3.7)–(3.9), one obtains

$$\Delta G_{mix} = kT[n_1 \ln \phi_1 + n_2 \ln \phi_2 + \chi n_1 \phi_2]. \tag{3.10}$$

The mixing of a pure solvent with a polymer solution creates an osmotic pressure, π, which can be expressed in terms of the polymer concentration c_2 and the volume fraction of the polymer,

$$\frac{\pi}{c_2} = RT \left[\frac{1}{M_2} + \left(\frac{v_2^2}{V_1} \right) \left(\frac{1}{2} - \chi \right) c_2 + \cdots \right], \tag{3.11}$$

where v_2 is the partial specific volume of the polymer ($v_2 = V_2/M_2$) and V_1 is the molar volume of the solvent.

The second term in equation (3.11) is the second virial coefficient B_2, i.e.

$$\frac{\pi}{c_2} = RT \left[\frac{1}{M_2} + B_2 + \cdots \right] \tag{3.12}$$

$$B_2 = \left(\frac{v_2^2}{V_1} \right) \left(\frac{1}{2} - \chi \right). \tag{3.13}$$

Note that $B_2 = 0$ when $\chi = 1/2$, i.e. the polymer behaves as ideal in mixing with the solvent. This condition was termed by Flory [2] as the θ-point. Under these conditions, the polymer chains in solution have no repulsion or attraction or they adopt their unperturbed dimension. Clearly when $\chi < 1/2$, B_2 is positive and mixing is nonideal leading to positive deviation (repulsion); this occurs when the polymer chains are in "good" solvent conditions. In contrast, when $\chi > 1/2$, B_2 is negative and mixing is nonideal leading to negative deviation (attraction); this occurs when the polymer chains are in "poor" solvent conditions (precipitation of the polymer may occur under these conditions). Since the polymer solvency depends on temperature, one can also define a theta temperature, θ, at which $\chi = 1/2$.

The function $[(1/2) - \chi]$ can also be expressed in terms of two mixing parameters, an enthalpy parameter κ_1 and an entropy parameter ψ_1, i.e.

$$\left(\frac{1}{2} - \chi\right) = \kappa_1 - \psi_1. \tag{3.14}$$

The θ-temperature can also be defined in terms of κ_1 and ψ_1:

$$\theta = \frac{\kappa_1 T}{\psi_1}. \tag{3.15}$$

Alternatively, one can write

$$\left(\frac{1}{2} - \chi\right) = \psi_1\left(1 - \frac{\theta}{T}\right). \tag{3.16}$$

The θ-temperature is an important parameter which describes a polymer-solvent system. At this temperature a polymer segment will not be able to tell whether it is in contact with another segment or a solvent molecule. The polymer will have the configuration it would have in its own liquid or it said to be in its "unperturbed dimension". In normal solvents the solvent quality increases as the temperature is raised (due to the larger thermal energy) and hence each polymer segment will have a tendency to be in contact with the solvent molecules rather in contact with the polymer's own segments. Thus, the polymer will expand its configuration. On the other hand, at temperatures below the θ-temperature, the polymer segments prefer to be in contact with other polymer segments rather than with the solvent molecules. Thus, the polymer will contract. The θ-temperature is also called the Flory temperature and the solvent or solvent mixture, at this temperature, is called a θ-solvent.

One can define an expansion or contraction parameter α that is given by

$$\alpha = \frac{R_G}{R_G^0}, \tag{3.17}$$

where R_G^0 is the radius of gyration at the θ-temperature.

Although the Flory–Huggins theory is sound in principle, several experimental results cannot be accounted for. For example, it was found that the χ parameter depends on the polymer concentration in solution. Most serious is the fact that many polymer solutions (such as PEO) show phase separation on heating, when the theory predicts that it should happen only on cooling. Another complication arises from specific interaction with the solvent, e.g. hydrogen bonding between polymer and solvent molecules (e.g. with PEO and PVA in water). Also aggregation in solution (lack of complete dissolution) may present another problem.

The Flory–Huggins equation was derived under the assumption that volume changes occurring upon mixing of polymer and solvent are negligible. As discussed above, the free volume concept must be considered. This predicts that near the critical point where phase separation occurs there are no bonds between the molecules constraining the separation of solvent molecules. They are, however, present for the segments of polymer molecules. Hence, upon heating a polymer solution, the increase in free volume for the solvent is large, and much larger than that for the polymer. This difference in free volume creates a large difference in the coefficient of expansion between the polymer and solvent and this leads to phase separation on heating.

Prigonine et al. [3] introduced the concept of the effect of free volume dissimilarity in their theory of solutions and they questioned the association of the χ parameter with heat of mixing only. This led Flory and co-workers [2, 4] to introduce some additional concepts. They pointed out that although a solution of a polymer in a chemically similar solvent would have no contact dissimilarities, it would have them due to the differences in length and size of the polymer chains. This means that the volume changes taking place during mixing, V_M, could not be neglected.

The Flory–Huggins theory only applies to ideal linear polymers. It fails to describe the properties of polymers with a structure having high monomer density, such as two- and three-dimensional branched polymers. It has been found that the theta temperature of star-branched polymers is generally lower than that of linear polymers and it also depends on the length of the arms [2, 4].

Another limitation of the Flory–Huggins theory is its applicability to polymers in aqueous solutions. Firstly, water produces strong specific interaction between the polymer and water molecules, mostly by hydrogen bonds. Secondly, complete dissolution of the polymer in water is questionable due to the aggregation of the molecules. For these reasons, the Flory–Huggins theory has been found not to be applicable for polymers such as poly(ethylene oxide) (PEO) or polyvinyl alcohol (PVA) which are known to form strong hydrogen bonds between the ethylene oxide or vinyl alcohol units and water. For example, colorimetric measurements showed that at low PEO concentrations each EO unit is hydrogen bonded to two water molecules, whereas at high PEO concentrations, the number of hydrogen bonds is significantly reduced to one water molecule per two EO units.

The solution properties of copolymers are much more complicated. This is due to the fact that the two copolymer components A and B behave differently in different solvents. Only when the two components are both soluble in the same solvent do they exhibit similar solution properties. This is the case for example for a nonpolar copolymer in a nonpolar solvent. Dilute solutions of copolymers in solvents that are good for both components exhibit similar behaviour to homopolymer chains, resulting from interactions with solvent molecules and each other. Two possible models for copolymers have been proposed. The first one, called the segregated model, assumes that there are only a few hetero-contacts, and the two different polymer components behave like homopolymer chains. The second model, referred to as the random structure model, takes into account some overlap between different blocks creating hetero-contacts between unlike segments. Several techniques can be applied to study the possible configurations of copolymer components in solution. Most studies have been carried out for block copolymers either in good solvents or θ-solvents for the blocks. For example, light scattering, viscosity and GPC studies were carried out for polystyrene-block-polyisoprene copolymers in methyl isobutyl ketone as the solvent. The solvent is a near θ-solvent for these polymers. The results showed segregated conformation for the block copolymer with a limited number of hetero-contacts. Small angle neutron scattering studies for polystyrene-block-poly(methylmethacrylate) in toluene showed that the poly(methylmethacrylate) block is in a tightly coiled conformation, surrounded by a slightly expanded polystyrene shell.

In a selective solvent, where the medium is a good solvent for one component, say A, and a poor solvent for the second component B, one part of the amphipathic block or graft separates as a distinct phase, while the other stays in solution. The insoluble portion of the amphipathic copolymer will aggregate reversibly to form micelles. It is believed that the polymeric micelles are spherical and have a narrow particle size distribution. Thus micelle formation in block and graft copolymers is analogous to small molecule surface active materials. For graft copolymers in selective solvents the formation of "molecular micelles" has been observed. These resemble particles with an insoluble swollen core surrounded by a sheath of soluble chains. Depending on the concentration of solution, the temperature and the nature of the solvent, multimolecular aggregates are observed. A schematic representation of monomolecular and multimolecular micelles is shown in Fig. 3.2.

The critical micelle concentration (cmc) of these block and graft copolymers is usually very low. Because of the relatively high molecular weight of block and graft copolymers (when compared with simple surfactants), the concentration of these materials has to be very low in order to precisely determine their cmc. Even at the cmc,

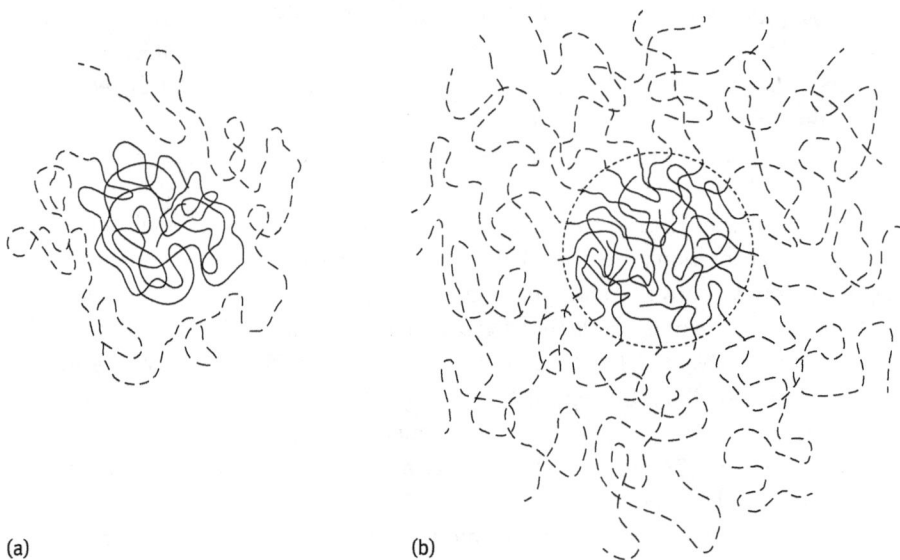

(a) (b)

Fig. 3.2: Schematic representation of monomolecular (a) and multimolecular (b) micelles.

the solution does not only contain micellar aggregates, but also single molecules over a very large range of concentrations. Similar to simple surfactants, the cmc can be determined using surface tension (γ) versus concentration measurements. From plots of γ versus log C one can obtain the cmc. γ decreases with increasing log C until the cmc is reached and then γ remains constant. Since with most block and graft copolymers, the molecular weight is not sufficiently narrow, calculation of C in $mol\,dm^{-3}$ is not straightforward and hence C is usually expressed in wt %.

Several methods may be applied to obtain the micellar size and shape of block and graft copolymers, of which light scattering, small angle X-ray and neutron scattering are probably the most direct. Dynamic light scattering (photon correlation spectroscopy) can also be applied to obtain the hydrodynamic radius of the micelle. This technique is relatively easy to perform when compared with static light scattering, since it does not require rigorous preparation of the samples.

3.3 Viscosity measurements for characterization of a polymer in solution

A convenient way to characterize a polymer in solution at low concentrations is to measure the viscosity using capillary viscometry. The most widely used capillary viscometer is the Ostwald type shown schematically in Fig. 3.3. A variant of the Ostwald viscometer is the Cannon–Fenske type, which is more convenient to use (Fig. 3.3).

In capillary viscometry, one measures the volumetric flow Q $(m^3\,s^{-1})$ and the viscosity η is calculated using Poiseuille's equation [5]:

$$\eta = \frac{\pi R^4 p}{8QL},\tag{3.18}$$

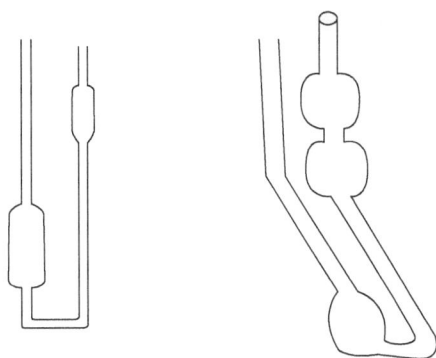

Ostwald viscometer   Cannon–Fenske viscometer

Fig. 3.3: Schematic pictures of Ostwald and Cannon–Fenske viscometers.

where R is the tube radius with length L; p is the pressure drop = hρg (where h is the liquid height with density ρ and g is the acceleration due to gravity).

One usually compares the viscosity of the liquid in question, η_2, with that of a liquid with known viscosity, η_1. In this way, one can measure the flow rates of the two liquids using the same viscometer with a bulb of volume V (the flow rate is simply given by V divided by the time taken for the liquid to flow between the two marks on the viscometer, t_1 and t_2, for the two liquids).

Using equation (3.18) one simply obtains

$$\frac{\eta_1}{\eta_2} = \frac{t_1 \rho_1}{t_2 \rho_2}, \tag{3.19}$$

where t_1 and t_2 are simply measured using a stopwatch (for automatic viscometers two fibre optics are used). Accurate temperature control is necessary (±0.01 °C). The flow time t must also be measured with an accuracy of ±0.01 s.

For measuring the intrinsic viscosity [η] of polymers, a measure that can be used to obtain the molecular weight and solvation of the polymer chains, one measures the relative viscosity η_r as a function of polymer concentration C (in the range 0.01–0.1 %):

$$\eta_r = \frac{\eta_s}{\eta_0}, \tag{3.20}$$

where η_s is the viscosity of the polymer solution and η_0 is that of the solvent.

From η_r one can obtain the reduced viscosity η_{red}:

$$\eta_{red} = (\eta_r - 1). \tag{3.21}$$

From η_{red} one can obtain the specific viscosity η_{sp}:

$$\eta_{sp} = \frac{\eta_{red}}{C}. \tag{3.22}$$

A plot of η_{sp} versus C gives a straight line that can be extrapolated to C = 0 to obtain the intrinsic viscosity [η]. This is illustrated in Fig. 3.4.

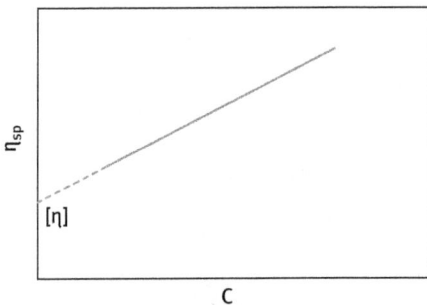

Fig. 3.4: Measurement of intrinsic viscosity of polymers.

From $[\eta]$ one can obtain the molecular weight M using the Mark–Houwink equation

$$[\eta] = KM^{\alpha}, \tag{3.23}$$

where K and α are constants for a particular polymer and solvent (values for many polymer-solvent systems are tabulated in the Polymer Handbook). α is related to the solvency of the medium for the polymer chain. In a good solvent $\alpha > 0.5$; it has values in the range 0.5–0.8. The higher the value of α, the better the solvent for the chain.

3.4 Phase separation of polymer solutions

When dissolving two liquids, the molecules are free to move and hence the entropy for a liquid mixture is large. In contrast, when dissolving a polymer in a solvent, a segment of the polymer is attached to several other segments and hence the entropy of a single segment with a free solvent molecule is much lower when compared with mixing two solvents [6]. Therefore, polymer solutions have lower total entropy and hence are less stable and more prone to phase separation when compared to mixtures of ordinary liquids. This was quantified by Flory and Huggins [2] as discussed above. The Flory–Huggins theory predicts that a solution with a higher molar mass is less stable towards phase when compared with a solution of the same polymer but with a lower molar mass. Hence, when a polymer phase-separates in solution, the high molar mass species will separate out first, leaving the lower molar mass species in solution. This phenomenon is used for fractionation of polymer samples with respect to molar mass.

The temperature at which phase separation occurs for 1 % solution is called the cloud point, due to the increased turbidity of the polymer solution as this temperature is reached. The highest, or lowest, temperature where a phase separation occurs is called the critical temperature and the corresponding polymer concentration is called the critical composition. The question of whether or not a polymer dissolves in a solvent is a matter of balance between the entropy and enthalpy of mixing. The entropy of mixing, which is low for polymers in solution, favours mixing. The enthalpy of mixing, which is a measure of the interaction energy between a segment and solvent molecule when compared with the interaction energy between segments and solvent molecule alone, is positive and it opposes mixing of the two components. In ordinary polymer/solvent systems, the stability with regard to phase separation will decrease with decreasing temperature. In this case, phase separation occurs when the temperature is sufficiently lowered and a concentrated polymer phase will be in equilibrium with a dilute polymer solution. Phase separation in this case can also occur by adding a nonsolvent to the polymer-solvent system.

However, some aqueous polymer solutions, where the polymer chain contains a polyoxyethylene oxide (PEO) chain, show phase separation when the temperature is increased. At very high temperatures, the two-phase region diminishes and the system is homogeneous again. This is due to the increased thermal energy that counteracts other forces at play. Thus, these polymer solutions show a low and upper critical solution temperature.

3.5 Solubility parameter concept for selecting the right solvent for a polymer

The solubility parameter is based on the assumption that "like dissolves like". A polymer is not soluble in certain liquids due to a large difference in the interaction between segments of the polymer and solvent molecules when compared to the interaction energy between segment-segment and solvent-solvent molecules. To achieve some solubility, the segment-solvent interaction energy should be as close as possible to the interaction energy between the segment-segment and solvent-solvent molecules. One method for measuring the interaction energies is to obtain the enthalpy of vaporization, ΔH_{vap}, which reflects the cohesive forces in the liquid. This concept was introduced by Hildebrand [7] who defined the cohesive energy ratio or solubility parameter δ by dividing ΔH_{vap} by the molar volume V:

$$\delta^2 = \frac{\Delta H_{vap}}{V}. \tag{3.24}$$

The units for the solubility parameter are $cal^{1/2} cm^{-3/2}$ or $J^{1/2} m^{-3/2}$ ($= MPa^{1/2}$). Values of solubility parameters for various polymer-solvent systems are given in the book by Barton [8]. In order to find a suitable solvent for a polymer one should first find the solubility parameter of the polymer and then select solvents that have solubility parameters that are close to that of the polymer. For example, polystyrene has a solubility parameter of $9.1 \, cal^{1/2} cm^{-3/2}$ and suitable solvents are cyclohexane ($\delta = 8.2$), benzene ($\delta = 9.2$) and methyl ethyl ketone ($\delta = 9.3$). In contrast, n-hexane ($\delta = 7.3$) and ethanol ($\delta = 12.7$) are nonsolvents, i.e. they do not dissolve polystyrene.

Hansen [9] subdivided δ^2 into three contributions: δ_d^2 (dispersion), δ_p^2 (polar) and δ_h^2 (hydrogen bonding):

$$\delta^2 = \delta_d^2 + \delta_p^2 + \delta_h^2. \tag{3.25}$$

Several theories and computations are available for calculating the above contributions for various polymer and solvent systems.

References

[1] Flory, P. J., "Statistical Mechanics of Chain Molecules", Interscience, NY (1969).

[2] Flory, P. J., "Principles of Polymer Chemistry", Cornell University Press, NY (1953).

[3] Prigogine, I., Trappeniers, N. and Mathos, V., Discussions Faraday Soc., 15, 93 (1953).

[4] Flory, P. J., Ellenson, J. L. and Eichinger, E., Macromolecules, 1, 279 (1968).

[5] Tadros, Th. F., "Rheology of Dispersions", Wiley-VCH, Germany (2010).

[6] Holmberg, K., Jonsson, B., Kronberg, B. and Lindman, B., "Surfactants and Polymers in Aqueous Solution", 2nd ed., John Wiley & Sons, USA (2003).

[7] Hildebrand, J. H., "Solubility of Non-Electrolytes", 2nd edition, Reinhold, New York (1936).

[8] Barton, A. F. M., "Handbook of Solubility Parameters and Other Cohesion Parameters", Boca Raton, Florida, CRC Press, Inc. (1983).

[9] Hansen, C. M., J. Paint Technol., 39, 505 (1967).

4 Adsorption and conformation of polymeric surfactants at interfaces

4.1 Introduction

As mentioned in Chapter 1, polymeric surfactants are essential materials for the preparation of many disperse systems, of which we mention dyestuffs, paper coatings, inks, agrochemicals, pharmaceuticals, personal care products, ceramics and detergents [1–3]. One of the most important applications of polymeric surfactants is in the preparation of oil-in-water (O/W) and water-in-oil (W/O) emulsions as well as solid/liquid dispersions [1–3]. In this case, the hydrophobic portion of the surfactant molecule should adsorb "strongly" at the O/W interface or become dissolved in the oil phase, leaving the hydrophilic components in the aqueous medium, where they become strongly solvated by the water molecules.

The other major application of surfactants is for the preparation of solid/liquid dispersions (usually referred to as suspensions). There are generally two methods for preparing suspensions, referred to as condensation and dispersions methods. In the first case, one starts with molecular units and builds up the particles by a process of nucleation and growth [1]. A typical example is the preparation of polymer lattices. In this case, the monomer (such as styrene or methylmethacrylate) is emulsified in water using an anionic or nonionic surfactant (such as sodium dodecyl sulphate or alcohol ethoxylate). An initiator such as potassium persulphate is added and when the temperature of the system is increased, initiation occurs resulting in the formation of the latex (polystyrene or polymethylmethacrylate). In the dispersion methods, preformed particles (usually powders) are dispersed in an aqueous solution containing a surfactant. The latter is essential for adequate wetting of the powder (both external and internal surfaces of the powder aggregates and agglomerates must be wetted) [1]. This is followed by dispersion of the powder using high speed stirrers and finally the dispersion is "milled" to reduce the particle size to the appropriate range.

For the stabilization of emulsions and suspensions against flocculation, coalescence and Ostwald ripening the following criteria must be satisfied:
(i) Complete coverage of the droplets or particles by the surfactant. Any bare patches may result in flocculation as a result of van der Waals attraction or bridging.
(ii) Strong adsorption (or "anchoring") of the surfactant molecule to the surface of droplet or particle.
(iii) Strong solvation (hydration) of the stabilizing chain to provide effective steric stabilization.
(iv) Reasonably thick adsorbed layer to prevent weak flocculation [1].

DOI 10.1515/9783110487282-005

Most of the above criteria for stability are best served by using a polymeric surfactant. In particular, molecules of the A–B, A–B–A blocks and BA_n (or AB_n) grafts (see Chapter 2) are the most efficient for stabilizing emulsions and suspensions. In this case, the B chain (referred to as the "anchoring" chain) is chosen to be highly insoluble in the medium and with a high affinity to the surface in the case of suspensions, or soluble in the oil in the case of emulsions. The A chain is chosen to be highly soluble in the medium and strongly solvated by its molecules. These block and graft copolymers are ideal for the preparation of concentrated emulsions and suspensions, which are needed in many industrial applications.

In this chapter, I will discuss in detail the adsorption of polymers at interfaces and their conformation. This is key to understanding how polymeric surfactants can be applied as stabilizers for disperse systems as will be discussed in Chapter 5.

4.2 Polymers at interfaces

As mentioned above understanding the adsorption and conformation of polymeric surfactants at interfaces is key to knowing how these molecules act as stabilizers. Most basic ideas on adsorption and conformation of polymers have been developed for the solid/liquid interface [4]. The first theories on polymer adsorption were developed in the 1950; and the 1960s, with extensive developments in the 1970s. The process of polymer adsorption is fairly complicated. In addition to the usual adsorption considerations such as polymer/surface, polymer/solvent and surface/solvent interactions, one of the principal problems to be resolved is the configuration (conformation) of the polymer at the solid/liquid interface. This was recognized by Jenkel and Rumbach in 1951 [5] who found that the amount of polymer adsorbed per unit area of the surface would correspond to a layer more than 10 molecules thick if all the segments of the chain are attached. They suggested a model in which each polymer molecule is attached in sequences separated by bridges which extend into solution. In other words, not all the segments of a macromolecule are in contact with the surface. The segments in direct contact with the surface are termed "trains"; those in between and extended into solution are termed "loops"; the free ends of the macromolecule also extending into solution are termed "tails". This is illustrated in Fig. 4.1 (a) for a homopolymer. Examples of homopolymers that are formed from the same repeating units are poly(ethylene oxide) or poly(vinylpyrrolidone). These homopolymers have little surface activity at the O/W interface, since the homopolymer segments (ethylene oxide or vinylpyrrolidone) are highly water soluble and have little affinity to the interface. However, such homopolymers may adsorb significantly at the S/L interface. Even if the adsorption energy per monomer segment to the surface is small (fraction of kT, where k is the Boltzmann constant and T is the absolute temperature), the total adsorption energy per molecule may be sufficient to overcome the unfavourable entropy loss of the molecule at the S/L interface.

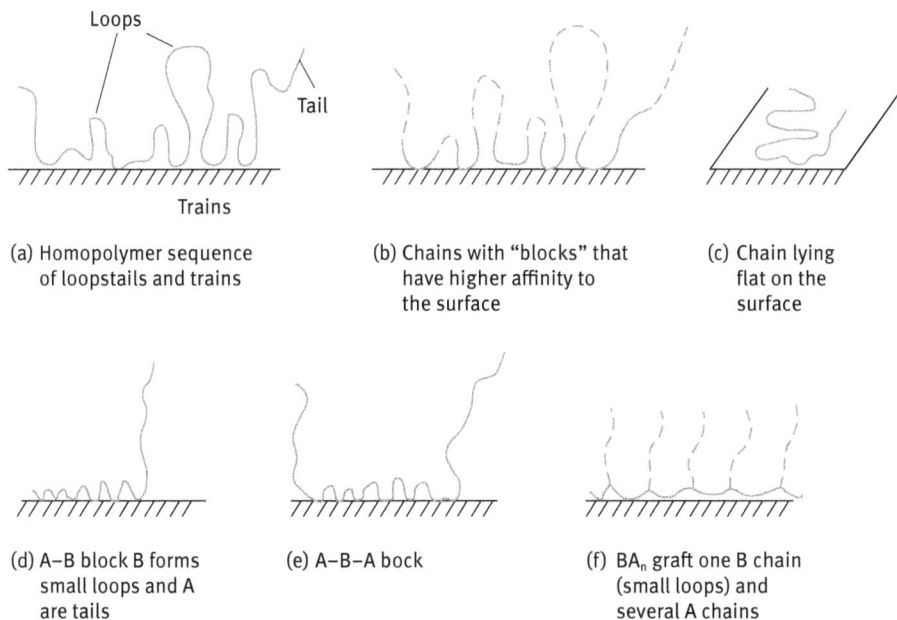

(a) Homopolymer sequence of loopstails and trains

(b) Chains with "blocks" that have higher affinity to the surface

(c) Chain lying flat on the surface

(d) A–B block B forms small loops and A are tails

(e) A–B–A bock

(f) BA$_n$ graft one B chain (small loops) and several A chains

Fig. 4.1: Various conformations of macromolecules on a plane surface.

Clearly, homopolymers are not the most suitable emulsifiers or dispersants. A small variant is to use polymers that contain specific groups that have high affinity to the surface. This is exemplified by partially hydrolysed poly(vinyl acetate) (PVAc), technically referred to as poly(vinyl alcohol) (PVA). The polymer is prepared by partial hydrolysis of PVAc, leaving some residual vinyl acetate groups. Most commercially available PVA molecules contain 4–12 % acetate groups. These acetate groups, which are hydrophobic, give the molecule its amphipathic character. On a hydrophobic surface such as polystyrene, the polymer adsorbs with preferential attachment of the acetate groups on the surface, leaving the more hydrophilic vinyl alcohol segments dangling in the aqueous medium. The configuration of such "blocky" copolymers is illustrated in Fig. 4.1 (b). Clearly, if the molecule is made fully from hydrophobic segments, the chain will adopt a flat configuration as is illustrated in Fig. 4.1 (c).

The most convenient polymeric surfactants are those of the block and graft copolymer type. A block copolymer is a linear arrangement of blocks of variable monomer composition. The nomenclature for a diblock is poly-A-block-poly-B and for a triblock is poly-A-block-poly-B-poly-A. An example of an A–B diblock is polystyrene block-polyethylene oxide and its conformation is represented in Fig. 4.1 (d). One of the most widely used triblock polymeric surfactants are the "Pluronics" (BASF, Germany) which consists of two poly-A blocks of poly(ethylene oxide) (PEO) and one block of poly(propylene oxide) (PPO). Several chain lengths of PEO and PPO are available. More recently, triblocks of PPO–PEO–PPO (inverse Pluronics) have become available for some specific applications. These polymeric triblocks can be applied as emulsifiers or dispersants, whereby the assumption is made that the hydrophobic PPO chain resides at the hydrophobic surface, leaving the two PEO chains dangling in aqueous solution and hence providing steric repulsion. Several other triblock copolymers have been synthesized, although these are of limited commercial availability. Typical examples are triblocks of poly(methyl methacrylate)-block poly(ethylene oxide)-block poly(methyl methacrylate). The conformation of these triblock copolymers is illustrated in Fig. 4.1 (e). An alternative (and perhaps more efficient) polymeric surfactant is the amphipathic graft copolymer consisting of a polymeric backbone B (polystyrene or polymethyl methacrylate) and several A chains ("teeth") such as polyethylene oxide. This graft copolymer is sometimes referred to as a "comb" stabilizer. Its configuration is illustrated in Fig. 4.1 (f).

The polymer/surface interaction is described in terms of adsorption energy per segment χ^s. The polymer/solvent interaction is described in terms of the Flory–Huggins interaction parameter χ. For adsorption to occur, a minimum energy of adsorption per segment χ^s is required. When a polymer molecule adsorbs on a surface, it loses configurational entropy and this must be compensated by an adsorption energy χ^s per segment. This is schematically shown in Fig. 4.2, where the adsorbed amount Γ is plotted versus χ^s The minimum value of χ^s can be very small ($< 0.1 \, \text{kT}$), since a large number of segments per molecule are adsorbed For a polymer with say 100 segments and 10 % of these are in trains, the adsorption energy per molecule now reaches 1 kT (with $\chi^s = 0.1 \, \text{kT}$). For 1000 segments, the adsorption energy per molecule is now 10 kT.

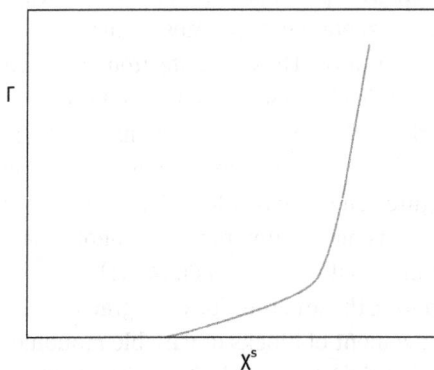

Fig. 4.2: Variation of adsorption amount Γ with adsorption energy per segment χ^s.

As mentioned above, homopolymers are not the most suitable for stabilizing dispersions. For strong adsorption, one needs the molecule to be "insoluble" in the medium and to have strong affinity ("anchoring") to the surface. For stabilization, one needs the molecule to be highly soluble in the medium and strongly solvated by its molecules; this requires a Flory–Huggins interaction parameter less than 0.5. The above opposing effects can be resolved by introducing "short" blocks into the molecule which are insoluble in the medium and have a strong affinity to the surface, as for example partially hydrolysed polyvinyl acetate (88 % hydrolysed, i.e. with 12 % acetate groups), usually referred to as polyvinyl alcohol (PVA),

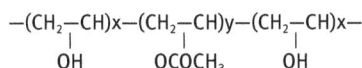

$$-(CH_2-CH)x-(CH_2-CH)y-(CH_2-CH)x-$$
$$\qquad | \qquad\qquad | \qquad\qquad |$$
$$\quad OH \qquad\quad OCOCH_3 \qquad OH$$

As mentioned above, these requirements are better satisfied using A–B, A–B–A and BA$_n$ graft copolymers. B is chosen to be highly insoluble in the medium and it should have high affinity to the surface. This is essential to ensure strong "anchoring" to the surface (irreversible adsorption). A is chosen to be highly soluble in the medium and strongly solvated by its molecules. The Flory–Huggins χ parameter can be applied in this case. For a polymer in a good solvent, χ has to be lower than 0.5; the smaller the χ, value the better the solvent for the polymer chains. Examples of B for hydrophobic particles in aqueous media are polystyrene and polymethylmethacrylate. Examples of A in aqueous media are polyethylene oxide, polyacrylic acid, polyvinyl pyrrolidone and polysaccharides. For nonaqueous media such as hydrocarbons, the A chain(s) could be poly(12-hydroxystearic acid).

For a full description of polymer adsorption one needs to obtain information on the following:

(i) The amount of polymer adsorbed Γ (in mg or mol) per unit area of the particles. It is essential to know the surface area of the particles in the suspension. Nitrogen adsorption on the powder surface may give such information (by application of the BET equation) provided that there will be no change in area on dispersing the particles in the medium. For many practical systems, a change in surface area may occur on dispersing the powder, in which case one has to use dye adsorption to measure the surface area (some assumptions have to be made in this case).

(ii) The fraction of segments in direct contact with the surface, i.e. the fraction of segments in trains p (p = (number of segments in direct contact with the surface)/total number).

(iii) The distribution of segments in loops and tails, $\rho(z)$, which extend in several layers from the surface. $\rho(z)$ is usually difficult to obtain experimentally although recently the application of small angle neutron scattering could obtain such information. An alternative and useful parameter for assessing "steric stabilization" is the hydrodynamic thickness, δ_h (thickness of the adsorbed or grafted polymer layer plus any contribution from the hydration layer). Several methods can be applied to measure δ_h as will be discussed below.

4.3 Theories of polymer adsorption

Two main approaches have been developed to treat the problem of polymer adsorption:

(i) Random walk approach. This is based on Flory's treatment of the polymer chain in solution; the surface was considered as a reflecting barrier.

(ii) Statistical mechanical approach. The polymer configuration was treated as being made of three types of structures, trains, loops and tails, each having a different energy state.

The random walk approach is based on the random walk concept which was originally applied to the problem of diffusion and later adopted by Flory to deduce the conformations of macromolecules in solution. The earliest analysis was by Simha, Frisch and Eirich [6] who neglected excluded volume effects and treated the polymer as a random walk. Basically, the solution was represented by a three-dimensional lattice and the surface by a two-dimensional lattice. The polymer was represented by a realization of a random walk on the lattice. The probabilities of performing steps in different directions were considered to be the same except at the interface which acts as a reflecting barrier. The polymer molecules were, therefore, effectively assumed to be adsorbed with large loops protruding into the solvent and with few segments actually attached to the surface, unless the segment-surface attractive forces were very high. This theory predicts an isotherm for flexible macromolecules that is considerable different from the Langmuir-type isotherm. The number of attached segments per chain is proportional to $n^{1/2}$, where n is the total number of segments. Increasing the molecular weight results in increased adsorption, except for strong chain interaction with the surface.

This approach has been criticized by Silberberg [7] and by Di Marzio [8]. One of the major problems was the use of a reflecting barrier as the boundary condition, which meant overcounting the number of distinguishable conformations. To overcome this problem, Di Marzio and McCrackin [9] used a Monte Carlo method to calculate the average number of contacts of the chain with the surface, the end-to-end length and distribution of segments $\rho(z)$ with respect to the distance z from the surface, as a function of chain length of the polymer and the attractive energy of the surface. The same method was also used by Clayfield and Lumb [10, 11].

The statistical mechanical approach is a more realistic model for the problem of polymer adsorption since it takes into account the various interactions involved. This approach was first used by Silberberg [12] who treated separately the surface layer which contains adsorbed units (trains) and the adjacent layer in solution (loops or tails). The units in each layer were considered to be in two different energy states and partition functions were used to describe the system. The units close to the surface are adsorbed with an internal partition function determined by the short-range forces between the segments and the surface, whereas the units in loops and tails were considered to have an internal partition function equivalent to the segments in the bulk. By equating the chemical potential of macromolecules in the adsorbed state and in bulk solution, the adsorption isotherm could be determined. In this treatment, Silberberg [12] assumed a narrow distribution of loop sizes and predicted small loops for all values of the adsorption energy. Later, the loop size distribution was introduced by Hoeve et al. [13–15] and this theory predicted large loops for small adsorption free energies and small loops and more units adsorbed for larger adsorption free energies when the chains are sufficiently flexible. Most of these theories considered the case of an isolated polymer molecule at an interface, i.e. under conditions of low surface coverage, θ. These theories were extended by Silberberg [16] and Hoeve [17, 18] to take into account the lateral interaction between the molecules on the surface, i.e. high surface coverage. These theories also considered the excluded volume effect, which reduces the number of configurations available for interacting chains near the surface. Excluded volume effects are strongly dependent on the solvent as is the case for chains in solution. Some progress has been made in the analysis of the problem of multilayer adsorption [17, 18].

One feature of an adsorbed layer that is important in the theory of steric stabilization is the actual segment distribution normal to the interface. Hoeve [17, 18] was the first to calculate this quantity for an adsorbed homopolymer of loops and tails, using random flight statistics. He showed that at a distance from the interface corresponding to the thickness of the trains, there was a discontinuity in the distribution. Beyond this the segment density falls exponentially with distance, as shown schematically in Fig. 4.3.

Similarly, Meier [19] developed an equation for the segment density distribution of a single terminally adsorbed tail. Hesselink [20, 21] has developed Meier's theory and given the segment density distribution for single tails, single loops, homopolymers and random copolymers, as illustrated in Fig. 4.4.

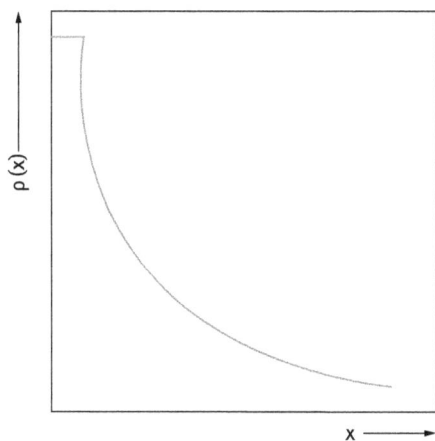

Fig. 4.3: Segment density-distance distribution.

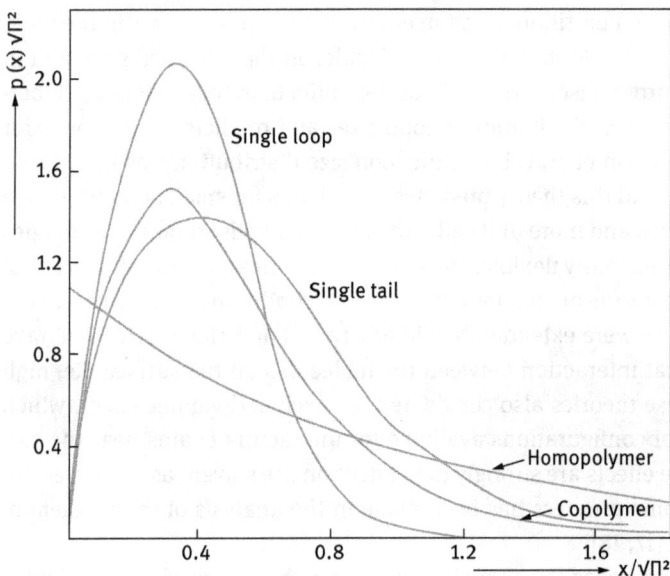

Fig. 4.4: Segment density distribution for single tails, single loops, homopolymers and random copolymers.

A useful model for treating polymer adsorption and configuration was suggested by Roe [22] and Scheutjens and Fleer (SF theory) [23–26] that is referred to as the step weighted random walk approach. In order to be able to describe all possible chain conformations, Scheutjens and Fleer [23–26] used a model of a quasi-crystalline lattice with lattice layers parallel to the surface. Starting from the surface the layers are numbered $I = 1, 2, 3, \ldots, M$, where M is a layer in bulk solution. All the lattice sites within one layer were considered to be energetically equivalent. The probability of finding any lattice site in layer I occupied by a segment was assumed to be equal to the volume fraction ϕ_I in this layer. The conformation probability and the free energy of mixing were calculated with the assumption of random mixing within each layer (the Brag–Williams or mean field approximation). The energy for any segment is only determined by the layer number, and each segment can be assigned a weighting or Boltzmann factor, p_i, which depends only on the layer number. The partition functions were derived for the mixture of free and adsorbed polymer molecules, as well as for the solvent molecules. As mentioned before, all chain conformations were described as step weighted random walks on a quasi-crystalline lattice which extends in parallel layers from the surface; this is schematically shown in Fig. 4.5.

The partition function is written in terms of a number of configurations. These were treated as connected sequences of segments. In each layer, random mixing between segments and solvent molecules was assumed (mean field approximation).

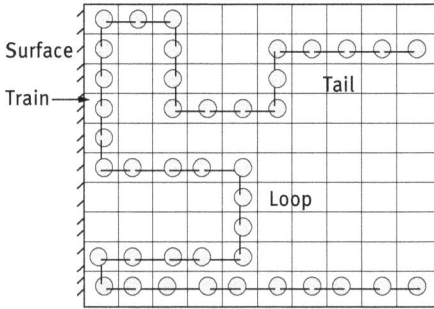

Fig. 4.5: Schematic representation of a polymer molecule adsorbing on a flat surface – quasi-crystalline lattice with segments filling layers that are parallel to the surface (random mixing of segments and solvent molecules in each layer is assumed).

Each step in the random walk was assigned a weighting factor, p_i, that consists of three contributions:

(i)   an adsorption energy, χ^s (which exists only for the segments that are near the surface);

(ii) configurational entropy of mixing (that exists in each layer);

(iii) segment-solvent interaction parameter χ (the Flory–Huggins interaction parameter; note that $\chi = 0$ for an athermal solvent; $\chi = 0.5$ for a θ-solvent).

The adsorption energy gives rise to a Boltzmann factor, $\exp \chi^s$, in the weighting factor for the first layer, provided χ^s is interpreted as the adsorption energy difference (in units of kT) between a segment and a solvent molecule. The configurational entropy for the segment, as a part of the chain, is accounted for in the matrix procedure in which all possible chain conformations are considered. However, the configurational entropy loss of the solvent molecule, going from a layer i with low solvent concentration to the bulk solution with a higher solvent concentration, has to be introduced in p_i. This entropy loss can be written as $\Delta s^0 = k \ln \phi_*^0/\phi_i^0$ per solvent molecule, where ϕ_i^0 and ϕ_*^0 are the solvent volume fractions in layer I and in bulk solution respectively. This change is equivalent to introducing a Boltzmann factor $\exp(-\Delta s^0/k) = \phi_i^0/\phi_*^0$ in the weighting factor p_i. The last contribution stems from the mixing energy of the exchange process. The transfer of a segment from the bulk solution to layer i is accompanied by an energy change (in units of kT) $\chi(\phi_i^0 - \phi_*^0)$, where χ is the Flory–Huggins segment solvent interaction parameter.

Figure 4.6 shows typical adsorption isotherms plotted as surface coverage (in equivalent monolayers) versus polymer volume fraction ϕ_* in bulk solution (ϕ_* was taken to vary between 0 and 10^{-3} which is the normal experimental range). The results in Fig. 4.6 show the effect of increasing the chain length r and the effect of solvency (athermal solvent with $\chi = 0$ and theta solvent with $\chi = 0.5$). The adsorption energy, χ^s, was taken to be the same and equal to 1 kT. When r = 1, θ is very small

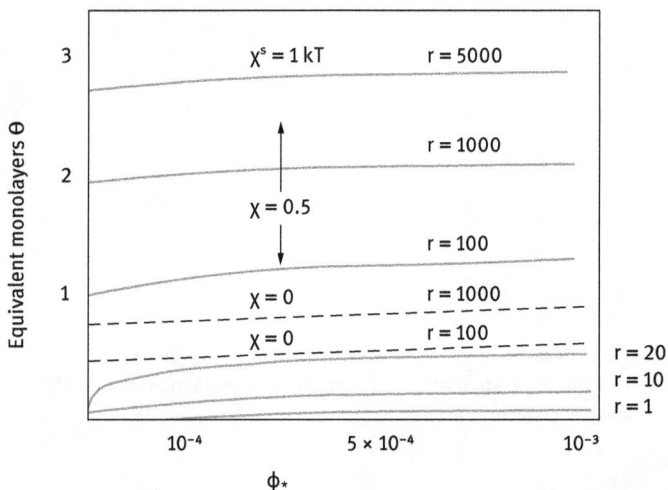

Fig. 4.6: Adsorption isotherms for oligomers and polymers in the dilute region based on SF theory. Full curves $\chi = 0.5$; dashed curves $\chi = 0$.

and the adsorption increases linearly with increasing ϕ^* (Henry's type isotherm). On the other hand, when $r = 10$, the isotherm deviates strongly from a straight line and approaches a Langmuirian type. However, when $r \geq 20$, high affinity isotherms are obtained. This implies that the first added polymer chains are completely adsorbed resulting in extremely low polymer concentration in solution (approaching zero). This explains the irreversibility of adsorption of polymeric surfactants with $r > 100$. The adsorption isotherms with $r = 100$ and above are typical of those observed experimentally for most polymers that are not too polydisperse, i.e. showing a steep rise followed by a nearly horizontal plateau (which only increases a few percent per decade increase of ϕ^*). In these dilute solutions, the effect of solvency is most clearly seen, with poor solvents giving the highest adsorbed amounts. In good solvents θ is much smaller and levels off for long chains to attain an adsorption plateau which is essentially independent of molecular weight.

Some general features of the adsorption isotherms over a wide concentration range can be illustrated by using logarithmic scales for both θ and ϕ_*, which highlight the behaviour in extremely dilute solutions. Such a presentation [25–27] is shown in Fig. 4.7.

These results show a linear Henry region followed by a pseudoplateau region. A transition concentration, ϕ_*^{1c}, can be defined by extrapolation of the two linear parts. ϕ_*^c decreases exponentially with increasing chain length and when $r = 50$, ϕ_*^c is so small (10^{-12}) that it does not appear within the scale shown in Fig. 4.7. With $r = 1000$, ϕ_*^c reaches the ridiculously low value of 10^{-235}. The region below ϕ_*^c is the Henry region where the adsorbed polymer molecules behave essentially as isolated molecules. The representation in Fig. 4.7 also answers the question of reversibility versus

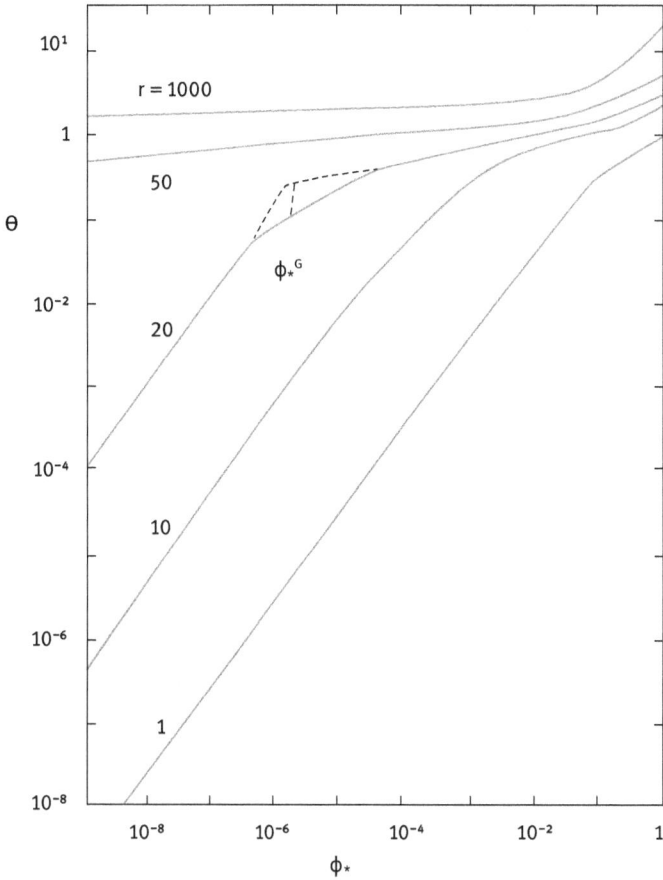

Fig. 4.7: log-log representation of adsorption isotherms of various r values, $\chi^s = 1$; $\chi = 0.5$. Hexagonal lattice.

irreversibility for polymer adsorption. When r > 50, the pseudoplateau region extends down to very low concentration ($\phi_*^c = 10^{-12}$) and this explains why one cannot easily detect any desorption upon dilution. Clearly if such an extremely low concentration can be reached, desorption of the polymer may take place. Thus, the lack of desorption (sometimes referred to as irreversible adsorption) is due to the fact that the equilibrium between adsorbed and free polymer is shifted far in favour of the surface because of the high possible number of possible attachments per chain.

Another point that emerges from the SF theory is the difference in shape between the experimental and theoretical adsorption isotherms in the low concentration region. The experimental isotherms are usually rounded, whereas those predicted from theory are flat. This is accounted for in terms of the molecular weight distribution (polydispersity) which is encountered with most practical polymers. This effect has been explained by Cohen-Stuart et al. [27]. With polydisperse polymers, the larger molecular weight fractions adsorb preferentially over the smaller ones. At low polymer concentrations, nearly all polymer fractions are adsorbed leaving a small fraction of the polymer with the lowest molecular weights in solution. As the polymer concentration is increased, the higher molecular weight fractions displace the lower ones on the surface, which are now released in solution, thus shifting the molecular weight distribution in solution to lower values. This process continues, with a further increase in polymer concentration leading to fractionation whereby the higher molecular weight fractions are adsorbed at the expense of the lower fractions which are released to the bulk solution. However, in very concentrated solutions, monomers adsorb preferentially with respect to polymers and short chains with respect to larger ones. This is due to the fact that in this region, the conformational entropy term predominates over the free energy, disfavouring the adsorption of long chains.

According to the SF theory, the bound fraction p and the direct surface coverage θ_1 depend on the chain length for the same volume fraction. This is illustrated in Fig. 4.8 which shows the adsorbed amount Γ (Fig. 4.8 (a)), surface coverage θ (Fig. 4.8 (b)) and fraction of adsorbed segments p = θ/Γ (Fig. 4.8 (c)) as a function of volume fraction ϕ_*.

In the Henry region ($\phi_* < \phi_*^c$), p is rather high and independent of chain length for $r \geq 20$. In this region the molecules lie nearly flat on the surface, with 87 % of segments in trains. At the other end of the concentration range ($\phi_* = 1$), p is proportional to $r^{-1/2}$. At intermediate concentrations, p is within these two extremes. With increasing polymer concentration, the adsorbed molecules become gradually more extended (lower p) until at very high ϕ_* values they become Gaussian at the interface. In better solvents the direct surface coverage is lower due to the stronger repulsion between the segments. This effect is more pronounced if the surface concentration differs strongly from the solution concentration. If the adsorption is small, the effect of the excluded volume effect (and therefore of χ) is rather weak; the same applies if both the concentrations in the bulk solution and near the surface are high. Both θ_1 and θ decrease with increasing solvent power (decreasing χ) but the effect is stronger for θ than for θ_1 resulting in a higher bound fraction (thus flatter chains) from better solvents at the same solution concentration.

The structure of the adsorbed layer is described in terms of the segment density distribution. As an illustration, Fig. 4.9 shows some calculations using SF theory for loops and tails with r = 1000, $\phi^* = 10^{-6}$ and $\chi = 0.5$. In this example, 38 % of the segments are in trains, 55.5 % in loops and 6.5 % in tails. This theory demonstrates the importance of tails which dominate the total distribution in the outer region.

▸ Fig. 4.8: Adsorbed amount Γ (a), surface coverage θ (b) and fraction of adsorbed segments p = θ/Γ (c) as a function of volume fraction ϕ_*. Full lines for a θ-solvent ($\chi = 0.5$), broken line for an athermal solvent ($\chi = 0$).

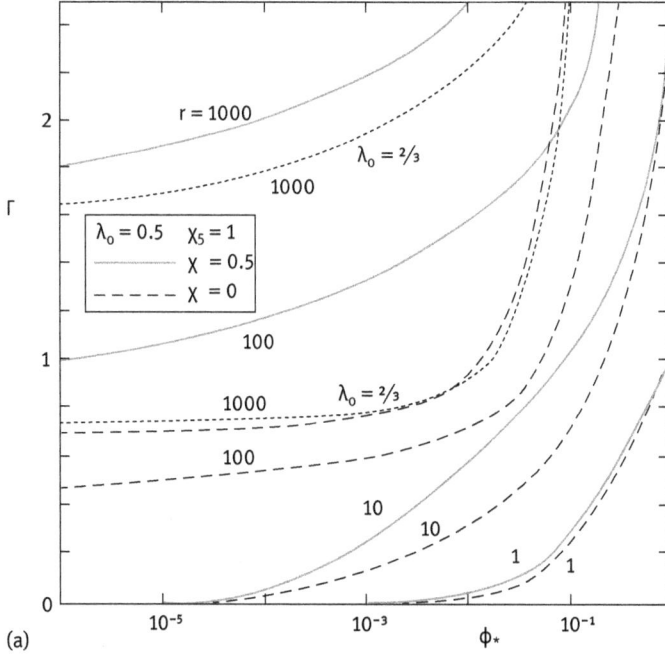

(a)

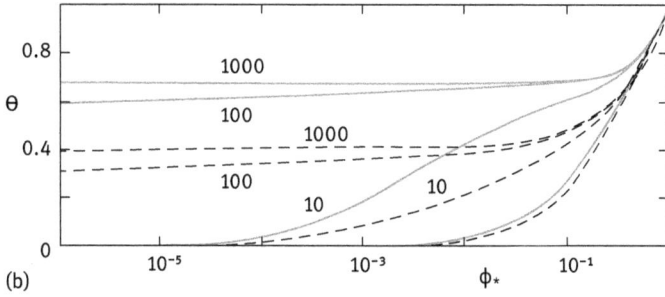

(b)

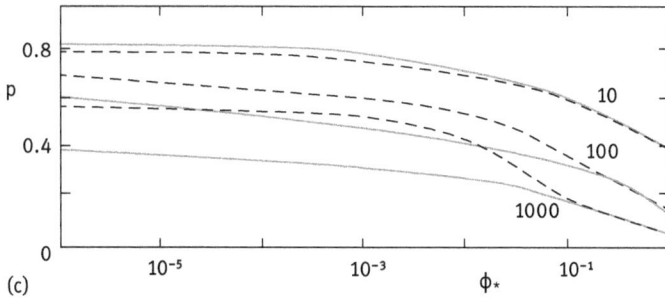

(c)

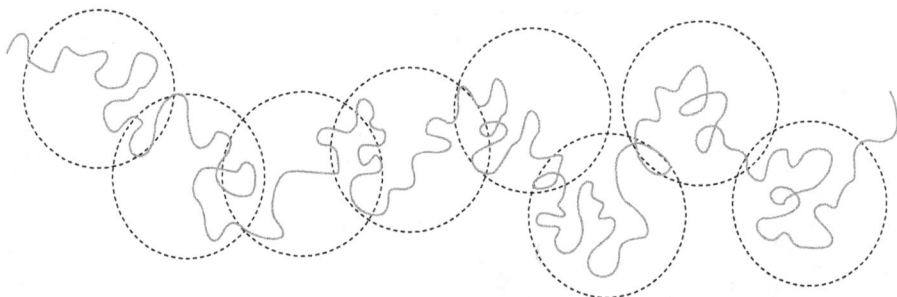

Fig. 4.9: Loop, tail and total segment profile according to the SF theory.

4.4 Scaling theory for polymer adsorption

De Gennes [28] introduced a simple theory for terminally attached polymer chains on a flat surface [29]. He considered the chain to be broken up into "blobs" as illustrated in Fig. 4.10.

Inside the blob the chain is self-avoiding, but the blobs themselves can overlap and are essentially ideal. For g monomers per blob, the blob size $\xi \approx g^{v/2} \approx g^{2/3}$. If the blobs can overlap, then $R \approx \xi^{0.5}$ and

$$R \approx n^{0.5} g^{0.1}. \tag{4.1}$$

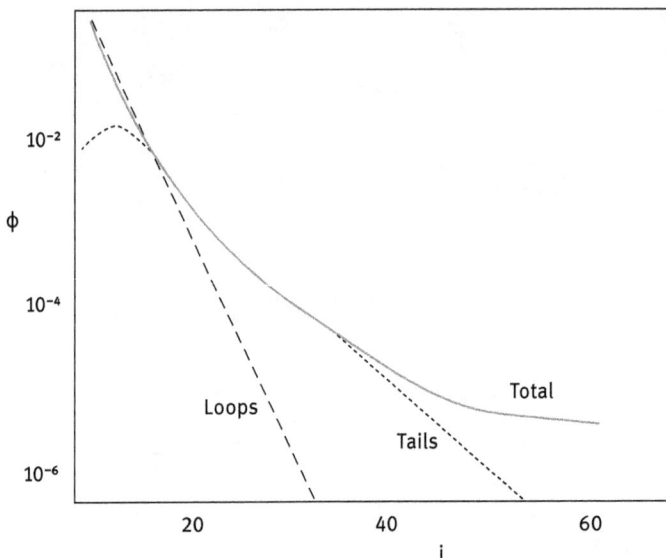

Fig. 4.10: Schematic representation of the blob model according to de Gennes [28].

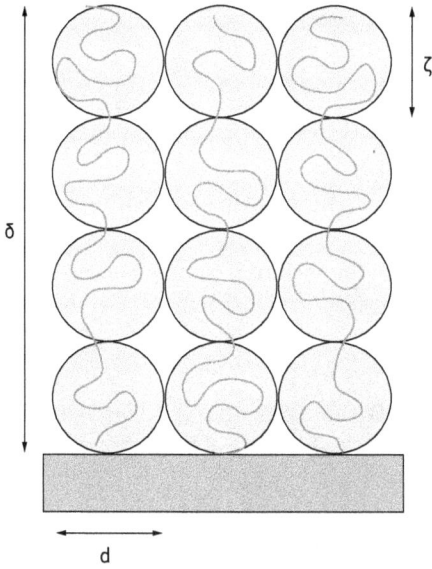

Fig. 4.11: Schematic representation of a terminally attached chain, d, by blobs of size ξ.

For an ideal chain, g = 1 and for a chain with full excluded volume n = g, which
corresponds to the two extreme cases of an ideal and a swollen chain.

A schematic representation of the blob presentation of terminally attached chains
is given in Fig. 4.11.

If each blob contains a self-avoiding walk of g monomers, then the blob size is given
by

$$\xi = g^{1/3}. \tag{4.2}$$

For N monomers the brush length δ is given by

$$\delta = \left(\frac{N}{g}\right)\xi. \tag{4.3}$$

The assumption is made that the blob size is directly related to the grafted amount σ
so that

$$\sigma = \frac{1}{\xi^2}. \tag{4.4}$$

Combining equations (4.2)–(4.4),

$$\delta = N\sigma^{1/3}. \tag{4.5}$$

The above analysis predicts that the brush length is linear in chain length. This is
clearly true for a rod normal to the surface but suggests that chains closely grafted together on a surface are very strongly stretched. A more sophisticated approach which
confirms this result predicts that the brush volume fraction profile is parabolic [30].

4.5 Experimental techniques for studying polymeric surfactant adsorption

As mentioned above, for full characterization of polymeric surfactant adsorption one needs to determine three parameters:

(i)   The adsorbed amount Γ (mg m^{-2} or mol m^{-2}) as a function of equilibrium concentration C_{eq}, i.e. the adsorption isotherm.

(ii) The fraction of segments in direct contact with the surface p (number of segments in trains relative to the total number of segments).

(iii) The segment density distribution $\rho(z)$ or the hydrodynamic adsorbed layer thickness δ_h.

It is important to obtain the adsorption parameters as a function of the important variables of the system:

(i)   Solvency of the medium for the chain, which can be affected by temperature, addition of salt or a nonsolvent. The Flory–Huggins interaction parameter, χ, could be measured separately.

(ii) The molecular weight of the adsorbed polymer.

(iii) The affinity of the polymer to the surface as measured by the value of χ^s, the segment-surface adsorption energy.

(iv) The structure of the polymer; this is particularly important for block and graft copolymers.

4.6 Measurement of the adsorption isotherm

This is by far the easiest to obtain. One measures the polymeric surfactant concentration before ($C_{initial}$, C_1) and after ($C_{equilibrium}$, C_2)

$$\Gamma = \frac{(C_1 - C_2)V}{A},\qquad(4.6)$$

where V is the total volume of the solution and A is the specific surface area (m^2 g^{-1}). It is necessary in this case to separate the particles from the polymer solution after adsorption. This could be carried out by centrifugation and/or filtration. One should make sure that all particles are removed. To obtain this isotherm, one must develop a sensitive analytical technique for determining the polymeric surfactant concentration in the ppm range. It is essential to follow the adsorption as a function of time to determine the time required to reach equilibrium. For some polymer molecules such as polyvinyl alcohol, PVA, and polyethylene oxide, PEO (or blocks containing PEO), analytical methods based on complexation with iodine/potassium iodide or iodine/boric acid potassium iodide have been established. For some polymers with specific functional groups spectroscopic methods may be applied, e.g. UV, IR or fluorescence spectroscopy. A possible method is to measure the change in refractive index of the polymer solution before and after adsorption. This requires very sensitive refractometers. High resolution NMR has been recently applied since the polymer molecules in the adsorbed state are in a different environment than those in the bulk. The chemical shift of functional groups within the chain is different in these two environments. This has the attraction of measuring the amount of adsorption without separating the particles.

4.7 Measurement of the fraction of segments p

The fraction of segments in direct contact with the surface can be directly measured using spectroscopic techniques:
(i)   IR if there is specific interaction between the segments in trains and the surface, e.g. polyethylene oxide on silica from nonaqueous solutions [31, 32].
(ii) Electron spin resonance (ESR); this requires labelling of the molecule [33].
(iii) NMR, pulse gradient or spin ECO NMR. This method is based on the fact that the segments in trains are "immobilized" and hence they have lower mobility than those in loops and tails [34].

An indirect method of determining p is to measure the heat of adsorption ΔH using microcalorimetry [35]. One should then determine the heat of adsorption of a monomer H_m (or molecule representing the monomer, e.g. ethylene glycol for PEO); p is then given by the equation

$$p = \frac{\Delta H}{H_m n},\qquad(4.7)$$

where n is the total number of segments in the molecule.

  The above indirect method is not very accurate and can only be used in a qualitative sense. It also requires very sensitive enthalpy measurements (e.g. using an LKB microcalorimeter).

4.8 Determination of the segment density distribution $\rho(z)$ and adsorbed layer thickness δ_h

The segment density distribution $\rho(z)$ is given by the number of segments parallel to the surface in the z-direction. Three direct methods can be applied for determining adsorbed layer thickness: ellipsometry, attenuated total reflection (ATR) and neutron scattering. Both ellipsometry and ATR [36] depend on the difference between refractive indices between the substrate, the adsorbed layer and bulk solution and they require a flat reflecting surface. Ellipsometry [36] is based on the principle that light undergoes a change in polarizability when it is reflected at a flat surface (whether covered or uncovered with a polymer layer).

The above limitations when using ellipsometry or ATR are overcome by the technique of neutron scattering, which can be applied to both flat surfaces as well as particulate dispersions. The basic principle of neutron scattering is to measure the scattering due to the adsorbed layer, when the scattering length density of the particle is matched to that of the medium (the so-called "contrast-matching" method). Contrast matching of particles and medium can be achieved by changing the isotopic composition of the system (using deuterated particles and mixture of D_2O and H_2O). It was used for measurement of the adsorbed layer thickness of polymers, e.g. PVA or poly(ethylene oxide) (PEO) on polystyrene latex [37]. Apart from obtaining δ, one can also determine the segment density distribution $\rho(z)$.

The above technique of neutron scattering gives clearly a quantitative picture of the adsorbed polymer layer. However, its application in practice is limited since one needs to prepare deuterated particles or polymers for the contrast matching procedure. The practical methods for determining the adsorbed layer thickness are mostly based on hydrodynamic methods. Several methods may be applied to determine the hydrodynamic thickness of adsorbed polymer layers of which viscosity, sedimentation coefficient (using an ultracentrifuge) and dynamic light scattering measurements are the most convenient. A less accurate method is from zeta potential measurements.

The viscosity method [38] depends on measuring the increase in the volume fraction of the particles as a result of the presence of an adsorbed layer of thickness δ_h. The volume fraction of the particles, ϕ, plus the contribution of the adsorbed layers is usually referred to as the effective volume fraction ϕ_{eff}. Assuming the particles behave as hard spheres, then the measured relative viscosity, η_r, is related to the effective volume fraction by the Einstein equation, i.e.

$$\eta_r = 1 + 2.5\phi_{eff}. \tag{4.8}$$

ϕ_{eff} and ϕ are related from simple geometry by

$$\phi_{eff} = \phi\left[1 + \left(\frac{\delta_h}{R}\right)\right]^3, \tag{4.9}$$

where R is the particle radius. Thus, from a knowledge of η_r and ϕ one can obtain δ_h using the equations (4.8) and (4.9).

The sedimentation method depends on measuring the sedimentation coefficient (using an ultracentrifuge) of the particles S_0' (extrapolated to zero concentration) in the presence of the polymer layer [39]. Assuming the particles obey Stokes' law, S_0' is given by the expression

$$S_0' = \frac{(4/3)\pi R^3(\rho - \rho_s) + (4/3)\pi[(R + \delta_h)^3 - R^3](\rho_s^{ads} - \rho_s)}{6\pi\eta(R + \delta_h)}, \tag{4.10}$$

where ρ and ρ_s are the mass density of the solid and solution phase respectively, and ρ^{ads} is the average mass density of the adsorbed layer which may be obtained from the average mass concentration of the polymer in the adsorbed layer.

In order to apply the above methods one should use a dispersion with monodisperse particles with a radius that is not much larger than δ_h. Small model particles of polystyrene may be used.

A relatively simple sedimentation method for determining δ_h is the slow speed centrifugation applied by Garvey et al. [39]. Basically, a stable monodisperse dispersion is slowly centrifuged at low g values ($< 50g$) to form a close-packed (hexagonal or cubic) lattice in the sediment. From a knowledge of ϕ and the packing fraction (0.74 for hexagonal packing), the distance of separation between the center of two particles, R_δ, may be obtained, i.e.

$$R_\delta = R + \delta_h = \left(\frac{0.74 V \rho_1 R^3}{W} \right), \tag{4.11}$$

where V is the sediment volume, ρ_1 is the density of the particles and W their weight.

The most rapid technique for measuring δ_h is photon correlation spectroscopy (PCS) (sometime referred to as quasi-elastic light scattering) which allows one to obtain the diffusion coefficient of the particles with and without the adsorbed layer (D_δ and D respectively). This is obtained from measuring the intensity fluctuation of scattered light as the particles undergo Brownian diffusion [40]. When a light beam (e.g. a monochromatic laser beam) passes through a dispersion, an oscillating dipole is induced in the particles, thus re-radiating the light. Due to the random arrangement of the particles (which are separated by a distance comparable to the wavelength of the light beam, i.e. the light is coherent with the interparticle distance), the intensity of the scattered light will, at any instant, appear as random diffraction or a "speckle" pattern. As the particles undergo Brownian motion, the random configuration of the speckle pattern changes. The intensity at any one point in the pattern will, therefore, fluctuate such that the time taken for an intensity maximum to become a minimum (i.e. the coherence time) corresponds approximately to the time required for a particle to move one wavelength. Using a photomultiplier of active area about the size of a diffraction maximum, i.e. approximately one coherence area, this intensity fluctuation can be measured. A digital correlator is used to measure the photocount or intensity correlation function of the scattered light. The photocount correlation function can be used to obtain the diffusion coefficient, D, of the particles. For monodisperse non-interacting particles (i.e. at sufficient dilution), the normalized correlation function $[g^{(1)}(\tau)]$ of the scattered electric field is given by the equation

$$[g^{(1)}(\tau)] = \exp -(\Gamma \tau) \tag{4.12}$$

where τ is the correlation delay time and Γ is the decay rate or inverse coherence time. Γ is related to D by the equation

$$\Gamma = DK^2, \tag{4.13}$$

where K is the magnitude of the scattering vector that is given by

$$K = \left(\frac{4n}{\lambda_0} \right) \sin \left(\frac{\theta}{2} \right), \tag{4.14}$$

where n is the refractive index of the solution, λ is the wavelength of light in vacuum and θ is the scattering angle.

From D, the particle radius R is calculated using the Stokes–Einstein equation,

$$D = \frac{kT}{6\pi\eta R},\qquad(4.15)$$

where k is the Boltzmann constant and T is the absolute temperature. For a polymer coated particle, R is denoted R_δ which is equal to $R + \delta_h$. Thus, by measuring D_δ and D, one can obtain δ_h. It should be mentioned that the accuracy of the PCS method depends on the ratio of δ_δ/R, since δ_h is determined by difference. Since the accuracy of the measurement is $\pm1\%$, δ_h should be at least 10 % of the particle radius. This method can only be used with small particles and reasonably thick adsorbed layers.

Electrophoretic mobility, u, measurements can also be applied to measure δ_h [41]. From u, the zeta potential ζ, i.e. the potential at the slipping (shear) plane of the particles can be calculated. Adsorption of a polymer causes a shift in the shear plane from its value in the absence of a polymer layer (which is close to the Stern plane) to a value that depends on the thickness of the adsorbed layer. Thus by measuring ζ in the presence (ζ_δ) and absence (ζ) of a polymer layer, one can estimate δ_h. Assuming that the thickness of the Stern plane is Δ, then ζ_δ may be related to the ζ (which may be assumed to be equal to the Stern potential ψ_d) by the equation

$$\tanh\left(\frac{e\psi_\delta}{4kT}\right) = \tanh\left(\frac{e\zeta}{4kT}\right)\exp[-\kappa(\delta_h - \Delta)],\qquad(4.16)$$

where κ is the Debye parameter that is related to electrolyte concentration and valency.

It should be mentioned that the value of δ_h calculated using the above simple equation shows a dependence on electrolyte concentration and hence the method cannot be used in a straightforward manner. Cohen-Stuart et al. [41] showed that the measured electrophoretic thickness δ_e approaches δ_h only at low electrolyte concentrations. Thus, to obtain δ_h from electrophoretic mobility measurements, results should be obtained at various electrolyte concentrations and δ_e should be plotted versus the Debye length $(1/\kappa)$ to obtain the limiting value at high $(1/\kappa)$ (i.e. low electrolyte concentration), which now corresponds to δ_h.

4.9 Examples of the adsorption isotherms of nonionic polymeric surfactants

Figure 4.12 shows the adsorption isotherms for PEO with different molecular weights on PS (at room temperature). It can be seen that the amount adsorbed in mg m^{-2} increases with increasing polymer molecular weight [42]. Figure 4.13 shows the variation of the hydrodynamic thickness δ_h with molecular weight M; δ_h shows a linear

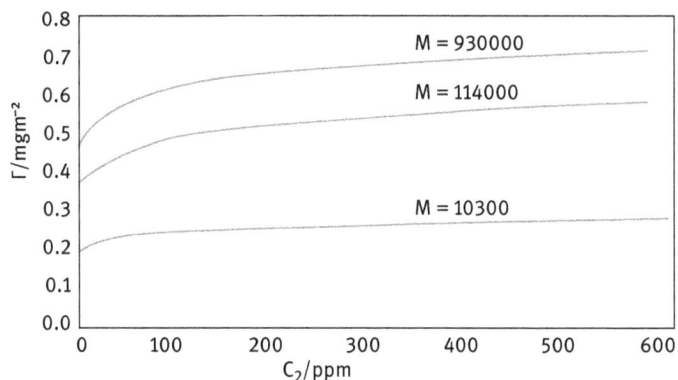

Fig. 4.12: Adsorption isotherms for PEO on PS.

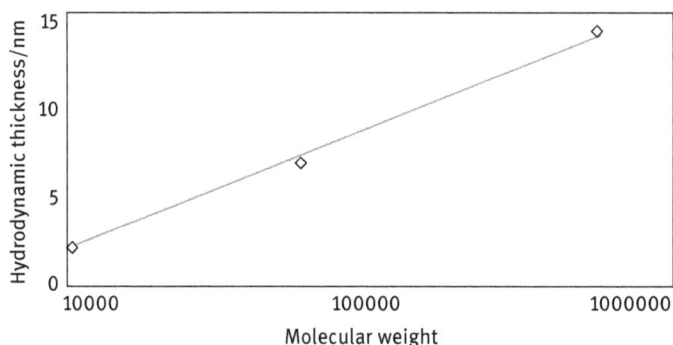

Fig. 4.13: Hydrodynamic thickness of PEO on PS as a function of the molecular weight.

increase with $\log M$; δ_h increases with n, the number of segments in the chain according to

$$\delta_h \approx n^{0.8}. \tag{4.17}$$

Figure 4.14 shows the adsorption isotherms of PVA with various molecular weights on PS latex (at 25 °C) [43]. The polymers were obtained by fractionation of a commercial sample of PVA with an average molecular weight of 45 000. The polymer also contained 12 % vinyl acetate groups. As with PEO, the amount of adsorption increases with increasing M. The isotherms are also of the high affinity type. Γ at the plateau increases linearly with $M^{1/2}$.

The hydrodynamic thickness was determined using PCS and the results are given below.

M	67 000	43 000	28 000	17 000	8 000
δ_h / nm	25.5	19.7	14.0	9.8	3.3

δ_h seems to increase linearly with increasing molecular weight.

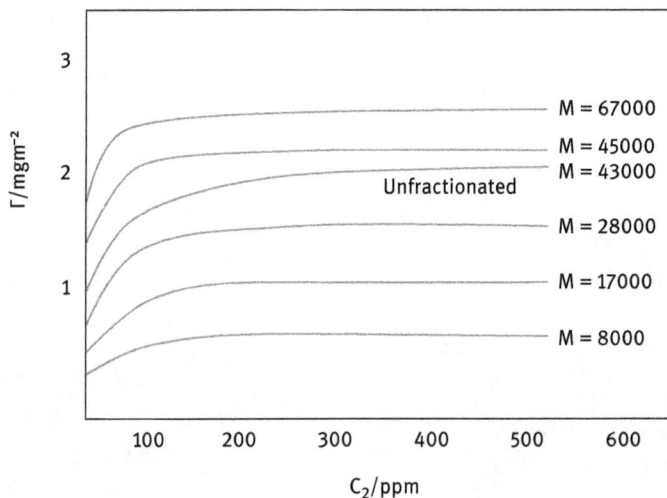

Fig. 4.14: Adsorption isotherms of PVA with different molecular weights on polystyrene latex at 25 °C.

The effect of solvency on adsorption was investigated by increasing the temperature (the PVA molecules are less soluble at higher temperature) or addition of electrolyte (KCl) [44, 45]. The results are shown in Fig. 4.15 and 4.16 for M = 65 100. As can be seen from Fig. 4.15, increasing temperature results in reduced solvency of the medium for the chain (due to breakdown of hydrogen bonds) and this results in an increase in the amount adsorbed. Addition of KCl (which reduces the solvency of the medium for the chain) results in an increase in adsorption (as predicted by theory).

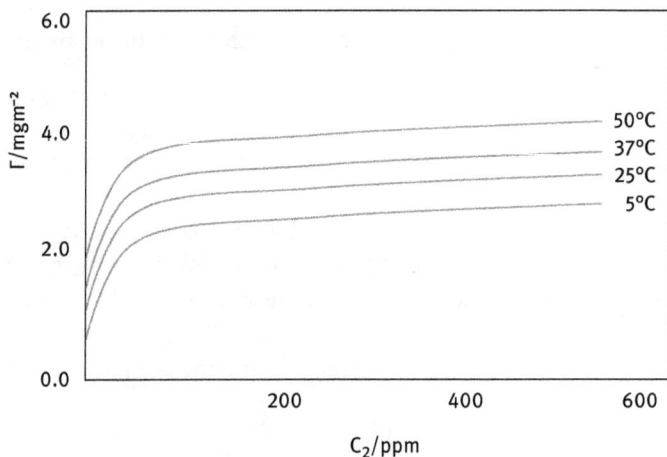

Fig. 4.15: Influence of temperature on adsorption.

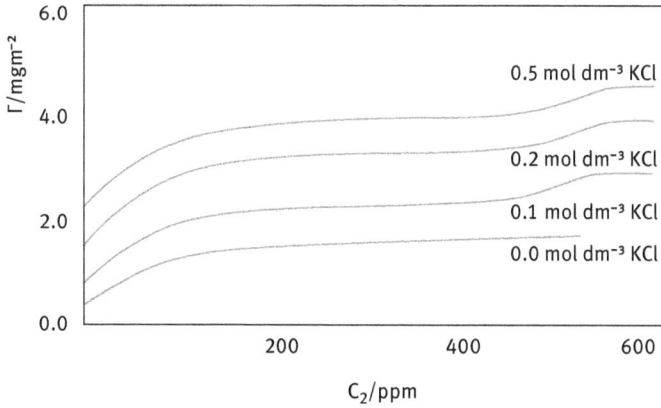

Fig. 4.16: Influence of addition of KCl on adsorption.

The adsorption of block and graft copolymers is more complex since the intimate structure of the chain determines the extent of adsorption [42]. Random copolymers adsorb in an intermediate way to that of the corresponding homopolymers. Block copolymers retain the adsorption preference of the individual blocks. The hydrophilic block (e.g. PEO), the buoy, extends away from the particle surface into the bulk solution, whereas the hydrophobic anchor block (e.g. PS or PPO) provides firm attachment to the surface. Figure 4.17 shows the theoretical prediction of diblock copolymer adsorption according to the Scheutjens and Fleer theory. The surface density σ is plotted versus the fraction of anchor segments v_A. The adsorption depends on the anchor/buoy composition.

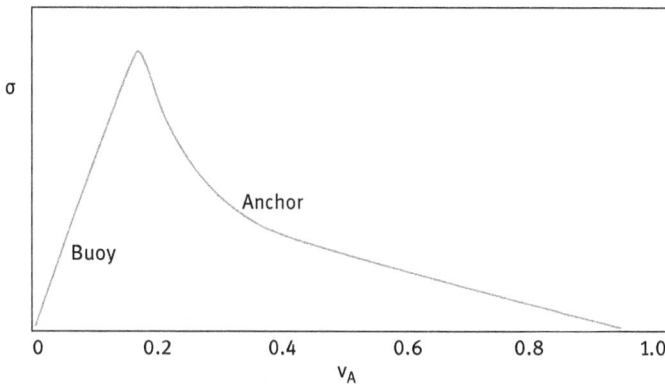

Fig. 4.17: Prediction of adsorption of diblock copolymer.

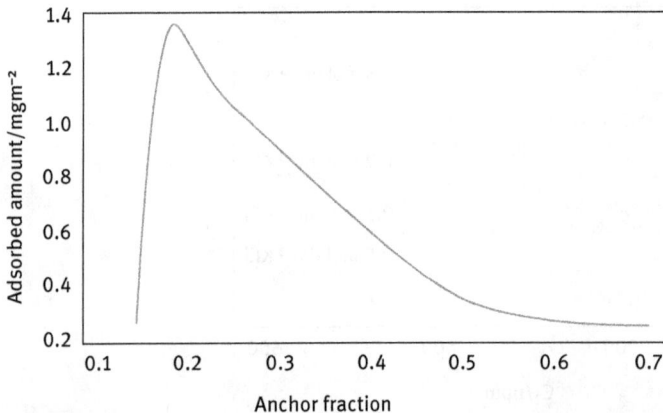

Fig. 4.18: Adsorbed amount (mg m^{-2}) versus fraction of anchor segment for an A–B–A triblock copolymer (PEO–PPO–PEO).

The amount of adsorption is higher than for homopolymers and the adsorbed layer thickness is more extended and dense. For a triblock copolymer A–B–A, with two buoy chains and one anchor chain, the behaviour is similar to that of diblock copolymers. This is shown in Fig. 4.18 for PEO–PPO–PEO block (Pluronic).

4.10 Adsorbed layer thickness results

Figure 4.19 shows a plot of $\rho(z)$ against z for PVA (M = 37 000) adsorbed on deuterated PS latex in D_2O/H_2O.

The results shows a monotonic decay of $\rho(z)$ with distance z from the surface and several regions may be distinguished. Close to the surface (0 < z < 3 nm), the decay in $\rho(z)$ is rapid and assuming a thickness of 1.3 nm for the bound layer, p was calculated to be 0.1, which is in close agreement with the results obtained using NMR measurements. In the middle region, $\rho(z)$ shows a shallow maximum followed by a slow decay which extends to 18 nm, i.e. close to the hydrodynamic layer thickness δ_h of the polymer chain (see below). δ_h is determined by the longest tails and is about 2.5 times the radius of gyration in bulk solution ($\approx$ 7.2 nm). This slow decay of $\rho(z)$ with z at long distances is in qualitative agreement with Scheutjens and Fleer's theory [23] which predicts the presence of long tails. The shallow maximum at intermediate distances suggests that the observed segment density distribution is a summation of a fast monotonic decay due to loops and trains together with the segment density for tails with a maximum density away from the surface. The latter maximum was clearly observed for a sample which had PEO grafted to a deuterated polystyrene latex [37] (where the configuration is represented by tails only).

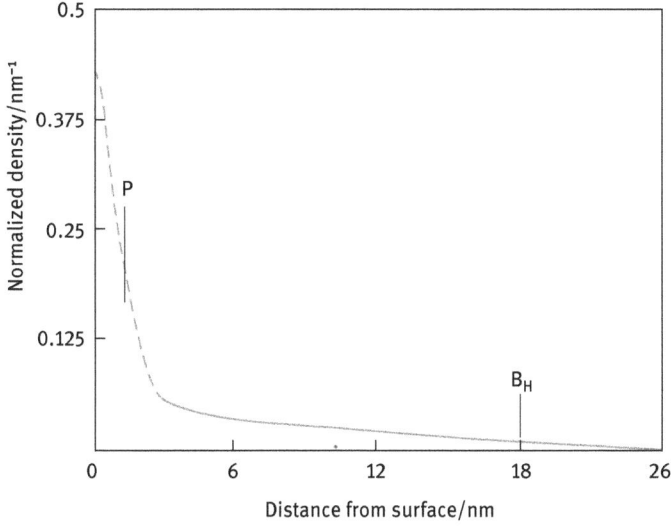

Fig. 4.19: Plot of $\rho(z)$ against z for PVA (M = 37 000) adsorbed on deuterated PS latex in D_2O/H_2O.

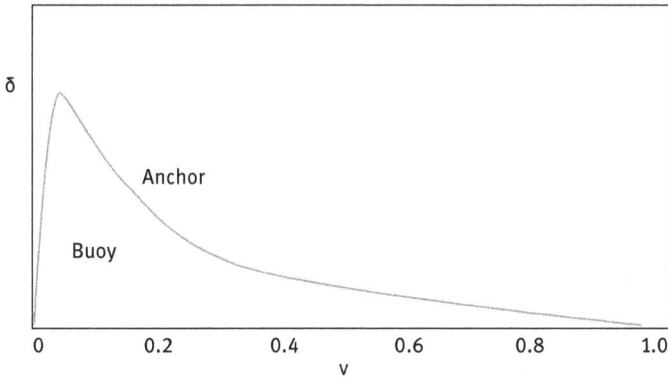

Fig. 4.20: Theoretical predictions of the adsorbed layer thickness for a diblock copolymer.

The hydrodynamic thickness of block copolymers shows different behaviour from that of homopolymers (or random copolymers). Figure 4.20 shows the theoretical prediction for the adsorbed layer thickness δ which is plotted as a function of v_A.

Figure 4.21 shows the hydrodynamic thickness versus fraction of anchor segment for an A–B–A block copolymer of (polyethylene oxide)-poly(propylene oxide)-poly(ethylene oxide) (PEO–PPO–PEO) [42] versus fraction of anchor segment. The theoretical (Scheutjens and Fleer) prediction of adsorbed amount and layer thickness versus fraction of anchor segment are shown in the inserts of Fig. 4.18. When there are two buoy blocks and a central anchor block, as in the above example, the A–B–A block shows similar behaviour to that of an A–B block. However, if there are two anchor blocks and a central buoy block, surface precipitation of the polymer molecule at the particle surface is observed and this is reflected in a continuous increase of adsorption with increasing polymer concentration as has been show for an A–B–A block of PPO–PEO–PPO [42].

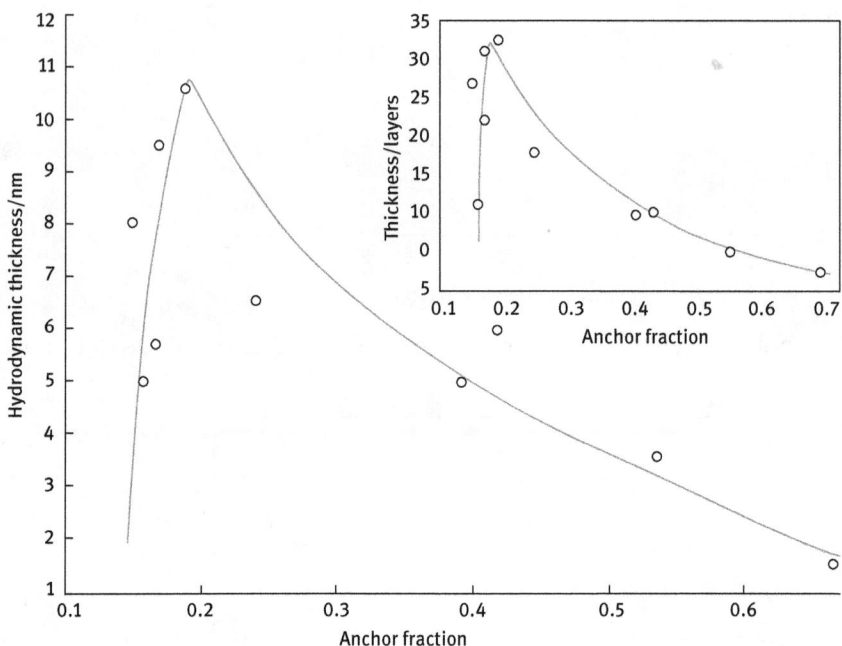

Fig. 4.21: Hydrodynamic thickness versus fraction of anchor segment v_A for PEO–PPO–PEO block copolymer onto polystyrene latex. Insert shows the mean field calculation of thickness versus anchor fraction using the SF theory.

4.11 Kinetics of polymer adsorption

The kinetics of polymer adsorption is a highly complex process. Several distinct processes can be distinguished, each with a characteristic timescale [42]. These processes may occur simultaneously and hence it is difficult to separate them. The first process is the mass transfer of the polymer to the surface, which may be either diffusion or convection. Having reached the surface, the polymer must then attach itself to a surface site, which depends on any local activation energy barrier. Finally, the polymer will undergo large-scale rearrangements as it changes from its solution conformation to a "tail-train-loop" conformation. Once the polymer has reached the surface, the amount of adsorption increases with time. The increase is rapid at the beginning but subsequently slows as the surface becomes saturated. The initial rate of adsorption is sensitive to the bulk polymer solution concentration and molecular weight as well as the solution viscosity. Nevertheless, all the polymer molecules arriving at the surface tend to adsorb immediately. The concentration of unadsorbed polymer around the periphery of the forming layer (the surface polymer solution) is zero and, therefore the concentration of polymer in the interfacial region is significantly greater than the bulk polymer concentration. Mass transport is found to dominate the kinetics of adsorption until 75 % of full surface coverage. At higher surface coverage, the rate of adsorption decreases since the polymer molecules arriving at the surface cannot immediately adsorb. Over time, equilibrium is set up between this interfacial concentration of polymer and the concentration of polymer in the bulk. Given that the adsorption isotherm is of the high affinity type, no significant change in adsorbed amount is expected, even over decades of polymer concentration. If the surface polymer concentration increases toward that of the bulk solution, the rate of adsorption decreases because the driving force for adsorption (the difference in concentration between the surface and bulk solutions) decreases. Adsorption processes tend to be very rapid and an equilibrated polymer layer can form within several 1000 s. However, desorption is a much slower process and this can take several years!

References

[1]   Tadros, Th. F., "Applied Surfactants", Wiley-VCH, Germany (2005).
[2]   Tadros, Th. F., in "Principles of Polymer Science and Technology in Cosmetics and Personal Care", E. D. Goddard and J. V. Gruber (eds.), Marcel Dekker, NY (1999).
[3]   Tadros, Th. F., in "Novel Surfactants", K. Holmberg (ed.), Marcel Dekker, NY (2003).
[4]   Tadros, Th. F., in "Polymer Colloids", R. Buscall, T. Corner and J. F. Stageman (eds.), Applied Sciences, Elsevier, London (1985) p. 105.
[5]   Jenkel E., and Rumbach, R., Z. Elekrochem., 55, 612 (1951).
[6]   Simha, R., Frisch, L. and Eirich, F. R., J. Phys. Chem., 57, 584 (1953).
[7]   Silberberg, A., J. Phys. Chem., 66, 1872 (1962).
[8]   Di Marzio, E. A., J. Chem. Phys., 42, 2101 (1965).
[9]   Di Marzio, E. A. and McCrakin, F. L., J. Chem. Phys., 43, 539 (1965).
[10] Clayfield, E. J. and Lumb, E. C., J. Colloid Interface Sci., 22, 269, 285 (1966).
[11] Clayfield, E. J. and Lumb, E. C., Macromolecules, 1, 133 (1968).
[12] Silberberg, A., J. Chem. Phys., 48, 2835 (1968).
[13] Hoeve, C. A., Di Marzio, E. A. and Peyser, P., J. Chem. Phys., 42, 2558 (1965).
[14] Hoeve, C. A. J., J. Chem. Phys., 44, 1505 (1965); 47, 3007 (1966).
[15] Hoeve, C. A., J. Polym. Sci., 30, 361 (1970); 34, 1 (1971).
[16] Silberberg, A., J. Colloid Interface Sci., 38, 217 (1972).
[17] Hoeve, C. A. J., J. Chem. Phys., 44, 1505 (1965).
[18] Hoeve, C. A. J., J. Chem. Phys., 47, 3007 (1966).
[19] Meier, D. J., J. Phys. Chem., 71, 1861 (1965).
[20] Hesselink, F. Th., J. Phys. Chem., 73, 3488 (1969).
[21] Hesselink, F. Th., J. Phys. Chem. 75, 65 (1971).

[22] Roe, R. J., J. Chem. Phys., 60, 4192 (1974).

[23] Scheutjens, J. M. H. M. and Fleer, G. J, J. Phys. Chem., 83, 1919 (1979).

[24] Scheutjens, J. M. H. M. and Fleer, G. J., J. Phys. Chem., 84, 178 (1980).

[25] Scheutjens, J. M. H. M. and Fleer, G. J., Adv. Colloid Interface Sci., 16, 341 (1982).

[26] Fleer, G. J., Cohen-Stuart, M. A., Scheutjens, J. M. H. M. Cosgrove, T. and Vincent, B., "Polymers at Interfaces", Chapman and Hall, London (1993).

[27] Cohen-Stuart, M. A., Scheutjens, J. M. H. M. and Fleer, G. J., J. Polym. Sci., Polym. Phys. Ed., 18, 559 (1980).

[28] de Gennes, P. G., "Scaling Concepts of Polymer Physics", Cornell University Press, Isthaca (1979).

[29] Cosgrove, T., "Polymers at Interfaces", Blackwell, Oxford (2005), Chapter 7.

[30] Milner, S. T., Science, 251, 905 (1991).

[31] Killmann, E., Eisenlauer, E. and Horn, M., J. Polymer Sci. Polymer Symposium, 61, 413 (1977).

[32] Fontana, B. J. and Thomas, J. R., J. Phys. Chem., 65, 480 (1961).

[33] Robb, I. D. and Smith, R. Eur. Polym. J., 10 1005 (1974).

[34] Barnett, K. G., Cosgrove, T., Vincent, B., Burgess, A., Crowley, T. L., Kims, J., Turner J. D. and Tadros, Th. F., Disc. Faraday Soc. 22, 283 (1981).

[35] Cohen-Stuart, M. A., Fleer, G. J. and Bijesterbosch, B., J. Colloid Interface Sci., 90, 321 (1982).

[36] Abeles, F., in "Ellipsometry in the Measurement of Surfaces and Thin Films", E. Passaglia, R. R. Stromberg and J. Kruger (eds.), Nat. Bur. Stand. Misc. Publ., 256, 41 (1964).

[37] Cosgrove, T., Crowley T. L. and Ryan, T., Macromolecules, 20, 2879 (1987).

[38] Einstein, A. "Investigations on the Theory of the Brownian Movement", Dover, NY (1906).

[39] Garvey, M. J., Tadros, Th. F. and Vincent, B., J. Colloid Interface Sci., 55, 440 (1976).

[40] Pusey, P. N., in "Industrial Polymers: Characterisation by Molecular Weights", J. H. S. Green and R. Dietz (eds.), London, Transcripta Books (1973).

[41] Cohen-Stuart, M. A., and Mulder, J. W., Colloids and Surfaces, 15, 49 (1985).

[42] Obey, T. M. and Griffiths, P. C., in "Principles of Polymer Science and Technology in Cosmetics and Personal Care", E. D. Goddard and J. V. Gruber (eds.), Marcel Dekker, NY (1999), Chapter 2.

[43] Garvey, M. J., Tadros, Th. F. and Vincent, B., J. Colloid Interface Sci., 49, 57 (1974).

[44] van den Boomgaard, Th., King, T. A., Tadros, Th. F., Tang, H. and Vincent, B., J. Colloid Interface Sci., 61, 68 (1978).

[45] Tadros, Th. F. and Vincent, B., J. Colloid Interface Sci., 72, 505 (1978).

5 Stabilization of disperse systems using polymeric surfactants

5.1 Introduction

The use of natural and synthetic polymers (referred to as polymeric surfactants) for the stabilization and destabilization of solid/liquid and liquid/liquid dispersions plays an important role in industrial applications, such as in paints, cosmetics, agrochemicals, ceramics, etc. Polymers are particularly important for the preparation of concentrated dispersions, i.e. at high volume fraction ϕ of the disperse phase,

$$\phi = \text{(volume of all particles)/(total volume of dispersion)} \qquad (5.1)$$

Polymers are also essential for the stabilization of nonaqueous dispersions, since in this case electrostatic stabilization is not possible (due to the low dielectric constant of the medium). To understand the role of polymers in dispersion stability, it is essential to consider the adsorption and conformation of the macromolecule at the solid/liquid interface and this was discussed in detail in Chapter 9. Polymers and polyelectrolytes are also used for destabilizing suspensions, e.g. for solid/liquid separation. In this chapter, I will cover the topic of interaction between particles containing adsorbed polymeric surfactants and the theory of steric stabilization [1–3].

5.2 Interaction between particles or droplets containing adsorbed polymer layers

When two particles, each with a radius R and containing an adsorbed polymer layer with a hydrodynamic thickness δ_h, approach each other to a surface-surface separation distance h that is smaller than $2\delta_h$, the polymer layers interact with each other resulting in two main situations [1–3]:
(i)   The polymer chains may overlap each other.
(ii) The polymer layer may undergo some compression.

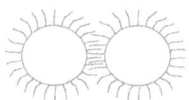

Interpenetration without   Compression without
compression   interpenetration

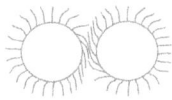

Fig. 5.1: Schematic representation of the interaction between particles containing adsorbed polymer layers.

DOI 10.1515/9783110487282-006

In both cases, there will be an increase in the local segment density of the polymer chains in the interaction region. This is schematically illustrated in Fig. 5.1. The real situation is perhaps in between the above two cases, i.e. the polymer chains may undergo some interpenetration and some compression.

Provided the dangling chains (the A chains in A–B, A–B–A block or BA_n graft copolymers) are in a good solvent, this local increase in segment density in the interaction zone will result in strong repulsion as a result of two main effects:

(i) An increase in the osmotic pressure in the overlap region as a result of the unfavourable mixing of the polymer chains, when these are in good solvent conditions. This is referred to as osmotic repulsion or mixing interaction and it is described by a free energy of interaction G_{mix}.

(ii) A reduction of the configurational entropy of the chains in the interaction zone; this entropy reduction results from the decrease in the volume available for the chains when these are either overlapped or compressed. This is referred to as volume restriction interaction, entropic or elastic interaction and it is described by a free energy of interaction G_{el}.

The combination of G_{mix} and G_{el} is usually referred to as the steric interaction free energy, G_s, i.e.

$$G_s = G_{mix} + G_{el}. \tag{5.2}$$

The sign of G_{mix} depends on the solvency of the medium for the chains. If in a good solvent, i.e. the Flory–Huggins interaction parameter χ is less than 0.5, then G_{mix} is positive and the mixing interaction leads to repulsion (see below). In contrast, if $\chi > 0.5$ (i.e. the chains are in a poor solvent condition), G_{mix} is negative and the mixing interaction becomes attractive. G_{el} is always positive and hence in some cases one can produce stable dispersions in a relatively poor solvent (enhanced steric stabilization).

5.2.1 Mixing interaction G_{mix}

This results from the unfavourable mixing of the polymer chains when these are in good solvent conditions. This is schematically shown in Fig. 5.2. Consider two spherical particles with the same radius and each containing an adsorbed polymer layer with thickness δ. Before overlap, one can define in each polymer layer a chemical potential for the solvent μ_i^α and a volume fraction for the polymer in the layer ϕ_2. In the overlap region (volume element dV), the chemical potential of the solvent is reduced to μ_i^β. This results from the increase in polymer segment concentration in this overlap region.

In the overlap region, the chemical potential of the polymer chains is now higher than in the rest of the layer (with no overlap). This amounts to an increase in the osmotic pressure in the overlap region; as a result solvent will diffuse from the bulk to the overlap region, thus separating the particles and hence a strong repulsive energy

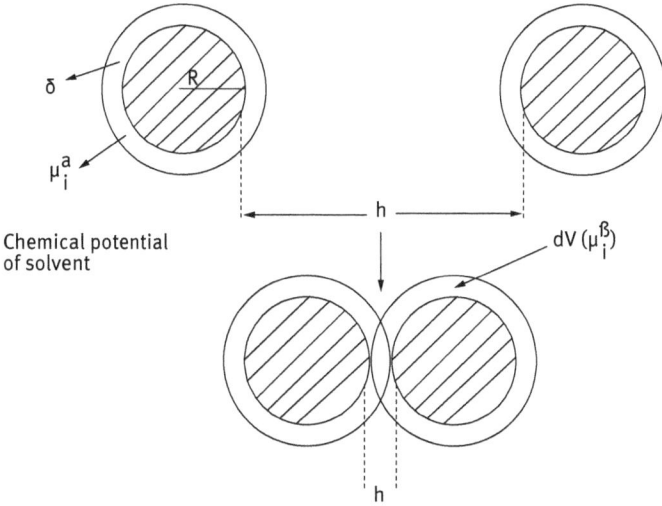

Fig. 5.2: Schematic representation of polymer layer overlap.

arises from this effect. The above repulsive energy can be calculated by considering the free energy of mixing of two polymer solutions, as for example treated by Flory and Krigbaum [4]. The free energy of mixing is given by two terms:

(i)   An entropy term that depends on the volume fraction of polymer and solvent.

(ii) An energy term that is determined by the Flory-Huggins interaction parameter χ:

$$\delta(G_{mix}) = kT(n_1 \ln \phi_1 + n_2 \ln \phi_2 + \chi n_1 \phi_2), \qquad (5.3)$$

where n_1 and n_2 are the number of moles of solvent and polymer with volume fractions ϕ_1 and ϕ_2, k is the Boltzmann constant and T is the absolute temperature.

The total change in free energy of mixing for the whole interaction zone, V, is obtained by summing over all the elements in V:

$$G_{mix} = \frac{2kTV_2^2}{V_1}v_2 \left(\frac{1}{2} - \chi\right) R_{mix}(h), \qquad (5.4)$$

where V_1 and V_2 are the molar volumes of solvent and polymer respectively, v_2 is the number of chains per unit area and $R_{mix}(h)$ is a geometric function that depends on the form of the segment density distribution of the chain normal to the surface, $\rho(z)$. k is the Boltzmann constant and T is the absolute temperature.

Using the above theory one can derive an expression for the free energy of mixing of two polymer layers (assuming a uniform segment density distribution in each layer) surrounding two spherical particles as a function of the separation distance h between the particles. The expression for G_{mix} is [5]

$$\frac{G_{mix}}{kT} = \left(\frac{2V_2^2}{V_1}\right) v_2^2 \left(\frac{1}{2} - \chi\right) \left(\delta - \frac{h}{2}\right)^2 \left(3R + 2\delta + \frac{h}{2}\right). \tag{5.5}$$

k is the Boltzmann constant, T is the absolute temperature, V_2 is the molar volume of polymer, V_1 is the molar volume of solvent and v_2 is the number of polymer chains per unit area.

The sign of G_{mix} depends on the value of the Flory–Huggins interaction parameter χ: if $\chi < 0.5$, G_{mix} is positive and the interaction is repulsive; if $\chi > 0.5$, G_{mix} is negative and the interaction is attractive; if $\chi = 0.5$, $G_{mix} = 0$ and this defines the θ-condition.

5.2.2 Elastic interaction G_{el}

This arises from the loss in configurational entropy of the chains on the approach of a second particle. As a result of this approach, the volume available for the chains becomes restricted, resulting in loss of the number of configurations. This can be illustrated by considering a simple molecule, represented by a rod that rotates freely in a hemisphere across a surface (Fig. 5.3). When the two surfaces are separated by an infinite distance ∞ the number of configurations of the rod is $\Omega(\infty)$, which is proportional to the volume of the hemisphere. When a second particle approaches to a distance h such that it cuts the hemisphere (losing some volume), the volume available to the chains is reduced and the number of configurations become $\Omega(h)$, which is less than $\Omega(\infty)$. For two flat plates, G_{el} is given by the following expression [6]:

$$\frac{G_{el}}{kT} = -2v_2 \ln\left[\frac{\Omega(h)}{\Omega(\infty)}\right] = -2v_2 R_{el}(h), \tag{5.6}$$

where $R_{el}(h)$ is a geometric function whose form depends on the segment density distribution. It should be stressed that G_{el} is always positive and could play a major role in steric stabilization. It becomes very strong when the separation distance between the particles becomes comparable to the adsorbed layer thickness δ.

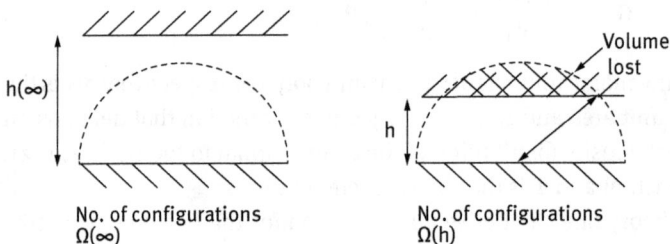

Fig. 5.3: Schematic representation of configurational entropy loss on approach of a second particle.

5.2.3 Total energy of interaction

Combining G_{mix} and G_{el} with G_A gives the total energy of interaction G_T (assuming there is no contribution from any residual electrostatic interaction) [7], i.e.,

$$G_T = G_{mix} + G_{el} + G_A. \tag{5.7}$$

A schematic representation of the variation of G_{mix}, G_{el}, G_A and G_T with surface-surface separation distance h is shown in Fig. 5.4. G_{mix} increases very sharply with decreasing h when $h < 2\delta$. G_{el} increases very sharply with decreasing h when $h < \delta$. G_T versus h shows a minimum, G_{min}, at separation distances comparable to 2δ. When $h < 2\delta$, G_T shows a rapid increase with decreasing h. The depth of the minimum depends on the Hamaker constant A, the particle radius R and adsorbed layer thickness δ. G_{min} increases with increasing A and R. At a given A and R, G_{min} increases with decreasing δ (i.e. with decreasing molecular weight, M_w, of the stabilizer). This is illustrated in Fig. 5.5 which shows the energy-distance curves as a function of δ/R. The larger the value of δ/R, the smaller the value of G_{min}. In this case the system may approach thermodynamic stability as is the case with nanodispersions.

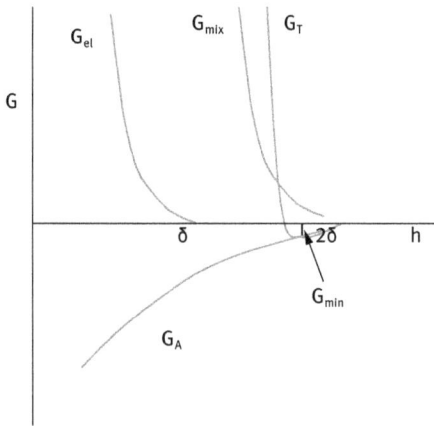

Fig. 5.4: Energy-distance curves for sterically stabilized systems.

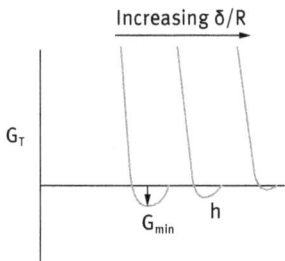

Fig. 5.5: Variation of G_{min} with δ/R.

5.2.4 Criteria for effective steric stabilization

(i) The particles should be completely covered by the polymer (the amount of polymer should correspond to the plateau value). Any bare patches may cause flocculation either by van der Waals attraction (between the bare patches) or by bridging flocculation (whereby a polymer molecule will become simultaneously adsorbed on two or more particles).

(ii) The polymer should be strongly "anchored" to the particle surfaces, to prevent any displacement during particle approach. This is particularly important for concentrated suspensions. For this purpose A–B, A–B–A block and BA$_n$ graft copolymers are the most suitable where the chain B is chosen to be highly insoluble in the medium and has a strong affinity to the surface. Examples of B groups for hydrophobic particles in aqueous media are polystyrene and polymethylmethacrylate.

(iii) The stabilizing chain A should be highly soluble in the medium and strongly solvated by its molecules. Examples of A chains in aqueous media are poly(ethylene oxide) and poly(vinyl alcohol).

(iv) δ should be sufficiently large (> 5 nm) to prevent weak flocculation.

5.3 Measurement of steric repulsion between adsorbed layers of polymeric surfactants

Two techniques can be applied for direct measurement of interaction forces between macroscopic bodies containing adsorbed layers of polymeric surfactants and these are summarized below.

5.3.1 Surface force methods

In this method, the energy E(D)-distance curves for adsorbed layers of polymeric surfactants (that are physically adsorbed on smooth mica sheets) are obtained using the surface force apparatus originally described by Israelachvili [8–11]. It consists of measuring the forces between mica sheets with molecularly smooth surfaces with a cross cylinder geometry. The mica sheets are partially silvered on the reverse side so that light interferometry can be used to measure the surface separation. The force between the surfaces is measured by monitoring the displacement of a leaf spring to which one of the sheets is attached. Initially measurements are made in the presence of electrolyte solution (10^{-2} mol dm^{-3} KNO$_3$) and then in the same electrolyte solution but with the mica sheets containing the adsorbed layers. In this manner one can subtract the double layer interaction from the total interaction to obtain the contribution from steric interaction. For this purpose a graft copolymer consisting of poly(methylmethacrylate) backbone to which several poly(ethylene oxide) are grafted, i.e. PMMA/MA(PEO)$_n$, is used.

The forces between the mica surfaces F(D) bearing the copolymer layers are converted to the interaction potential energy E(D) between flat surfaces using the Deryaguin approximation for cross cylinders [11]:

$$E(D) = \frac{F(D)}{2\pi a},\qquad(5.8)$$

where D is the surface separation distance and a is the cylinder radius. Figure 5.6 shows the energy-distance curve for mica sheets covered by the graft copolymer PMMA/MA(PEO)$_n$. The figure shows a monotonic and approximately exponential decrease of E(D) with increasing separation distance D. The exponential decay makes it difficult to assess precisely the point at which the interaction begins. It falls below the detection limit of the instrument at ≈ 25 nm.

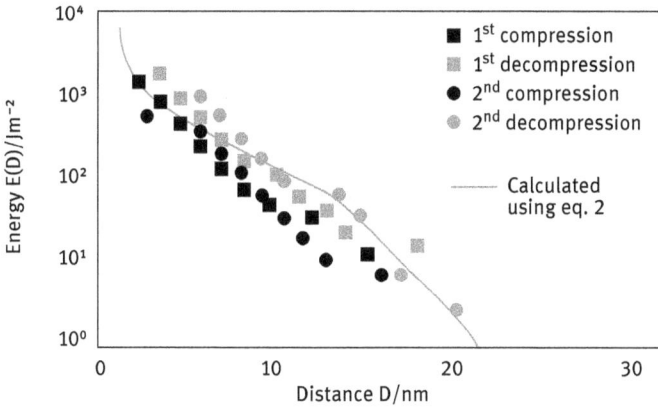

Fig. 5.6: Interaction energy E(D) versus separation distance D.

The energy of interaction between polymer layers can be calculated using de Gennes scaling theory [12],

$$E(D) = \frac{\beta kT}{s^3}\left[\frac{(2L)^{2.25}}{1.25(D)^{1.25}} + \frac{D^{1.75}}{1.75(2L)^{0.75}}\right] - \left[\frac{2L}{1.25} + \frac{2L}{1.75}\right],\qquad(5.9)$$

where L is the stabilizer thickness on each surface (taken to be 12.5 nm, i.e. half the separation distance D at which E(D) begins to increase with a further decrease in D). S is the distance between side chains, k is the Boltzmann constant and T is the absolute temperature. The solid line in Fig. 5.6 shows the theoretical calculations based on equation (5.9). Agreement between theory and experiment is satisfactory.

The high frequency modulus of latex dispersions containing adsorbed layers of PMMA/MA(PEO)$_n$ graft copolymer was also measured. For this purpose latex dispersions were prepared using surfactant-free emulsion polymerization. The particle diameter was 330 ± 9 nm. Latex dispersions containing adsorbed PMMA/MA(PEO)$_n$ graft copolymer were prepared at various latex volume fractions.

The storage modulus G$'$ of each dispersion was determined at low amplitudes (within the linear viscoelastic region) as a function of frequency. These measurements were obtained using dynamic (oscillatory) techniques. A plot of G$'$ versus frequency ω (rad s^{-1}) allows one to obtain the plateau value G$'_\infty$, i.e. the high frequency modulus which can be related to potential of mean force V(R) as illustrated below.

The relationship between G$_\infty$ and V(R) is given by the following expression [13–15]:

$$G'_\infty = NkT + \frac{2\pi N^2}{15} \int_0^\infty g(R) \frac{d}{dr} \left[R^4 \left(\frac{dV(R)}{dR} \right) \right] dR. \tag{5.10}$$

N is the number density of particles and g(R) is the radius distribution function. The assumption made in the derivation of the above expression that the particle interactions involve only central pairwise additive potentials applies if the particles slip over each other without contact friction. Both short-range and long-range order have been observed in dispersions of monodisperse particles and it is likely that at least short-range order is retained where the system is under oscillatory shear with a low strain amplitude. If within the short-range domain a perfect lattice arrangement exists and g(R) can be represented by a delta function centred at the nearest neighbour spacing, R. Evans and Lips showed that under these conditions equation (5.10) reduces to [15]:

$$G'_\infty = NkT + \frac{\phi_m n}{5\pi R^2} \left[4 \frac{dV(R)}{dR} + R \frac{d^2 V(R)}{dR^2} \right], \tag{5.11}$$

where ϕ_m is the maximum packing fraction and n is the coordination number. Equation (5.8) can be expressed in terms of interaction force F:

$$F = -\frac{dV(R)}{dR}. \tag{5.12}$$

The data of Fig. 5.6 are given as interaction energy between flat plates. They can be converted to the force between two spheres E(D) using the Deryaguin approximation. This leads to the following expression for G$'_\infty$:

$$G'_\infty = NkT - \frac{\phi_m na}{5R^2} \left[4E(D) + R \frac{dE(D)}{dD} \right]. \tag{5.13}$$

Using equation (5.9) for E(D) and assuming a reasonable value for L (12.5 nm) and β (7×10^{-3}, the value giving the best fit to equation (5.9), it is possible to calculate G$'_\infty$ as a function of the volume fraction ϕ of the latex, from the energy-distance curve. The results of these calculations are shown in Fig. 5.7 together with the measured values of G$'_\infty$.

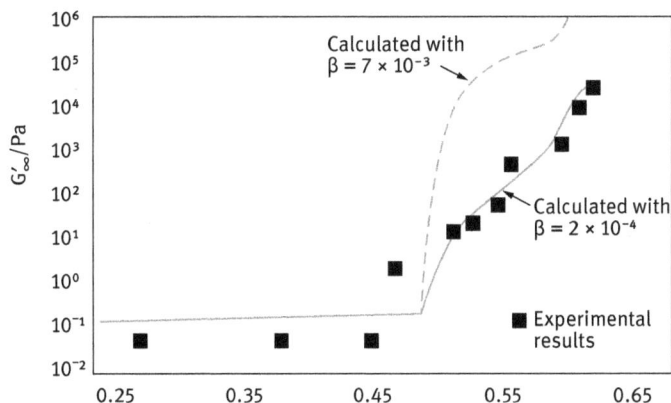

Fig. 5.7: G'_∞ versus ϕ.

It can be seen from Fig. 5.7 that the form of G'_∞ versus ϕ is correctly predicted. However, the calculated moduli values are about two orders of magnitude higher than the experimental values when the correct numerical prefactor β in the de Gennes expression has been used. By adjusting this numerical prefactor (solid line in Fig. 5.7) agreement between theoretical and experimental values of G'_∞ may be obtained.

5.3.2 Atomic Force Microscopy (AFM) measurements

In this case, the force between hydrophobic glass spheres and hydrophobic glass plate, both containing an adsorbed polymeric surfactant layer, is measured as a function of distance of separation both in water and in the presence of various electrolyte concentrations. The AFM is capable of measuring picoNewton surface forces at nanometre length scales. The interaction between glass spheres that were attached to the AFM cantilever and glass plates (both hydrophobized using dichlorodimethylsilane) and containing adsorbed layers of a graft copolymer of hydrophobically modified inulin (INUTEC® SP1, see Chapter 3) was measured as a function of INUTEC® SP1 concentration in water and at various Na_2SO_4 concentrations [16].

Measurements were initially carried out as a function of time (2, 5 and 24 h) at 2×10^{-4} mol dm^{-3} INUTEC® SP1. The force-separation distance curves showed that after 2 and 5 h equilibration time, the force-separation distance curve showed some residual attraction on withdrawal. By increasing the equilibration time (24 h) this residual attraction on withdrawal disappeared and the approach and withdrawal curves were very close, indicating full coverage of the surfaces with polymer within this time. All subsequent measurements were carried out after 24 h to ensure complete adsorption. Measurements were carried out at 5 different concentrations of INUTEC® SP1: 6.6×10^{-6}, 1×10^{-5}, 6×10^{-5}, 1.6×10^{-4} and 2×10^{-4} mol dm^{-3}.

At concentrations $< 1.6 \times 10^{-4}$ mol dm^{-3}, the withdrawal curve showed some residual attraction as is illustrated in Fig. 5.8 for 6×10^{-5} mol dm^{-3}. At concentrations $> 1.6 \times 10^{-4}$ mol dm^{-3}, the approach and withdrawal curves were very close to each other as is illustrated in Fig. 5.9 for 2×10^{-4} mol dm^{-3}. These results indicate full coverage of the surfaces by the polymer when the INUTEC® SP1 concentration becomes equal to or higher than 1.6×10^{-4} mol dm^{-3}. The results at full coverage give an adsorbed layer thickness of the order of 9 nm which indicates strong hydration of the loops and tails of inulin [17].

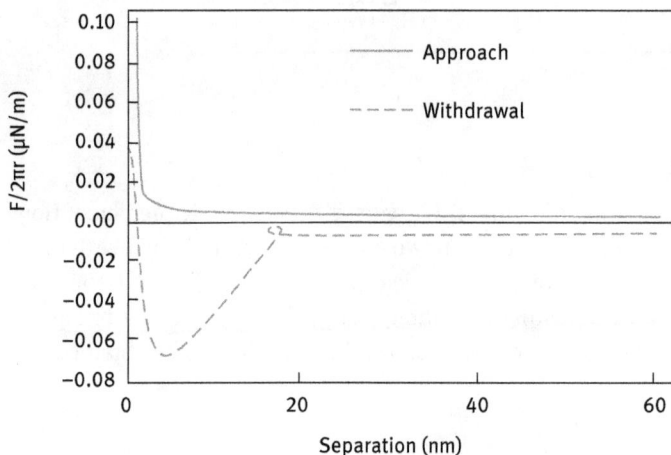

Fig. 5.8: Force-distance curves at 6×10^{-5} mol dm^{-3} INUTEC® SP1.

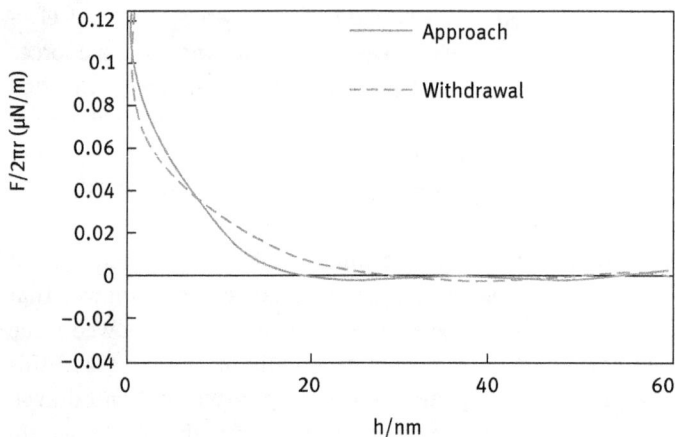

Fig. 5.9: Force-distance curves at 2×10^{-4} mol dm^{-3} INUTEC® SP1.

Several investigations of the stability of emulsions and suspensions stabilized using INUTEC® SP1 showed absence of flocculation in the presence of high electrolyte concentrations (up to 4 mol dm^{-3} NaCl and 1.5 mol dm^{-3} MgSO$_4$) [17]. This high stability in the presence of high electrolyte concentrations is attributed to the strong hydration of inulin (polyfructose) loops and tails. This strong hydration was confirmed by measuring the cloud point of inulin in the presence of such high electrolyte concentrations (the cloud point exceeded 100 °C up to 4 mol dm^{-3} NaCl and 1.0 mol dm^{-3} MgSO$_4$). Evidence of such strong repulsion was obtained from the force-distance curves in the presence of different concentrations of Na$_2$SO$_4$ as shown in Fig. 5.10.

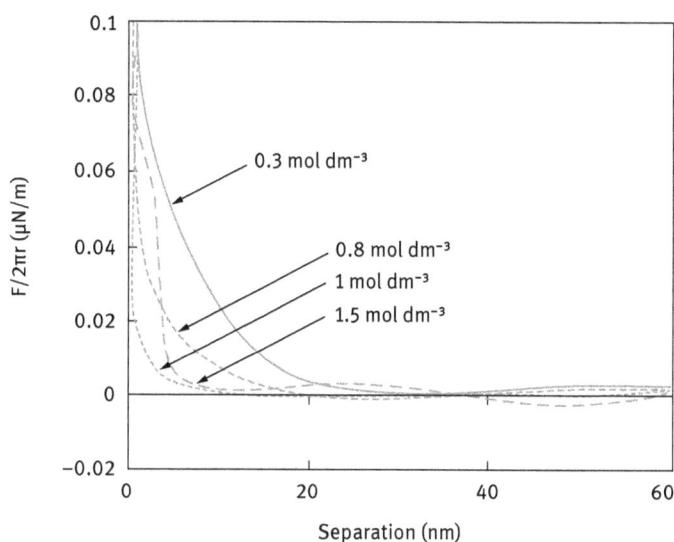

Fig. 5.10: Force-distance curves for hydrophobized glass surfaces containing adsorbed INUTEC® SP1 at various Na$_2$SO$_4$ concentrations.

The force-distance curves clearly show that the interaction remains repulsive up to the highest Na$_2$SO$_4$ concentration (1.5 mol dm^{-3}) studied. The adsorbed layer thickness decreases from approximately 9 nm at 0.3 mol dm^{-3} to about 3 nm at 1.5 mol dm^{-3} Na$_2$SO$_4$. This reduction in hydrodynamic thickness in the presence of high electrolyte concentrations is probably due to the change in the conformation of polyfructose loops and tails. It is highly unlikely that dehydration of the chains occurs, since cloud point measurements have shown absence of any cloud point up to 100 °C. Even at such low adsorbed layer thickness, strong repulsive interaction is observed indicating a high elastic repulsive term.

The interaction between INUTEC® SP1 adsorbed layers was investigated using rheological measurements. Steady state shear stress – shear rate curves were obtained at various volume fractions ϕ of polystyrene latex (PS) dispersions (in the range $\phi =$

0.1–0.42) that contained adsorbed layers of INUTEC® SP1. The results showed a shear thinning behaviour with the viscosity decreasing with applied shear rate and eventually reaching a plateau value when the shear rate exceeded $100\,s^{-1}$. Figure 5.11 shows the variation in the relative viscosity η_r (measured at a shear rate of $1000\,s^{-1}$) as a function of the latex core volume fraction ϕ (with particle radius of 161 nm). It can be clearly seen that η_r increases gradually with increasing ϕ, but when the latter increases above 0.3 there is a rapid increase in η_r with any further increase in ϕ. This trend is typical for concentrated dispersions.

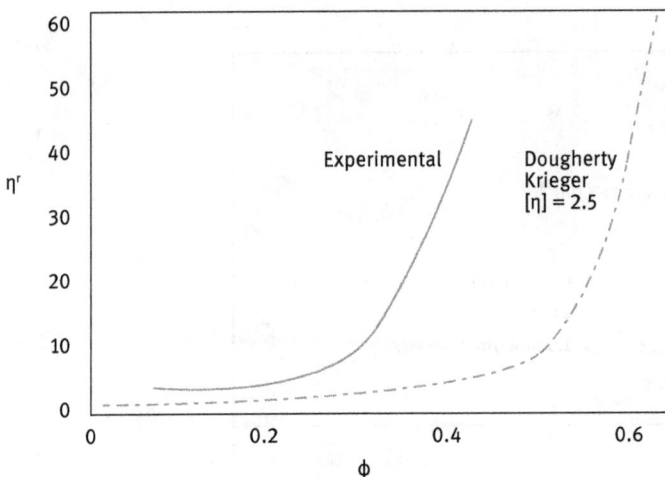

Fig. 5.11: Variation of η_r with ϕ for polystyrene latex dispersions.

The η_r–ϕ curve was calculated on the basis of Dougherty–Krieger equation [18]:

$$\eta_r = \left[1 - \frac{\phi}{\phi_p}\right]^{-[\eta]\phi_p}. \tag{5.14}$$

$[\eta]$ is the intrinsic viscosity and was taken to be equal to 2.5 (for hard spheres) and ϕ_p was obtained from a plot of $1/(\eta_r)^{1/2}$ versus ϕ (straight lines were obtained) and extrapolation of the results to $1/(\eta_r)^{1/2} = 0$. This gave a value of $\phi_p = 0.51$ for PS latex dispersions containing adsorbed INUTEC® SP1 layers.

It can be seen from Fig. 5.11 that the measured viscosity of the latex dispersions is significantly higher than the value calculated using the Dougherty–Krieger equation. This is due to the presence of adsorbed polymer layers on the latex particles.

From the measured η_r values one can obtain the effective volume fraction ϕ_{eff} of the latex dispersion, which can be used to obtain the adsorbed layer thickness δ:

$$\phi_{eff} = \phi \left[1 + \left(\frac{\delta}{R} \right) \right]^3 . \tag{5.15}$$

This gave a value of $\delta = 9.6$ nm, which is close to the value obtained using the AFM measurements, again indicating the strong hydration of the polyfructose loops and tails.

Dynamic (oscillatory) measurements were obtained at various volume fractions ($\phi = 0.1$–0.42) of PS latex dispersions containing adsorbed layers of INUTEC® SP1. Initially, the frequency was fixed at 1 Hz (6.28 rad s^{-1}) and the storage modulus G' and the loss modulus G'' were measured as a function of applied stress (or strain) to obtain the linear viscoelastic region (where G' and G'' are independent of applied stress or strain). Measurements were then carried out as a function of frequency (0.01–10 Hz) while keeping the strain or stress in the linear region.

Figure 5.12 shows the variation of G' and G'' with ϕ (at a stress amplitude in the linear region and frequency of 1 Hz). At $\phi < 0.2$, $G'' > G'$, whereas at $\phi > 0.2$, $G' > G''$ with $G' = G''$ at $\phi = 0.2$, corresponding to an effective volume fraction ϕ_{eff} of 0.24. This low value of ϕ_{eff} indicates very strong repulsive interaction beginning at long separation distance (long-range interaction) between the hydrated loops and tails of polyfructose chains.

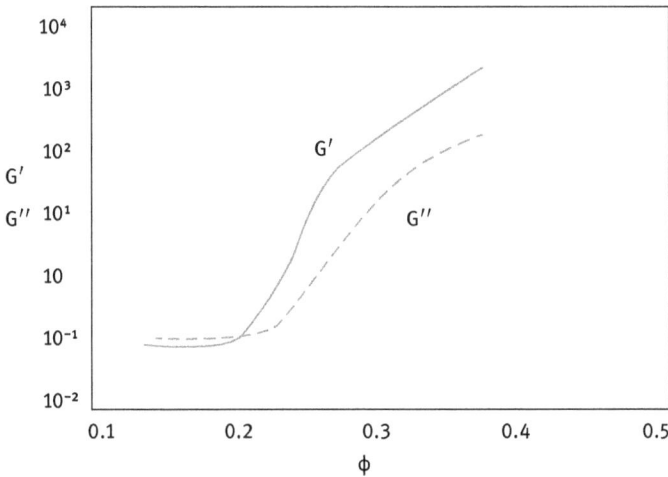

Fig. 5.12: Variation of G' and G'' (at 1 Hz) with volume fraction of PS latex containing adsorbed INUTEC® SP1 layers.

References

[1]   Tadros, Th. F., "Polymer Adsorption and Dispersion Stability", in "The Effect of Polymers on Dispersion Properties", Th. F. Tadros (ed.), Academic Press, London (1981).

[2]   Napper, D. H., "Polymeric Stabilisation of Colloidal Dispersions", Academic Press, London (1981).

[3]   Tadros, Th. F., "Polymeric Surfactants", in "Encyclopedia of Colloid and Interface Science", Th. F. Tadros (ed.), Springer, Germany (2013).

[4]   Flory, P. J. and Krigbaum, W. R., J. Chem. Phys. 18, 1086 (1950).

[5]   Fischer, E. W., Kolloid Z. 160, 120 (1958).

[6]   Mackor, E. L. and van der Waals, J. H., J. Colloid Sci., 7, 535 (1951).

[7]   Hesselink, F. Th., Vrij, A. and Overbeek, J. Th. G., J. Phys. Chem. 75, 2094 (1971).

[8]   Israelachvili, J. N. and Adams, G. E., J.Chem. Soc., Faraday Trans. I, 74, 975 (1978).

[9]   Luckham, P. F., Powder Technol., 58, 75 (1989).

[10] De L. Costello, B. A., Luckham, P. F. and Tadros, Th. F., Colloids and Surfaces, 34, 301 (1988/1989).

[11] De L. Costello, B. A., Luckham, P. F. and Tadros, Th. F., J. Colloid Interface Sci., 152, 237 (1992).

[12] de Gennes, P. G., Advances Colloid Interface Sci., 27, 189 (1987).

[13] White, L. R., J. Colloid Interface Sci., 95, 286 (1983).

[14] Zwanzig, R. and Mountain, R. D., J. Chem. Phys., 43, 4464 (1965).

[15] Evans, I. D. and Lips, A., J. Chem. Soc. Faraday Trans., 86, 3413 (1990).

[16] Nestor, J., Esquena, J., Solans, C., Luckham, P. F., Levecke, B. and Tadros, Th. F., J. Colloid Interface Sci., 311, 430 (2007).

[17] Nestor, J., Esquena, J., Solans, C., Levecke, B., Booten, K. and Tadros, Th. F., Langmuir, 21, 4837 (2005).

[18] Krieger, I. M., Advances Colloid Interface Sci., 3, 111 (1972).

6 Flocculation of disperse systems containing adsorbed polymeric surfactants

6.1 Introduction

Disperse systems (emulsions and suspensions) that are stabilized by adsorbed polymers or polyelectrolytes experience several flocculation mechanisms, depending on the conditions and/or addition of other materials [1–3]. Flocculation can be weak and reversible when for example the minimum in the energy-distance curves becomes deep enough (several kT units, where k is the Boltzmann constant and T is the absolute temperature) for sufficient attraction to occur [1–3]. Flocculation can be strong and irreversible when the minimum becomes very deep. This irreversible flocculation occurs when the solvent for the chains becomes worse than a θ-solvent or the Flory–Huggins interaction parameter χ becomes higher than 0.5. Another mechanism of flocculation occurs when a "free", nonadsorbing polymer is added to the continuous medium. This phenomenon is referred to as depletion flocculation [1–3]. A fourth mechanism of flocculation occurs under conditions of incomplete coverage of the particle surface by the adsorbed polymer. Under these conditions, the polymer chain may become simultaneously adsorbed on two or more particles. This phenomenon is referred to as bridging flocculation [1–3]. A specific type of flocculation occurs in the presence of polyelectrolytes which are partially and strongly adsorbed on an oppositely charged surface (in the absence of complete coverage of the surface by the polyelectrolyte chains). Under these conditions the partially covered patches become attracted to the partially uncovered and oppositely charged surfaces [1–3]. A description of each flocculation mechanism is given below.

6.2 Weak (reversible) flocculation

This occurs when the thickness of the adsorbed layer is small (usually < 5 nm), particularly when the particle radius and Hamaker constant are large [1–3]. As mentioned above, the minimum depth may become sufficiently large for flocculation to take place. For a given particle radius and Hamaker constant, the minimum depth can be controlled by controlling the adsorbed layer thickness. To illustrate this, polystyrene latex dispersions were prepared using different molar mass of partially hydrolysed poly(vinyl acetate), to be referred to as PVA. A commercial sample of PVA (88 % hydrolysed poly(vinyl acetate), i.e. containing 12 % acetate group) was fractionated into narrow molar mass ranges using gel permeation chromatography [4]. The weight-average molar mass of each fraction was determined by measuring the sedimentation coefficient using an ultracentrifuge [4]. The molecular dimensions in solution were also determined from intrinsic viscosity measurements. The latter were also used to

DOI 10.1515/9783110487282-007

determine the Flory–Huggins interaction parameter χ at 25, 37 and 50 °C. The χ values obtained are 0.464, 0.478 and 0.485 at 25, 37 and 50 °C respectively, indicating that water is a good solvent for PVA (note that χ increases slightly with increasing temperature due to the possible dehydration of the polymer at higher temperature, but still $\chi < 0.5$ at all temperatures studied). The adsorption isotherms of the various molecular weight fractions were determined at 25 °C and the results are shown in Fig. 6.1.

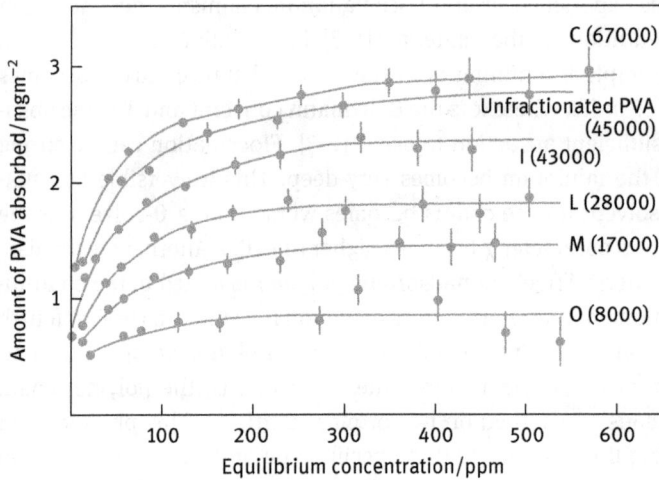

Fig. 6.1: Adsorption isotherms of PVA with various molar mass on polystyrene latex at 25 °C [4].

The isotherms shown in Fig. 6.1 are of the high affinity type (irreversible adsorption) and, as predicted, the plateau adsorption Γ (mg m^{-2}) increases with increasing molar mass. The hydrodynamic adsorbed layer thickness, δ_h, was determined using three different techniques, namely, ultracentrifugation, photon correlation spectroscopy and slow-speed centrifugation [5]. The results obtained using the three methods agreed well with each other. A comparison was also made between δ_h and the hydrodynamic radius R_h. A summary of the adsorption, adsorbed layer thickness and hydrodynamic radius for the various PVA fractions is given in Tab. 6.1.

It can be seen from Tab. 6.1 that δ_h is significantly larger than $2R_h$. Since the volume occupied by the polymer chain at the surface is the same as the hydrodynamic volume of the chain in solution, the larger value of δ_h may be due to some deformation of the random coil on adsorption at the interface.

Using the above values of δ_h it is possible to calculate the energy-distance curves for polystyrene latex particles containing adsorbed PVA layers of various molecular weights, using for example the theory developed by Hesselink et al. [6]. The results of these calculations are shown in Fig. 6.2. It is clear that in the high molecular weight fractions (M, L and I with molar mass 17 000, 28 000 and 43 000 respectively) δ_h is

Tab. 6.1: Adsorption parameters for the various PVA fractions

Fraction	M_w	Γ / mg m^{-2}	δ_h / nm	R_h / nm
unfractionated	45 000	2.79 ± 0.15	22.6 ± 2.6	5.5
C	67 000	2.91 ± 0.17	25.5 ± 3.2	6.5
I	43 000	2.38 ± 0.10	19.7 ± 2.2	5.4
L	28 000	1.82 ± 0.16	14.0 ± 1.7	4.5
M	17 000	1.47 ± 0.14	9.8 ± 2.4	3.6
O	8 000	0.89 ± 0.07	3.3 ± 1.9	2.1

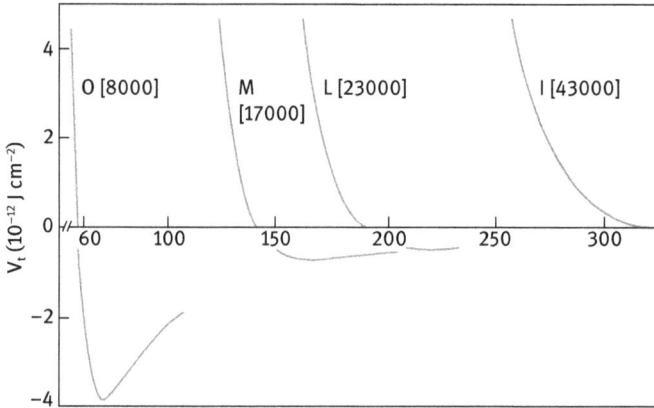

Fig. 6.2: Total energy of interaction versus separation distance for polystyrene latex with adsorbed layers of PVA of various molecular weights.

large (> 9.8 nm) and the energy minimum is too small for flocculation to occur. In contrast, with PVA with molar mass 8000 and $\delta_h \approx 3.3$ nm an appreciable attraction prevails at a separation distance in the region of $2\delta_h$. In this case a weakly flocculated open structure could be produced. To illustrate this point, dispersions stabilized with PVA of various M_w were slowly centrifuged at 50g, and the sediment was freeze-dried and examined by scanning electron microscopy [7]. The results are shown in Fig. 6.3 which clearly shows the close-packed sediment (absence of flocculation) with the high molar mass fractions C and L and the weakly flocculated open-structure obtained with the low molar mass fraction O.

The minimum depth required for causing weak flocculation depends on the volume fraction of the suspension. The higher the volume fraction, the lower the minimum depth required for weak flocculation. This can be understood if one considers the free energy of flocculation that consists of two terms, an energy term determined by the depth of the minimum (G_{min}) and an entropy term that is determined by reduction in configurational entropy on aggregation of particles:

$$\Delta G_{flocc} = \Delta H_{flocc} - T\Delta S_{flocc}. \tag{6.1}$$

Fig. 6.3: Scanning electron micrographs of polystyrene sediments stabilized with PVA fractions: (C) M_w = 43 000; (L) M_w = 28 000; (O) M_w = 8000.

With dilute suspensions, the entropy loss on flocculation is larger than with concentrated suspensions. Hence for flocculation of a dilute suspension, a higher energy minimum is required when compared to the case with concentrated suspensions.

The above flocculation is weak and reversible, i.e. on shaking the container redispersion of the suspension occurs. On standing, the dispersed particles aggregate to form a weak "gel". This process (referred to as sol ↔ gel transformation) leads to reversible time dependence of viscosity (thixotropy) [8].

On shearing the suspension, the viscosity decreases and when the shear is removed, the viscosity is recovered. This phenomenon is applied in paints. On application of the paint (by a brush or roller), the gel is fluidized, allowing uniform coating of the paint. When shearing is stopped, the paint film recovers its viscosity and this avoids any dripping [9].

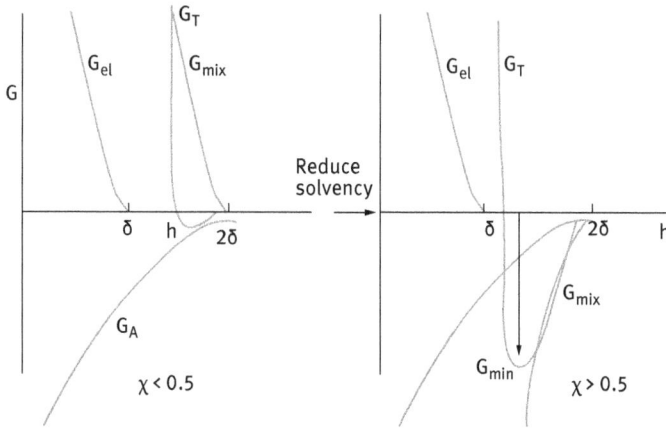

Fig. 6.4: Influence of reduction in solvency on the energy-distance curve.

6.3 Incipient flocculation

This occurs when the solvency of the medium is reduced to become worse than θ-solvent (i.e. $\chi > 0.5$). This reduction in solvency can be induced by temperature changes [10, 11] or addition of a nonsolvent [10, 11] for the stabilizing chain. When the solvency is reduced, the dispersion often exhibits a sharp transition from long-term stability to fast flocculation. This process of incipient flocculation is, for example, observed when a dispersion stabilized by poly(ethylene oxide) moieties is heated. Over a few degrees temperature rise, the turbidity of the dispersion rises sharply indicating excessive flocculation. Flocculation can also occur by the addition of a nonsolvent, e.g. by addition of ethanol to polymethylmethacrylate dispersion stabilized by poly(hydroxystearic) acid in a hydrocarbon solvent [12, 13]. The critical point at which flocculation is first observed is referred to as the critical flocculation temperature (CFT) or critical flocculation concentration of the added nonsolvent (CFV).

An illustration of incipient flocculation is given in Fig. 6.4 where χ was increased from < 0.5 (good solvent) to > 0.5 (poor solvent). One of the characteristic features of sterically stabilized systems, which distinguish them from electrostatically stabilized dispersions, is the temperature dependence of stability. Indeed, some dispersions flocculate on heating [14, 15]; others flocculate on cooling [14, 15]. Furthermore, in some cases dispersions can be produced which do not flocculate at any accessible temperature, whilst some sterically stabilized systems have been found to flocculate both on heating and cooling. This temperature dependence led Napper [10, 11] to describe stability in terms of the thermodynamic process that governs stabilization. Thus, the temperature dependence of the Gibbs free energy of interaction (ΔG_R) for two sterically stabilized particles or droplets is given by

$$\frac{\partial \Delta G_R}{\partial T} = -\Delta S_R, \tag{6.2}$$

where ΔS_R is the corresponding entropy change.

In passing from the stability to the instability domain, ΔG_R must change sign, i.e. from being positive to being negative. It is convenient to split ΔG_R into its enthalpy and entropy contributions:

$$\Delta G_R = \Delta H_R - T\Delta S_R. \tag{6.3}$$

Thus the sign of ΔG_R will depend on the signs and relative magnitudes of ΔH_R and ΔS_R as summarized in Tab. 6.2.

Tab. 6.2: Types of steric stabilization

ΔH_R	ΔS_R	$\Delta H_R/T\Delta S_R$	ΔG_R	Type	Flocculation
+	+	> 1	+	Enthalpic	On heating
−	−	< 1	+	Entropic	On cooling
+	−	≥	+	Combined	Not accessible
		<	+	Enthalpic = entropic	

As a result of extensive investigations on model sterically stabilized dispersions, it has been demonstrated that a strong correlation exists between the critical flocculation point and the θ-point of the stabilizing moieties in free solution. As mentioned before, the θ-point is that at which $\chi = 0.5$, i.e. the point at which the second virial coefficient of the polymer chain is equal to zero. The absolute methods for determining the θ-point include light scattering and osmotic pressure measurements. Less sound methods for determining the θ-point depend on establishing the phase diagrams of polymer solutions.

Using lattices with terminally-anchored polymer chains of various kinds, it has been established that the CFT is independent of the molar mass of the chain, the size of the particle core and the nature of the disperse phase [12]. The CFT correlates strongly with the θ-temperature. Similar correlations have been found between the CFV and the θ-point [12]. However, such correlations are only obtained if the surface is fully covered by the polymer chains. Under conditions of incomplete coverage, flocculation occurs in dispersion media that are better than θ-solvents. This may be due to lateral movement of the stabilizer, desorption or even bridging flocculation (see below).

The correlation between the critical flocculation point and the θ-point implies that G_{mix} dominates the steric interaction. It has been argued that the contribution from G_{el} can be neglected until $h < \delta$, i.e. the polymer layer from one particle comes into direct contact with the second interface. With the high molar mass chains, the contribution from G_A to the total interaction is also negligible. This means that G_T is approximately equal to G_{mix} and it shows that χ is the main parameter controlling the stability. This is clearly illustrated in Fig. 6.4 which shows a significant value of G_{min} when $\chi > 0.5$.

Thus by measuring the θ-point (CFT or CFV) for the polymer chains (A) in the medium under investigation (which could be obtained from light scattering or viscosity measurements) one can establish the stability conditions for a dispersion, before its preparation. This procedure helps also in designing effective steric stabilizers such as block and graft copolymers.

6.4 Depletion flocculation

Depletion flocculation is produced by adding "free" nonadsorbing polymer [16]. In this case, the polymer coils cannot approach the particles to a distance Δ (that is determined by the radius of gyration of free polymer R_G), since the reduction of entropy on close approach of the polymer coils is not compensated by an adsorption energy. The suspension particles will be surrounded by a depletion zone with thickness Δ. Above a critical volume fraction of the free polymer, ϕ_p^+, the polymer coils are "squeezed out" from between the particles and the depletion zones begin to interact. The interstices between the particles are now free from polymer coils and hence an osmotic pressure is exerted outside the particle surface (the osmotic pressure outside is higher than in between the particles) resulting in weak flocculation [16]. A schematic representation of depletion flocculation is shown in Fig. 6.5.

The magnitude of the depletion attraction free energy, G_{dep}, is proportional to the osmotic pressure of the polymer solution, which in turn is determined by ϕ_p and molecular weight M. The range of depletion attraction is proportional to the thickness of the depletion zone, Δ, which is roughly equal to the radius of gyration, R_G, of the free polymer. A simple expression for G_{dep} is [16]

$$G_{dep} = \frac{2\pi R \Delta^2}{V_1}(\mu_1 - \mu_1^0)\left(1 + \frac{2\Delta}{R}\right), \tag{6.4}$$

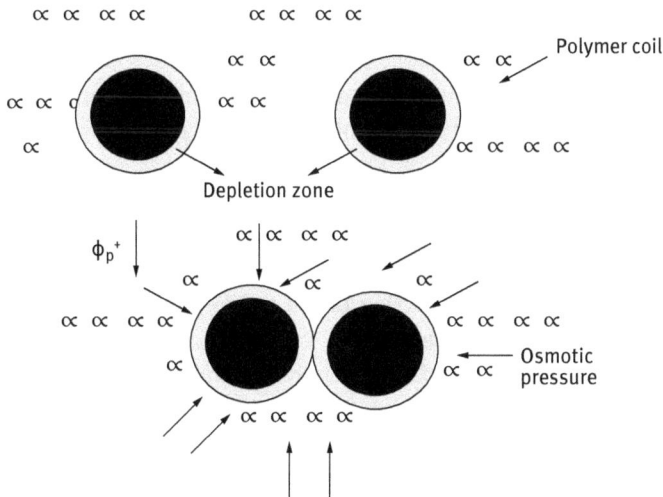

Fig. 6.5: Schematic representation of depletion flocculation.

where V_1 is the molar volume of the solvent, μ_1 is the chemical potential of the solvent in the presence of free polymer with volume fraction ϕ_p and μ_1^0 is the chemical potential of the solvent in the absence of free polymer. $(\mu_1 - \mu_1^0)$ is proportional to the osmotic pressure of the polymer solution.

6.5 Bridging flocculation by polymers and polyelectrolytes

Certain long-chain polymers may adsorb in such a way that different segments of the same polymer chain are adsorbed on different particles, thus binding or "bridging" the particles together, despite the electrical repulsion [17, 18]. With polyelectrolytes of opposite charge to the particles, another possibility exists; the particle charge may be partly or completely neutralized by the adsorbed polyelectrolyte, thus reducing or eliminating the electrical repulsion and destabilizing the particles.

Effective flocculants are usually linear polymers, often of high molecular weight, which may be nonionic, anionic or cationic in character. Ionic polymers should be strictly referred to as polyelectrolytes. The most important properties are molecular weight and charge density. There are several polymeric flocculants that are based on natural products, e.g. starch and alginates, but the most commonly used flocculants are synthetic polymers and polyelectrolytes, e.g. polyacrylamide and copolymers of acrylamide and a suitable cationic monomer such as dimethylaminoethyl acrylate or methacrylate. Other synthetic polymeric flocculants are poly(vinyl alcohol), poly(ethylene oxide) (nonionic), sodium polystyrene sulphonate (anionic) and polyethyleneimine (cationic).

As mentioned above, bridging flocculation occurs because segments of a polymer chain adsorb simultaneously on different particles thus linking them together. Adsorption is an essential step and this requires favourable interaction between the polymer segments and the particles. Several types of interactions are responsible for adsorption that is irreversible in nature:

(i) Electrostatic interaction when a polyelectrolyte adsorbs on a surface bearing oppositely charged ionic groups, e.g. adsorption of a cationic polyelectrolyte on a negative oxide surface such as silica.

(ii) Hydrophobic bonding that is responsible for adsorption of nonpolar segments on a hydrophobic surface, e.g. partially hydrolysed poly(vinyl acetate) (PVA) on a hydrophobic surface such as polystyrene.

(iii) Hydrogen bonding as for example interaction of the amide group of polyacrylamide with hydroxyl groups on an oxide surface.

(iv) Ion binding as is the case of adsorption of anionic polyacrylamide on a negatively charged surface in the presence of Ca^{2+}.

Effective bridging flocculation requires that the adsorbed polymer extends far enough from the particle surface to attach to other particles and that there is sufficient free surface available for adsorption of these segments of extended chains. When excess polymer is adsorbed, the particles can be restabilized, either because of surface saturation or by steric stabilization as discussed before. This is one explanation of the fact that an "optimum dosage" of flocculant is often found; at low concentration there is insufficient polymer to provide adequate links and with larger amounts restabilization may occur. A schematic picture of bridging flocculation and restabilization by adsorbed polymer is given in Fig. 6.6.

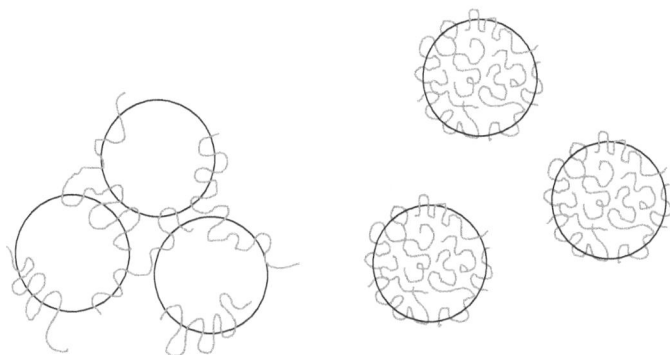

Fig. 6.6: Schematic illustration of bridging flocculation (left) and restabilization (right) by adsorbed polymer.

If the fraction of particle surface covered by polymer is θ, then the fraction of uncovered surface is $(1 - \theta)$ and the successful bridging encounters between the particles should be proportional to $\theta(1 - \theta)$, which has its maximum when $\theta = 0.5$. This is the well-known condition of "half-surface-coverage" that has been suggested as giving the optimum flocculation.

An important condition for bridging flocculation with charged particles is the role of electrolyte concentration. The latter determines the extension ("thickness") of the double layer which can reach values as high as 100 nm (in 10^{-5} mol dm^{-3} 1:1 electrolyte such as NaCl). For bridging flocculation to occur, the adsorbed polymer must extend far enough from the surface to a distance over which electrostatic repulsion occurs (> 100 nm in the above example).

This means that at low electrolyte concentrations quite high molecular weight polymers are needed for bridging to occur. As the ionic strength is increased, the range of electrical repulsion is reduced and lower molecular weight polymers should be effective.

In many practical applications, it has been found that the most effective flocculants are polyelectrolytes with a charge opposite to that of the particles. In aqueous media most particles are negatively charged, and cationic polyelectrolytes such as polyethyleneimine are often necessary. With oppositely charged polyelectrolytes it is likely that adsorption occurs to give a rather flat configuration of the adsorbed chain, due to the strong electrostatic attraction between the positive ionic groups on the polymer and the negative charged sites on the particle surface. This would probably reduce the probability of bridging contacts with other particles, especially with fairly low molecular weight polyelectrolytes with high charge density. However, the adsorption of a cationic polyelectrolyte on a negatively charged particle will reduce the surface charge of the latter, and this charge neutralization could be an important factor in destabilizing the particles. Another mechanism for destabilization has been suggested by Gregory [8] who proposed an "electrostatic-patch" model. This applied to cases where the particles have a fairly low density of immobile charges and the polyelectrolyte has a fairly high charge density. Under these conditions, it is not physically possible for each surface site to be neutralized by a charged segment on the polymer chain, even though the particle may have sufficient adsorbed polyelectrolyte to achieve overall neutrality. There are then "patches" of excess positive charge, corresponding to the adsorbed polyelectrolyte chains (probably in a rather flat configuration), surrounded by areas of negative charge, representing the original particle surface. Particles which have this "patchy" or "mosaic" type of surface charge distribution may interact in such a way that the positive and negative "patches" come into contact, giving quite strong attraction (although not as strong as in the case of bridging flocculation). A schematic illustration of this type of interaction is given in Fig. 6.7. The electrostatic-patch concept (which can be regarded as another form of "bridging") can explain a number of features of flocculation of negatively charged particles with positive polyelectrolytes. These include the rather small effect of increasing the molecular weight and the effect of ionic strength on the breadth of the flocculation dosage range and the rate of flocculation at optimum dosage.

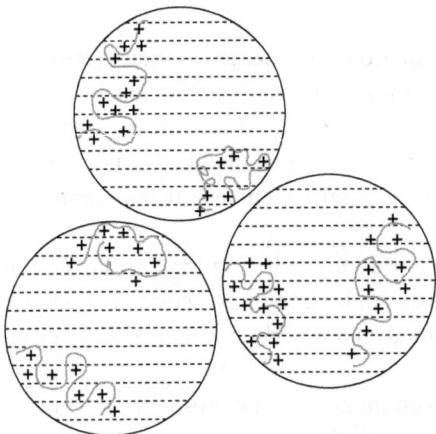

Fig. 6.7: "Electrostatic-patch" model for the interaction of negatively charged particles with adsorbed cationic polyelectrolytes.

References

[1]   Tadros, Th. F., "Interfacial Phenomena and Colloid Stability", Vol. 1, De Gruyter, Germany (2015).
[2]   Tadros, Th. F., "Formulation of Disperse Systems", Wiley-VCH, Germany (2014).
[3]   Tadros, Th. F., "Applied Surfactants", Wiley-VCH, Germany (2005).
[4]   Garvey, M. J., Tadros, Th. F. and Vincent, B., J. Colloid Interface Sci., 49, 57 (1974).
[5]   Garvey, M. J., Tadros, Th. F. and Vincent, B., J. Colloid Interface Sci., 55, 440 (1976).
[6]   Hesselink, F. Th., Vrij, A. and Overbeek, J. Th. G., J. Phys. Chem. 75, 2094 (1971).
[7]   Tadros, Th. F., Advances in Colloid and Interface Science, 12, 141 (1980).
[8]   Tadros, Th. F., "Rheology of Dispersions", Wiley-VCH, Germany (2010).
[9]   Tadros, Th. F., "Colloids in Paints", Wiley-VCH, Germany (2010).
[10] Napper, D. H., "Polymeric Stabilisation of Colloidal Dispersions", Academic Press, London (1983).
[11] Napper, D. H., Trans. Faraday Soc., 64, 1701 (1968); Napper, D. H., J. Colloid Sci., 29, 168 (1969).
[12] Napper, D. H., Kolloid Z., Z. Polym., 234, 1149 (1969).
[13] Napper, D. H., Ind. Eng. Chem. Prod. Res. Develop., 9, 467 (1970).
[14] Napper, D. H. and Netschey, A., J. Colloid Interface Sci., 37, 528 (1971).
[15] Napper, D. H., J. Colloid Interface Sci., 58, 390 (1977).
[16] Asakura, S. and Oosawa, F., J. Phys. Chem., 22, 1255 (1954); Asakura, S. and Oosawa, F., J. Polym. Sci., 33, 183 (1958).
[17] Gregory, J., in "Solid/Liquid Dispersions", Th. F. Tadros (ed.), Academic Press, London (1987).
[18] Gregory, J., "Flocculation Fundamentals", in "Encyclopedia of Colloid and Interface Science", Th. F. Tadros (ed.), Springer, Germany (2013).

7 Polymeric surfactants for stabilization of emulsions and nanoemulsions

7.1 Introduction

Emulsions are a class of disperse systems consisting of two immiscible liquids [1–4]. The liquid droplets (the disperse phase) are dispersed in a liquid medium (the continuous phase). Several classes may be distinguished: Oil-in-Water (O/W); Water-in-Oil (W/O); Oil-in-Oil (O/O). The latter class may be exemplified by an emulsion consisting of a polar oil (e.g. propylene glycol) dispersed in a nonpolar oil (paraffinic oil) and vice versa. To disperse two immiscible liquids one needs a third component, namely the emulsifier. The choice of the emulsifier is crucial in formation of the emulsion and its long-term stability [1–4].

Several breakdown processes may occur on storage depending on: particle size distribution and density difference between the droplets and the medium; magnitude of the attractive versus repulsive forces which determines flocculation; solubility of the disperse droplets and the particle size distribution which determines Ostwald ripening; stability of the liquid film between the droplets that determines coalescence; and phase inversion, where the two phases exchange, e.g. an O/W emulsion inverting to a W/O emulsion and vice versa. Phase inversion can be catastrophic as is the case when the oil phase in an O/W emulsion exceeds a critical value. The inversion can be transient when for example the emulsion is subjected to a temperature increase.

The various breakdown processes are illustrated in the Fig. 7.1. The physical phenomena involved in each breakdown process are not simple and require analysis of the various surface forces involved. In addition, the above processes may take place simultaneously rather than consecutively and this complicates the analysis. Model emulsions, with monodisperse droplets cannot be easily produced and hence any theoretical treatment must take into account the effect of droplet size distribution. Theories that take into account the polydispersity of the system are complex and in many cases only numerical solutions are possible. In addition, measuring surfactant and polymer adsorption in an emulsion is not easy and one has to extract such information from measurements at a planer interface.

Polymeric surfactants enhance the stability of emulsions against flocculation, Ostwald ripening and coalescence. Flocculation is prevented by the presence of strong repulsive energy as discussed in detail in Chapter 5. Ostwald ripening is significantly reduced when using strongly adsorbed polymeric surfactants by enhancing the Gibbs dilational elasticity as will be discussed below. Emulsion coalescence is prevented by enhancing the film stability between emulsion droplets. This enhanced stability is due to the high disjoining pressure and absence of film rupture on approach of the droplets as will be discussed below.

DOI 10.1515/9783110487282-008

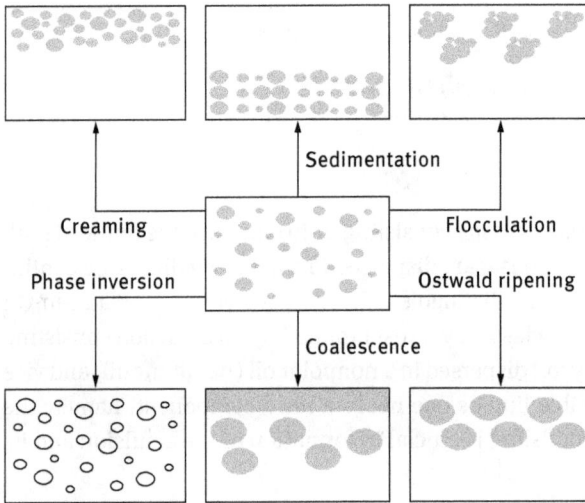

Fig. 7.1: Schematic representation of the various breakdown processes in emulsions.

Nanoemulsions are transparent or translucent systems mostly covering the size range 50–200 nm [5, 6]. Nanoemulsions were also referred to as mini-emulsions [7, 8]. Unlike microemulsions (which are also transparent or translucent and thermodynamically stable) nanoemulsions are only kinetically stable. However, the long-term physical stability of nanoemulsions (with no apparent flocculation or coalescence) makes them unique and they are sometimes referred to as "approaching thermodynamic stability". The inherently high colloid stability of nanoemulsions can be well understood from a consideration of their steric stabilization when using polymeric surfactants and how this is affected by the ratio of the adsorbed layer thickness to droplet radius as will be discussed below. Unless adequately prepared (to control the droplet size distribution) and stabilized against Ostwald ripening (that occurs when the oil has some finite solubility in the continuous medium), nanoemulsions may lose their transparency with time as a result of an increase in droplet size.

7.2 Polymeric surfactants for prevention of emulsion and nanoemulsion flocculation

The driving force for the prevention of flocculation is the strong repulsive energy between two emulsion or nanoemulsion droplets that can overcome the van der Waals attraction between these droplets. This was discussed in detail in Chapter 5 and only a summary is given here [9–11].

When two droplets, each containing an adsorbed layer of thickness δ, approach to a distance of separation h, whereby h becomes less than 2δ, repulsion occurs as result of two main effects:

(i) Unfavourable mixing of the stabilizing chains A of the adsorbed layers, when these are in good solvent conditions. This is referred to as the mixing (osmotic interaction), G_{mix}, and is given by the following expression:

$$\frac{G_{mix}}{kT} = \frac{4\pi}{3V_1}\phi_2^2 \left(\frac{1}{2} - \chi\right)\left(3a + 2\delta + \frac{h}{2}\right), \tag{7.1}$$

where k is the Boltzmann constant, T is the absolute temperature, V_1 is the molar volume of the solvent, ϕ_2 is the volume fraction of the polymer (the A chains) in the adsorbed layer and χ is the Flory–Huggins (polymer-solvent interaction) parameter.

It can be seen that G_{mix} depends on three main parameters: the volume fraction of the A chains in the adsorbed layer (the denser the layer is, the higher the value of G_{mix}); the Flory–Huggins interaction parameter χ (for G_{mix} to remain positive, i.e. repulsive, χ should be lower than 1/2); and the adsorbed layer thickness δ.

(ii) Reduction in configurational entropy of the chains on significant overlap. This is referred to as elastic (entropic) interaction and is given by the expression

$$G_{el} = 2v_2 \ln\left[\frac{\Omega(h)}{\Omega(\infty)}\right], \tag{7.2}$$

where v_2 is the number of chains per unit area, $\Omega(h)$ is the configurational entropy of the chains at a separation distance h and $\Omega(\infty)$ is the configurational entropy at infinite distance of separation.

Combining G_{mix} and G_{el} with the van der Waals attraction G_A gives the total energy of interaction G_T:

$$G_T = G_{mix} + G_{el} + G_A. \tag{7.3}$$

Figure 7.2 gives a schematic representation of the variation of G_{mix}, G_{el}, G_A and G_T with h. As can be seen from Fig. 7.2, G_{mix} increases very rapidly with decreasing h as soon as $< 2\delta$ and G_{el} increases very rapidly with decreasing h when $h < \delta$. G_T shows one minimum, G_{min}, and it increases very rapidly with decreasing h when $h < 2\delta$.

The magnitude of G_{min} depends on the following parameters: the particle radius R; the Hamaker constant A, the adsorbed layer thickness δ. As an illustration, Fig. 7.3 shows the variation of G_T with h at various ratios of δ/R. It can be seen from Fig. 7.3 that the depth of the minimum decreases with increasing δ/R. This is the basis of the high kinetic stability of nanoemulsions. If nanoemulsions have a radius in the region of 50 nm and an adsorbed layer thickness of say 10 nm, the value of δ/R is 0.2. This high value (when compared to the situation with macroemulsions where δ/R is at least an order of magnitude lower) results in a very shallow minimum (which could be less than kT).

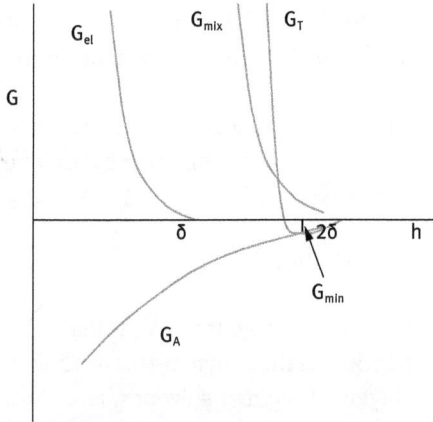

Fig. 7.2: Variation of G_{mix}, G_{el}, G_A and G_T with h.

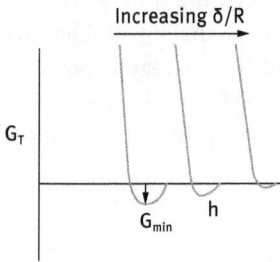

Fig. 7.3: Variation of G_T with h with increasing δ/R.

The above situation results in very high stability with no flocculation (weak or strong). In addition, the very small size of the droplets and the dense adsorbed layers ensure lack of deformation of the interface, lack of thinning and disruption of the liquid film between the droplets and hence coalescence is also prevented.

7.3 Polymeric surfactants for reduction of Ostwald ripening

The driving force of Ostwald ripening is the difference in solubility between the smaller and larger droplets [12]. The small particles or droplets with radius r_1 will have higher solubility than the larger particle or droplet with radius r_2. This can be easily recognized from the Kelvin equation [13] which relates the solubility of a particle or droplet $S(r)$ to that of a particle or droplet with infinite radius $S(\infty)$:

$$S(r) = S(\infty) \exp\left(\frac{2\gamma V_m}{rRT}\right), \tag{7.4}$$

where γ is the solid/liquid or liquid/liquid interfacial tension, V_m is the molar volume of the disperse phase, R is the gas constant and T is the absolute temperature. The quantity $(2\gamma V_m/RT)$ has the dimension of length and it termed the characteristic length with an order of $\approx 1\,nm$.

Kelvin Equation

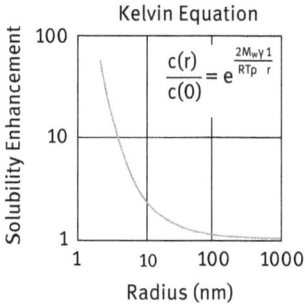

$$\frac{c(r)}{c(0)} = e^{\frac{2M_w\gamma}{RT\rho}\frac{1}{r}}$$

Fig. 7.4: Solubility enhancement with decreasing droplet radius.

A schematic representation of the enhancement of the solubility $c(r)/c(0)$ with decreasing droplet size according to the Kelvin equation is shown in Fig. 7.4.

It can be seen from Fig. 7.4 that the solubility of nanodispersion droplets increases very rapidly with decreasing radius, particularly when $r < 100$ nm. This means that a particle with a radius of say 4 nm will have about 10 times solubility enhancement compared say with a droplet with 10 nm radius, which has a solubility enhancement of only 2 times. Thus with time molecular diffusion will occur between the smaller and larger droplet, with the ultimate disappearance of most of the small droplets. This results in a shift in the droplet size distribution to larger values on storage of the nanoemulsion. This could lead to the formation of an emulsion with average droplet size in the μm range. This instability can cause severe problems, such as creaming or sedimentation, flocculation and even coalescence of the nanoemulsion.

For two droplets with radii r_1 and r_2 ($r_1 < r_2$),

$$\frac{RT}{V_m} \ln\left[\frac{S(r_1)}{S(r_2)}\right] = 2\gamma\left[\frac{1}{r_1} - \frac{1}{r_2}\right] \tag{7.5}$$

Equation (7.5) is sometimes referred to as the Ostwald equation and it shows that the larger the difference between r_1 and r_2, the higher the rate of Ostwald ripening. That is why in the preparation of nanoemulsions, one aims at producing a narrow size distribution.

The kinetics of Ostwald ripening is described in terms of the theory developed by Lifshitz and Slesov [14] and by Wagner [15] (referred to as LSW theory). The LSW theory assumes that:

(i)   The mass transport is due to molecular diffusion through the continuous phase.
(ii) The dispersed phase droplets are spherical and fixed in space.
(iii) There are no interactions between neighbouring droplets (the droplets are separated by a distance much larger than the diameter of the droplets).
(iv) The concentration of the molecularly dissolved species is constant except adjacent to the droplet boundaries.

The rate of Ostwald ripening, ω, is given by

$$\omega = \frac{d}{dr}(r_c^3) = \left(\frac{8\gamma DS(\infty)V_m}{9RT}\right) f(\phi) = \left(\frac{4DS(\infty)\alpha}{9}\right) f(\phi), \qquad (7.6)$$

where r_c is the radius of a droplet that is neither growing nor decreasing in size, D is the diffusion coefficient of the disperse phase in the continuous phase, $f(\phi)$ is a factor that reflects the dependence of ω on the disperse volume fraction and α is the characteristic length scale ($= 2\gamma V_m/RT$).

Droplets with $r > r_c$ grow at the expense of smaller ones, while droplets with $r < r_c$ tend to disappear. The validity of the LSW theory was tested by Kabalnov et al. [16] who used 1,2 dichloroethane-in-water emulsions in which the droplets were fixed to the surface of a microscope slide to prevent their coalescence. The evolution of the droplet size distribution was followed as a function of time by microscopic investigations.

LSW theory predicts that the droplet growth over time will be proportional to r_c^3. This is illustrated in Fig. 7.5 for dichloroethane-in-water emulsions.

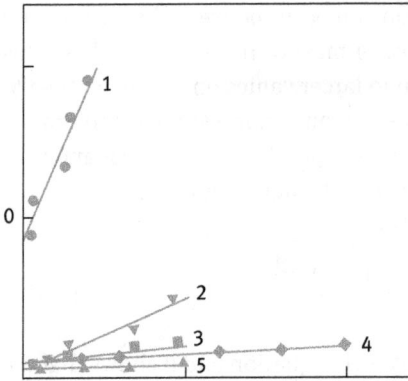

Fig. 7.5: Variation of average cube radius with time during Ostwald ripening in emulsions of: (1) 1,2 dichloroethane; (2) benzene; (3) nitrobenzene; (4) toluene; (5) p-xylene.

Another consequence of LSW theory is the prediction that the size distribution function $g(u)$ for the normalized droplet radius $u = r/r_c$ adopts a time independent form given by

$$g(u) = \frac{81eu^2 \exp[1/(2u/3 - 1)]}{3^{2^{1/3}}(u+3)^{7/3}(1.5 - u)^{11/3}} \quad \text{for } 0 < u \leq 1.5 \qquad (7.7)$$

and

$$g(u) = 0 \quad \text{for} \quad u > 1.5. \qquad (7.8)$$

A characteristic feature of the size distribution is the cut-off at $u > 1.5$.

A comparison of the experimentally determined size distribution (dichloroethane-in-water emulsions) with the theoretical calculations based on LSW theory is shown in Fig. 7.6.

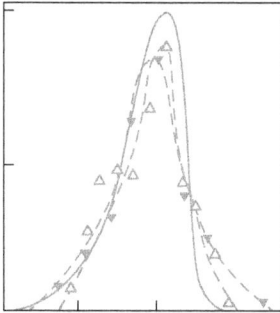

Fig. 7.6: Comparison between theoretical function g(u) (full line) and experimentally determined functions obtained for 1,2 dichloroethane droplets at time 0 (open triangles) and 300 s (inverted solid triangles).

The influence of the alkyl chain length of the hydrocarbon on the Ostwald ripening rate of nanoemulsions was systematically investigated by Kabalanov et al. [6]. An increase in the alkyl chain length of the hydrocarbon used for the emulsion results in a decrease in oil solubility. According to LSW theory, this reduction in solubility should result in a decrease in the Ostwald ripening rate. This was confirmed by the results of Kabalnov et al. [17] who showed that the Ostwald ripening rate decreases with increasing the alkyl chain length from C_9–C_{16}. Table 7.1 shows the solubility of the hydrocarbon, the experimentally determined rate, ω_e, the theoretical values, ω_t, and the ratio of ω_e/ω_t.

Tab. 7.1: Influence of alkyl chain length on the Ostwald ripening rate.

Hydrocarbon	$c(\infty)$ / ml ml^{-1} [a]	ω_e / cm^{-3} s^{-1}	ω_t / cm^{-3} s^{-1} [b]	$\omega_r = \omega_e/\omega_t$
C_9H_{20}	3.1×10^{-7}	6.8×10^{-19}	2.9×10^{-19}	2.3
$C_{10}H_{22}$	7.1×10^{-8}	2.3×10^{-19}	0.7×10^{-19}	3.3
$C_{11}H_{24}$	2.0×10^{-8}	5.6×10^{-20}	2.2×10^{-20}	2.5
$C_{12}H_{26}$	5.2×10^{-9}	1.7×10^{-20}	0.5×10^{-20}	3.4
$C_{13}H_{28}$	1.4×10^{-9}	4.1×10^{-21}	1.6×10^{-21}	2.6
$C_{14}H_{30}$	3.7×10^{-10}	1.0×10^{-21}	0.4×10^{-21}	2.5
$C_{15}H_{32}$	9.8×10^{-11}	2.3×10^{-22}	1.4×10^{-22}	1.6
$C_{16}H_{34}$	2.7×10^{-11}	8.7×10^{-23}	2.2×10^{-23}	4.0

a Molecular solubilities of hydrocarbons in water taken from: C. McAuliffe, *J. Phys. Chem.*, 1966, 1267.
b For theoretical calculations, the diffusion coefficients were estimated according to the Hayduk–Laudie equation (W. Hayduk and H. Laudie, *AIChE J.*, 1974, **20**, 611) and the correction coefficient f(ϕ) assumed to be equal to 1.75 for $\phi = 0.1$ (P. W. Voorhees. *J. Stat. Phys.*, 1985, **38**, 231).

Although the results showed the linear dependence of the cube of the droplet radius with time in accordance with LSW theory, the experimental rates were ≈ 2–3 times higher than the theoretical values. The deviation between theory and experiment has been ascribed to the effect of Brownian motion [17]. LSW theory assumes that the droplets are fixed in space and that molecular diffusion is the only mechanism of mass transfer. For droplets undergoing Brownian motion, one must take into account the contributions of molecular and convective diffusion as predicted by the Peclet number

$$Pe = \frac{rv}{D},$$ (7.9)

where v is the velocity of the droplets that is approximately given by

$$v = \left(\frac{3kT}{M}\right)^{1/2},$$ (7.10)

where k is the Boltzmann constant, T is the absolute temperature and M is the mass of the droplet. For r = 100 nm, Pe = 8, indicating that the mass transfer will be accelerated with respect to that predicted by LSW theory.

LSW theory assumes that there are no interactions between the droplets and it is limited to low oil volume fractions. At higher volume fractions the rate of ripening depends on the interaction between diffusion spheres of neighbouring droplets. It is expected that emulsions with higher volume fractions of oil will have broader droplet size distribution and faster absolute growth rates than those predicted by LSW theory. However, experimental results using high surfactant concentrations (5 %) showed the rate to be independent of the volume fraction in the range $0.01 \leq \phi \leq 0.3$. It has been suggested that the emulsion droplets may have been screened from one another by surfactant micelles [18]. A strong dependency on volume fraction has been observed for fluorocarbon-in-water emulsions. A threefold increase in ω was found when ϕ was increased from 0.08 to 0.52.

Two main processes have been suggested for reducing Ostwald ripening:

(i) Addition of a small proportion of highly insoluble oil. Huguchi and Misra [19] suggested that addition of a second disperse phase that is virtually insoluble in the continuous phase, such as squalane, can significantly reduce the Ostwald ripening rate. In this case, significant partitioning between different droplets is predicted, with the component having the low solubility in the continuous phase (e.g. squalane) being expected to be concentrated in the smaller droplets. During Ostwald ripening in a two-component disperse system, equilibrium is established when the difference in chemical potential between different sized droplets, which results from curvature effects, is balanced by the difference in chemical potential resulting from the partitioning of the two components. Huguchi and Misra [19] derived the following expression for the equilibrium condition, wherein the excess chemical potential of the medium soluble component, $\Delta\mu_1$, is equal for all of the droplets in a polydisperse medium:

$$\frac{\Delta\mu_i}{RT} = \left(\frac{a_1}{r_{eq}}\right) + \ln(1 - X_{eq2}) = \left(\frac{a_1}{r_{eq}}\right) - X_{02}\left(\frac{r_0}{r_{eq}}\right)^3 = \text{const},$$ (7.11)

where $\Delta\mu_1 = \mu_1 - \mu_1^*$ is the excess chemical potential of the first component with respect to the state μ_1^* when the radius r = ∞ and $X_{02} = 0$, r_0 and r_{eq} are the radii of an arbitrary drop under initial and equilibrium conditions respectively, X_{02} and X_{eq2} are the initial and equilibrium mole fractions of the medium insoluble component 2, a_1 is the characteristic length scale of the medium soluble component 1.

The equilibrium determined by equation (7.11) is stable if the derivative $\partial\Delta\mu_1/\partial r_{eq}$ is greater than zero for all the droplets in a polydisperse system. Based on this analysis, Kabalanov et al. [20] derived the following criterion:

$$X_{02} > \frac{2a_1}{3d_0}, \tag{7.12}$$

where d_0 is the initial droplet diameter. If the stability criterion is met for all droplets, two patterns of growth will result, depending on the solubility characteristic of the secondary component. If the secondary component has zero solubility in the continuous phase, then the size distribution will not deviate significantly from the initial one, and the growth rate will be equal to zero. In the case of limited solubility of the secondary component, the distribution is governed by rules similar to the LSW theory, i.e. the distribution function is time variant. In this case, the Ostwald ripening rate, ω_{mix}, will be a mixture growth rate that is approximately given by the following equation [21]:

$$\omega_{mix} = \left(\frac{\phi_1}{\omega_1} + \frac{\phi_2}{\omega_2}\right)^{-1}, \tag{7.13}$$

where ϕ_1 is the volume fraction of the medium soluble component and ϕ_2 is the volume fraction of the medium insoluble component respectively.

If the stability criterion is not met, a bimodal size distribution is predicted to emerge from the initially monomodal one. Since the chemical potential of the soluble component is predicted to be constant for all the droplets, it is also possible to derive the following equation for the quasi-equilibrium component 1:

$$X_{02} + \frac{2a_1}{d} = \text{const}, \tag{7.14}$$

where d is the diameter at time t.

Kabalanov et al. [21] studied the effect of addition of hexadecane to a hexane-in-water nanoemulsion. Hexadecane, which is less soluble than hexane, was studied at three levels, $X_{02} = 0.001$, 0.01 and 0.1. For the higher mole fraction of hexadecane, namely 0.01 and 0.1, the emulsion had a physical appearance similar to that of an emulsion containing only hexadecane and the Ostwald ripening rate was reliably predicted by equation (7.13). However, the emulsion with $X_{02} = 0.001$ quickly separated into two layers, a sedimented layer with a droplet size of ca. 5 µm and a dispersed population of submicron droplets (i.e. a bimodal distribution). Since the stability criterion was not met for this low volume fraction of hexadecane, the observed bimodal distribution of droplets is predictable.

(ii) Modification of the interfacial layer for reduction of Ostwald ripening. According to LSW theory, the Ostwald ripening rate, ω, is directly proportional to the interfacial tension, γ. Thus, by reducing γ, ω is reduced. This could be confirmed by measuring ω as a function of SDS concentration for decane-in-water emulsion [12] below the critical micelle concentration (cmc). Below the cmc, γ shows a linear decrease with increasing log[SDS] concentration. The results are summarized in Tab. 7.2.

Tab. 7.2: Variation of Ostwald ripening rate with SDS concentration for decane-in-water emulsions.

[SDS] Concentration / mol dm^{-3}	ω / cm^3 s^{-1}
0.0	2.50×10^{-18}
1.0×10^{-4}	4.62×10^{-19}
5.0×10^{-4}	4.17×10^{-19}
1.0×10^{-3}	3.68×10^{-19}
5.0×10^{-3}	2.13×10^{-19}

cmc of SDS = 8.0×10^{-3}

Several other mechanisms have been suggested to account for the reduction in the Ostwald ripening rate by a modification of the interfacial layer. For example, Walstra [22] suggested that emulsions could be effectively stabilized against Ostwald ripening by the use of polymeric surfactants that are strongly adsorbed at the interface and which do not desorb during the Ostwald ripening process. In this case the increase in interfacial dilational modulus, ε, and decreases in interfacial tension, γ, would be observed for the shrinking droplets. Eventually the difference in ε and γ between droplets would balance the difference in capillary pressure (i.e. curvature effects) leading to a quasi-equilibrium state. In this case, emulsifiers with low solubilities in the continuous phase such as proteins would be preferred. Long-chain phospholipids with a very low solubility (cmc $\approx 10^{-10}$ mol dm^{-3}) are also effective in reducing Ostwald ripening of some emulsions. The phospholipid would have to have a solubility in water about three orders of magnitude lower than the oil [23].

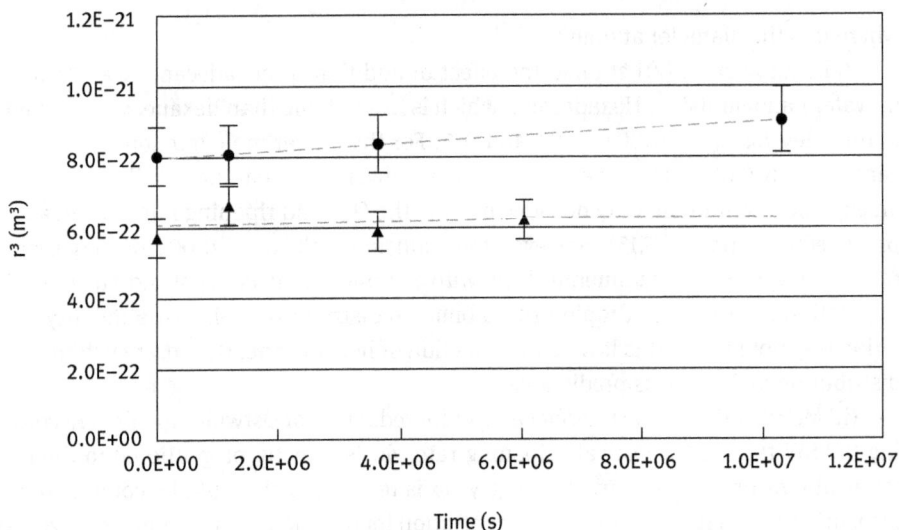

Fig. 7.7: Nanoemulsions stabilized with INUTEC® SP1 (top curve 1.6 % and bottom curve 2.4 % INUTEC® SP1).

As an illustration, Fig. 7.7 shows the variation of the cube radius with time for silicone oil/water nanoemulsions stabilized with hydrophobically modified inulin (INUTEC® SP1) at two different polymeric surfactant concentrations [24].

The rate of Ostwald ripening is 1.1×10^{-29} and $2.4 \times 10^{-30} \, \text{m}^3 \, \text{s}^{-1}$ for the 1.6% and 2.4% INUTEC® SP1 respectively, which is ≈ 3 orders of magnitude lower than those obtained using a nonionic surfactant.

7.4 Polymeric surfactants for reducing (or eliminating) coalescence

When two emulsion droplets come into close contact in a floc or creamed layer or during Brownian diffusion, a thin liquid film or lamella forms between them [25]. This is illustrated in Fig. 7.8. Coalescence results from the rupture of this film as illustrated in Fig. 7.8 (c) at the top. If the film cannot be ruptured, adhesion (Fig. 7.8 (c) middle) or engulfment (Fig. 7.8 (c) bottom) may occur. Film rupture usually commences at a specified "spot" in the lamella, arising from thinning in that region. This is illustrated in Fig. 7.9, where the liquid surfaces undergo some fluctuations forming surface waves. The surface waves may grow in amplitude and the apices may join as a result of the strong van der Waals attraction (at the apex, the film thickness is the smallest). The same applies if the film thins to a small value (critical thickness for coalescence). In order to understand the behaviour of these films, one has to consider two aspects of their physics: (i) the nature of the forces acting across the film; these determine whether the film is thermodynamically stable, metastable or unstable and (ii) the kinetic aspects associated with local (thermal or mechanical) fluctuations in film thickness.

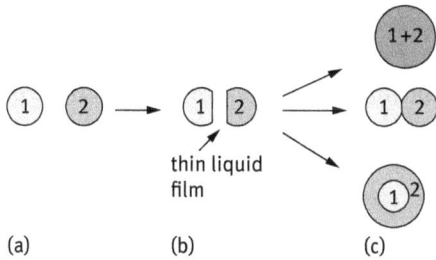

Fig. 7.8: Droplet coalescence, adhesion and engulfment.

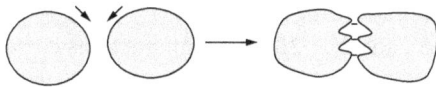

Fig. 7.9: Schematic representation of surface fluctuations.

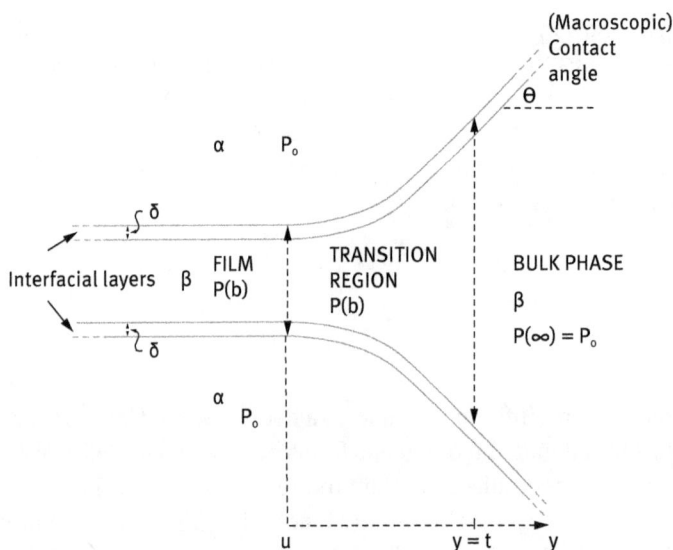

Fig. 7.10: Schematic representation of the thin film and border regions between two liquid droplets (α) in a continuous phase (β).

To understand the process of coalescence, one must consider the surface forces across the liquid films. Figure 7.10 shows the general features of the lamella between two droplets of phase α in a continuous phase β. The film consists of two flat parallel interfaces separated by a distance h. At the end of the film there is a border or transition region where the interfaces have a high curvature, i.e. compared to the curvature of the droplets themselves. Eventually, at larger values of b (effectively beyond the range of forces operating across the film) the curvature decreases to that of the droplets themselves, i.e. becomes effectively flat, on the scale considered here, for droplets in the 1 μm region. One may define a macroscopic contact angle θ as shown in Fig. 7.10.

In considering the forces acting across the film, two regions of separation are of interest:

(i) $h > 2\delta$, where δ is the film thickness, where the forces acting are long-range van der Waals forces and electrical double layer interactions as described by the DLVO (Deryaguin–Landau–Verwey–Overbeek) theory [26, 27], schematically shown in Fig. 7.11. This shows a secondary minimum, an energy maximum and a primary minimum. When the film is sitting in either the primary or secondary minimum, the net force on the film is zero (i.e. $dG/dh = 0$). These two metastable states correspond with the so-called Newton black films, respectively.

(ii) When the film is in the primary minimum and $h < 2\delta$, steric interactions come into play as discussed in Chapter 5. In this case, G increases very sharply with decreasing h and for film rupture to occur these steric interactions must break down.

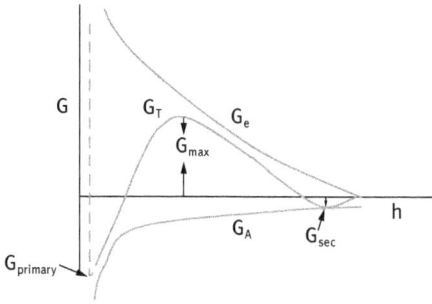

Fig. 7.11: Energy-distance curves according to DLVO theory [2, 3].

Two main approaches were considered to analyse the stability of thin films in terms of the relevant interactions. The first approach was considered by Deryaguin [28] who introduced the concept of disjoining pressure. The second approach considered the interfacial tension of the film that could be related to the tangential pressure across the interface [29].

(i) Disjoining pressure approach: Deryaguin [28] suggested that a "disjoining pressure", $\pi(h)$, is produced in the film which balances the excess normal pressure:

$$\pi(h) = P(h) - P_0, \tag{7.15}$$

where $P(h)$ is the pressure of a film with thickness h and P_0 is the pressure of a sufficiently thick film such that the net interaction free energy is zero.

$\pi(h)$ may be equated to the net force (or energy) per unit area acting across the film:

$$\pi(h) = -\frac{dG_T}{dh}, \tag{7.16}$$

where G_T is the total interaction energy in the film.

$\pi(h)$ is made up of three contributions due to electrostatic repulsion (π_E), steric repulsion (π_s) and van der Waals attraction (π_A):

$$\pi(h) = \pi_E + \pi_s + \pi_A. \tag{7.17}$$

To produce a stable film, $\pi_E + \pi_s > \pi_A$ and this is the driving force for preventing coalescence which can be achieved by two mechanisms and their combination:
(a) increased repulsion, both electrostatic and steric;
(b) dampening fluctuation by enhancing Gibbs elasticity.

In general, smaller droplets are less susceptible to surface fluctuations and hence coalescence is reduced.

(ii) Interfacial tension approach: The interfacial tension $\gamma(h)$ can be related to the variation in the tangential pressure tensor p_t across an interface. There will be a similar variation in P_t across the liquid film at some thickness h as shown schematically

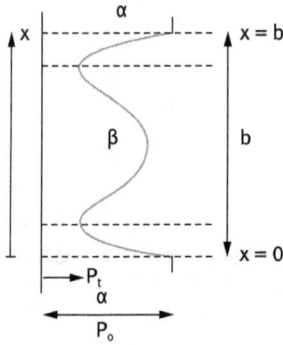

Fig. 7.12: The variation in tangential pressure P_t across a thin film.

in Fig. 7.12. One can define an interfacial tension for the whole film

$$\gamma(h) = \int_0^h (P_t - P_0)\, dx \tag{7.18}$$

by choosing some dividing plane in the middle of the film (conveniently at $x = b/2$, for a symmetrical film), one can divide $\gamma(h)$ into two contributions, one from the upper interface and one from the lower interface:

$$\gamma(h) = \gamma^{\alpha\beta}(h) + \gamma^{\beta\alpha}(h) = 2\gamma^{\alpha\beta}(h). \tag{7.19}$$

Note that $\gamma^{\alpha\beta}(h) \neq \gamma^{\alpha\beta}(\infty)$, the interfacial tension of an isolated $\alpha\beta$ interface, i.e. between the bulk liquids α or β, for an emulsion, in the region of the interface of the droplet far away from the contact zone. $\gamma(h)$ is related to $G_i(h)$ and $\pi(h)$ through the relations

$$\gamma(h) = \gamma(\infty) - G_i(h) \tag{7.20}$$

and

$$\gamma(h) = \gamma(\infty) + h. \tag{7.21}$$

Equation (7.21) is obtained by combining equations (7.20) and (7.18).

Film rupture is a nonequilibrium effect and is associated with local thermal or mechanical fluctuations in the film thickness h. A necessary condition for rupture to occur, i.e. for a spontaneous fluctuation to occur, is that

$$\frac{d\pi_A}{dh} > \frac{d\pi_E}{dh}. \tag{7.22}$$

However, this would assume that, at a given value of h, there are no changes in $\gamma(h)$ fluctuations. This is not so, since a local fluctuation is necessarily accompanied by a local increase in the interfacial area resulting in a decrease in surfactant or polymer adsorption in that region, and, therefore, a local rise in the interfacial tension.

This effect (referred to as the Gibbs–Marangoni effect) opposes the fluctuation. Equation (7.22) has to be modified by including a term at the right-hand side to take into account the fluctuation effect in the local interfacial tension:

$$\frac{d\pi_A}{dh} > \frac{d\pi_E}{dh} + \frac{d\pi_\gamma}{dh}. \tag{7.23}$$

As a film thins locally due to fluctuations, if the conditions of equation (7.23) are met at a critical thickness, h_{cr}, then the film becomes unstable and the fluctuation "grows", leading to rupture. Scheludko [29] introduce the concept of a critical thickness and he derived the following equation for the critical thickness:

$$h_{cr} = \left(\frac{A\pi}{32K^2\gamma_0} \right)^{1/4}, \tag{7.24}$$

where A is the net Hamaker constant of the film, K is the wave number of the fluctuation and γ_0 [= $\gamma(\infty)$] the interfacial tension of the isolated liquid/liquid interface. K depends on the radius of the (assumed) circular film zone.

Vrij [30] derived alternative expressions for b_{cr}. For large thicknesses where $\pi_A \ll \pi_\gamma$

$$h_{cr} = 0.268 \left(\frac{A^2 R^2}{\gamma_0 \pi_\gamma f} \right). \tag{7.25}$$

For small thicknesses, where $\pi_\gamma \ll \pi_{\gamma A}$

$$h_{cr} = 0.22 \left(\frac{AR^2}{\gamma f} \right)^{1/4}, \tag{7.26}$$

where f is a factor that depends on b.

Scheludko's and Vrij's equations (7.24) and (7.26) have the same form at small thicknesses. Equation (7.26) predicts that when $\gamma \to 0$, the film should spontaneously rupture at large h values. This is certainly not observed since when $\gamma \to 0$ the emulsion becomes highly stable. Also equation (7.26) predicts that as $R \to 0$, $h_{cr} \to 0$, i.e., very small droplets should never rupture. Experiments on aqueous foam films suggest that one observes finite values for b_{cr} as $R \to 0$. Experiments on emulsion droplets showed no changes in b_{cr} with change in the size of the contact area. This is because the lamella formed between two oil droplets, at nonequilibrium separations, do not have the idealized, planer interface depicted in Fig. 7.10. Rather they have a "dimple" structure as illustrated in Fig. 7.13.

The "dimple" structure in Fig. 7.13 does not arise from fluctuations, but rather by an effect produced by the draining of the solution from the film region and associated with hydrodynamic effects. The thinnest region of the film occurs at the periphery of the contact zone, and rupture tends to occur here. With polymer-stabilized films, dimpling is far less marked due to the increased rigidity of the interface. The dimpling effect also accounts for the fact that the interfacial tension γ_0 seems to have little effect on film rupture in emulsion systems.

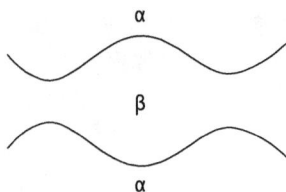

Fig. 7.13: Schematic representation of the "dimple" between two emulsion droplets.

A systematic study illustrating the use of polymeric surfactants in reducing or eliminating coalescence was carried out using isoparaffinic oil/water and silicone oil/water emulsions stabilized with a copolymer of linear polyfructose to which several C_{12} alkyl chains have been grafted on the polyfructose backbone (INUTEC® SP1, see Chapter 2) [31]. This graft copolymer is expected to be strongly adsorbed at the oil-water (O/W) interface by multipoint attachment with several alkyl chains, leaving several loops and tails of strongly hydrated polyfructose chains. A schematic representation of the adsorption and conformation of the polymeric surfactant at the O/W interface is shown in Fig. 7.14.

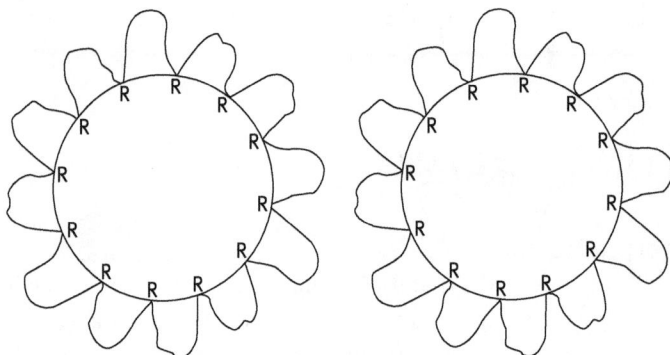

Fig. 7.14: Schematic representation of adsorption and conformation of INUTEC® SP1 at the O/W interface.

50/50 (v/v) ratio oil (Isopar M or cyclomethicone) in water emulsions were prepared using INUTEC® SP1 at concentrations from 0.25 to 2 (w/v)%. The emulsions were prepared using a high speed stirrer (Ultra-Turrax) and their quality was assessed using optical microscopy. The final emulsions were stored at room temperature and at 50 °C and all emulsions were diluted and their droplet size determined by optical microscopy and light diffraction techniques (using a Malvern Mastersizer). All emulsions containing INUTEC® SP1 at > 1 (w/v)% showed no increase in droplet size both at room temperature and at 50 °C. All emulsions were then prepared using 2 (w/v) INUTEC® SP1.

Emulsions were then prepared at 0.5, 1.0 and 2.0 mol dm^{-3} NaCl as well as in the presence of 0.5, 1.0, 1.5 and 2.0 mol dm^{-3} MgSO$_4$. All emulsions containing NaCl did not show any coalescence up to 50 °C for more than 1 year storage. With MgSO$_4$, the emulsions were stable up to 1.0 mol dm^{-3}, but those containing 1.5 and 2.0 MgSO$_4$ were only stable at room temperature and they showed coalescence at 50 °C. As an illustration, Fig. 7.15 shows optical micrographs of diluted Isopar M/water emulsions containing 2 (w/v) % INUTEC® SP1 that were stored for a period of 1.5 and 14 weeks at 50 °C.

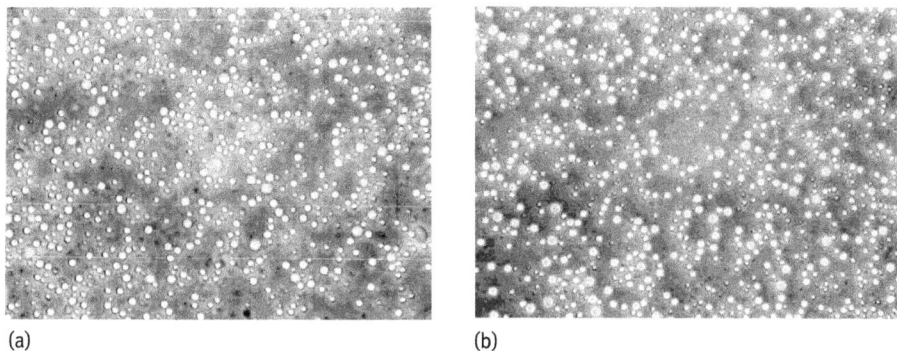

(a) (b)

Fig. 7.15: Optical micrographs of O/W emulsions stabilized with INUTEC® SP1 stored at 50 °C for 1.5 weeks (a) and 14 weeks (b).

The above stability in high electrolyte concentrations is not observed with polymeric surfactants containing poly(ethylene oxide) (PEO) as the stabilizing chain. The difference between the inulin and PEO chains can be understood if one considers the repulsive energy obtained when using polymeric surfactants. A shown in equation (7.1), the mixing free energy G$_{mix}$ for two droplets, stabilized by A chains with thickness δ, depends on the value of $(1/2 - \chi)$. When the latter is positive, i.e. $\chi < 1/2$, G$_{mix}$ is positive and the net interaction is repulsive. If $\chi > 1/2$, G$_{mix}$ is negative and this leads to incipient flocculation that is normally accompanied by coalescence.

The Flory–Huggins interaction parameter χ is related to the solvency of the medium for the chains. In water at room temperature, both inulin and PEO are strongly hydrated by water molecules and hence $\chi < 1/2$ under these conditions. On increasing the temperature, the H-bonds between the chains and the water molecules will be broken and the χ parameter will increase. With both inulin and PEO this will happen at much higher temperature than experienced on storage. The χ parameter is less than 1/2 below 80 °C for PEO and it remains $< 1/2$ for inulin up to 100 °C. Thus, in the absence of electrolyte, emulsions based on polymeric surfactants with PEO chains or inulin chains remain stable at temperatures > 80 °C.

On addition of electrolyte, dehydration of the A chains takes place and at a given electrolyte concentration and type, the χ parameter will change from $< 1/2$ to $> 1/2$ at a critical temperature and G_{mix} will change sign from being positive to being negative. This critical temperature is defined as the critical flocculation temperature (CFT). With inulin-based polymeric surfactant the CFT is higher than 80 °C up to 4 mol dm^{-3} NaCl and this explains the high stability of the emulsion in the presence of such high electrolyte concentration, prepared using inulin-based polymeric surfactants (INUTEC® SP1). In contrast with polymeric surfactants based on PEO chains, the CFT decreases significantly on addition of high NaCl concentrations reaching 40 °C at 4 mol dm^{-3} NaCl. This clearly shows that inulin (linear polyfructose) remains hydrated in the presence of high electrolyte concentrations and high temperature. In contrast, PEO becomes dehydrated at high temperature when high NaCl concentrations are added.

Evidence for the difference in hydration between inulin and PEO is obtained using cloud point measurements, as illustrated in Fig. 7.16. For inulin (INUTEC® N25) the solutions remain clear up to 100 °C in the presence of NaCl concentrations up to 4 mol dm^{-3} NaCl. In contrast, PEO shows a systematic reduction in the cloud point on addition of NaCl reaching 40 °C at 4 mol dm^{-3} NaCl.

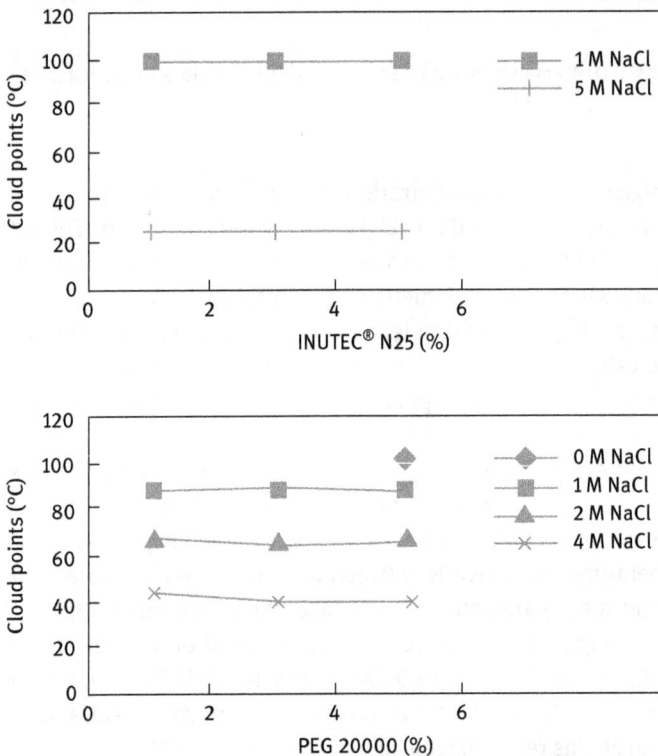

Fig. 7.16: Cloud points of INUTEC® N25 and PEO at various NaCl concentrations.

Direct evidence for the high emulsion stability when using INUTEC® SP1 was obtained from disjoining pressure measurements [32, 33]. These were obtained using the Scheludko–Exerowa cells that were described in detail by Exerowa and Kruglykov [34] for foam films. The setup is illustrated in Fig. 7.17 whereby the film between two oil droplets is investigated at constant capillary pressure (Fig. 7.17 (a)) using the so-called "pressure balance technique" (Fig. 7.17 (b)) that allows the direct measurement of disjoining pressure as a function of the thickness of the emulsion film [32, 33]. In both cases the film thickness is monitored by measuring the film reflectivity in monochromatic light using the microinterferometric method [34].

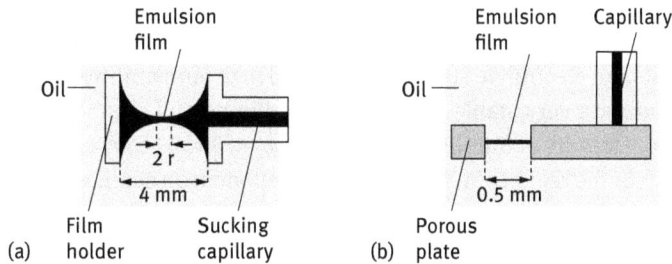

Fig. 7.17: Thin film cells: (a) Scheludko–Exerowa tube cell. (b) Exerowa–Scheludko porous plate cell.

Figure 7.18 shows the variation of the equivalent film thickness h_w with NaCl concentration at 22 °C at constant INUTEC® SP1 concentration $C_s = 2 \times 10^{-5}$ mol dm^{-3}, constant capillary pressure $P_c = 36$ Pa and film radius $r = 100$ µm. The equivalent film thickness is interferometrically determined with a film refractive index equal to that of the INUTEC® SP1 solution and 1.44 for the refractive index of the oil phase. As can be seen from Fig. 7.18, h_w decreases with increasing NaCl concentration to about 5×10^{-2} mol dm^{-3}, after which a constant film thickness value of 11 nm is maintained. The relatively large film thickness at low NaCl concentration ($< 5 \times 10^{-2}$ mol dm^{-3}) can be attributed to the contribution from electrostatic repulsion. The electrolyte concen-

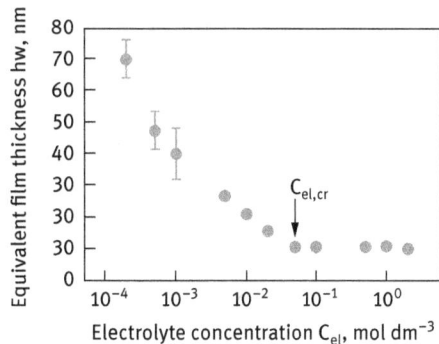

Fig. 7.18: Effect of NaCl concentration, C_e, on the equivalent film thickness, h_w, of O/W emulsion films from 2×10^{-5} mol dm^{-3} INUTEC® SP1 aqueous solutions.

tration at which a plateau in the h_w versus C_{el} curve is reached is called the "critical electrolyte concentration", $C_{el,cr}$. This concentration can be considered as the point of transition between DLVO and non-DLVO surface forces.

Figure 7.19 shows a plot of disjoining pressure versus equivalent film thickness between two emulsion droplets at various electrolyte concentrations. The results show that by increasing the capillary pressure a stable Newton Black Film (NBF) is obtained at a film thickness of ≈ 7 nm. The lack of rupture of the film at the highest pressure applied of 4.5×10^4 Pa indicates the high stability of the film in water and in high electrolyte concentrations (up to 2.0 mol dm^{-3} NaCl).

The lack of rupture of the NBF up to the highest pressure applied, namely 4.5×10^4 Pa, clearly indicates the high stability of the liquid film in the presence of high NaCl concentrations (up to 2 mol dm^{-3}). This result is consistent with the high emulsion stability obtained at high electrolyte concentrations and high temperature. Emulsions of Isopar M-in-water are very stable under such conditions and this could be accounted for by the high stability of the NBF. The droplet size of 50 : 50 O/W emulsed using 2 % INUTEC® SP1 is in the region of 1–10 μm. This corresponds to a capillary pressure of $\approx 3 \times 10^4$ Pa for the 1 μm drops and $\approx 3 \times 10^3$ Pa for the 10 μm drops. These capillary pressures are lower than those to which the NBF have been subjected and this clearly indicates the high stability obtained against coalescence in these emulsions.

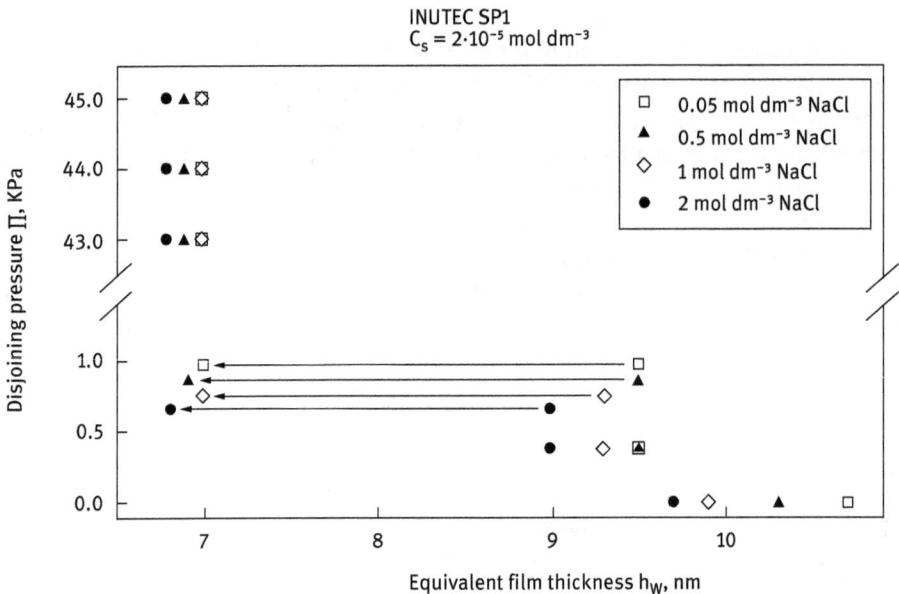

Fig. 7.19: Variation of disjoining pressure with equivalent film thickness at various NaCl concentrations.

References

[1] Tadros, Th. F. and Vincent, B., in "Encyclopedia of Emulsion Technology", P. Becher (ed.), Marcel Dekker, NY (1983).

[2] Binks, B. P. (ed.), "Modern Aspects of Emulsion Science", The Royal Society of Chemistry Publication (1998).

[3] Tadros, Th. F., "Applied Surfactants", Wiley-VCH, Germany (2005).

[4] Tadros, Th. F., in "Emulsion Formation and Stability", Th. F. Tadros (ed.), Wiley-VCH, Germany (2013), Chapter 1.

[5] Nakajima, H., Tomomossa, S. and Okabe, M., First Emulsion Conference, Paris (1993).

[6] Nakajima, H., in "Industrial Applications of Microemulsions", C. Solans and H. Konieda (eds.), Marcel Dekker (1997).

[7] Ugelstadt, J., El-Aassar, M. S. and Vanderhoff, J. W., J. Polym. Sci., 11, 503 (1973).

[8] El-Aasser, M., in "Polymeric Dispersions", J. M. Asua (ed.), Kluwer Academic Publications, The Netherlands (1997).

[9] Napper, D. H., "Polymeric Stabilisation of Colloidal Dispersions", Academic Press, London (1981).

[10] Tadros, Th. F., "Polymeric Surfactants", in "Encyclopedia of Colloid and Interface Science", Th. F. Tadros (ed.), Springer, Germany (2013).

[11] Tadros, Th. F., "Interfacial Phenomena and Colloid Stability", Vol. 1, De Gruyter, Germany (2015).

[12] Weers, J. G., "Molecular Diffusion in Emulsions and Emulsion Mixtures", in "Modern Aspects of Emulsion Science", B. P. Binks (ed.), Royal Society of Chemistry Publicaton, Cambridge, UK (2016).

[13] Thompson, W. (Lord Kelvin), Phil. Mag., 42, 448 (1871).

[14] Lifshitz I. M. and Slesov, V. V., Sov. Phys. JETP, 35, 331 (1959).

[15] Wagner, C., Z. Electrochem., 35, 581 (1961).

[16] Kabalnov, A. S. and Schukin, E. D., Adv. Colloid Interface Sci., 38, 69 (1992).

[17] Kabalnov, A. S., Makarov, K. N., Pertsov, A. V. and Schukin, E. D., J. Colloid Interface Sci., 138, 98 (1990).

[18] Taylor, P., Colloids and Surfaces A, 99, 175 (1995).

[19] Higuchi, W. I. and Misra, J., J. Pharm. Sci., 51, 459 (1962).

[20] Kabalnov, A. S., Pertsov, A. V. and Schukin, E. D., Colloids and Surfaces, 24, 19 (1987).

[21] Kabalnov, A. S., Pertsov, A. V., Aprosin, Yu. D. and Schukin, E. D., Kolloid Zh., 47, 1048 (1095).

[22] Walstra, P. in "Encyclopedia of Emulsion Technology", Vol. 4, P. Becher (ed.), Marcel Dekker, NY (1996).

[23] Kabalnov, A. S., Weers, J., Arlauskas, P. and Tarara, T., Langmuir, 11, 2966 (1995).

[24] Tadros, Th. F., "Nanodispersions", De Gruyter, Germany (2016).

[25] Tadros, Th. F., "Emulsions", De Gruyter, Germany (2016).

[26] Deryaguin, B. V. and Landau, L., Acta Physicochem. USSR, 14, 633 (1941).

[27] Verwey, E. J. W. and Overbeek, J. Th. G., "Theory of Stability of Lyophobic Colloids", Elsevier, Amsterdam (1948).

[28] Deryaguin, B. V. and Scherbaker, R. L., Kolloid Zh., 23, 33 (1961).

[29] Scheludko, A., Advances Colloid Interface Sci., 1, 391 (1967).

[30] Vrij, A., Discussion Faraday Soc., 42, 23 (1966).

[31] Tadros, Th. F., Vandamme, A., Levecke, B., Booten, K. and Stevens, C. V., Advances Colloid Interface Sci., 108–109, 207 (2004).

[32] Exerowa, D., Kolarev, T., Pigov, I, Levecke, B. and Tadros, Th. F., Langmuir, 22, 5013 (2006).

[33] Exerowa, D., Gotchev, G., Kolarev, T., Khristov, Khr., Levecke, B. and Tadros, Th. F., Langmuir, 23, 1711 (2007).

[34] Exerowa, D. and Kroglyakov, P. M., "Foam and Foam Films", Elsevier, Amsterdam (1998).

8 Polymeric surfactants for stabilizing suspensions

8.1 Introduction

The use of polymeric surfactants, both natural and synthetic, for stabilizing solid/liquid disperse systems (suspensions) plays an important role in industrial applications [1, 2], such as in paints, cosmetics, agrochemicals, ceramics, etc. Polymeric surfactants are particularly important for the preparation of concentrated suspensions, i.e. at high volume fraction ϕ of the disperse phase:

$$\phi = \frac{\text{volume of all particles}}{\text{total volume of suspension}}. \qquad (8.1)$$

Polymeric surfactants are also essential for stabilizing nonaqueous suspensions [2], since in this case electrostatic stabilization is not possible (due to the low dielectric constant of the medium). To understand the role of polymeric surfactants in suspension stability, it is essential to consider the adsorption and conformation of the polymeric surfactant at the solid/liquid interface [3] and this was discussed in detail in Chapter 4. With block polymeric surfactants used for the stabilization of aqueous suspensions, that may be represented as A–B or A–B–A structure, the hydrophobic "anchor" chain B becomes adsorbed on the hydrophobic particle leaving the strongly hydrated A chains dangling in solution [4]. The latter provides the steric repulsion [5], as discussed in Chapter 5, and it gives a hydrodynamic thickness, δ, that is determined by the number of monomer units in the A chains. With A–B–A type, the hydrophobic B chain (such as poly(propylene oxide), polystyrene or poly(methylmethacrylate) forms the "anchor" chain (by strong adsorption on the hydrophobic particle) and the A chains consist of hydrophilic components (such as ethylene oxide, EO, groups) and this provides the effective steric repulsion. In some cases a graft copolymer of one B chain with several A chains attached to the B backbone, i.e. BA_n is used. Again, the hydrodynamic thickness of the layer δ is determined by the number of monomer units in the A chains.

Block and graft copolymers based on poly(12-hydroxystearic acid) are used for nonaqueous suspensions. The poly(12-hydroxystearic acid) chains (the A chains of an A–B–A block copolymer or the A side chains of a graft copolymer), which are of low molecular weight (≈ 1000 Da), provide steric stabilization analogous to how PEO behaves in aqueous solution. The B anchor chains are chosen to be highly insoluble in the nonaqueous medium and have some specific interaction with the surface of the particle.

In this chapter, I will start with a list of the most commonly used polymeric surfactants for stabilizing aqueous suspensions. These systems were described in detail in Chapter 2 and only a summary is given here. This is followed by a summary highlighting the criteria for effective steric stabilization. Examples of suspensions that are stabilized by polymeric surfactants are given to illustrate the stabilization mechanism.

DOI 10.1515/9783110487282-009

Two main systems are described, namely polystyrene dispersions and aqueous suspensions of hydrophobic organic particles that are commonly used in agrochemical suspensions concentrates. Finally, the methods that may be applied to assess the stability/flocculation of suspensions are briefly discussed.

8.2 Examples of polymeric surfactants for aqueous suspensions

An example of a homopolymer that is used in many pharmaceutical formulations is poly(vinyl pyrrolidone) that is formed from repeating units of vinyl pyrrolidone.

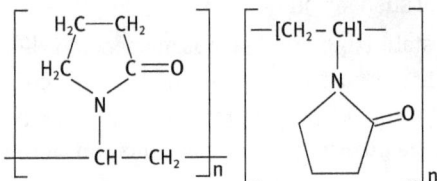

Such a homopolymer may adsorb significantly at the S/L interface. Even if the adsorption energy per monomer segment to the surface is small (fraction of kT, where k is the Boltzmann constant and T is the absolute temperature), the total adsorption energy per molecule may be sufficient to overcome the unfavourable entropy loss of the molecule at the S/L interface.

In general, homopolymers are not the most suitable stabilizers for suspensions, since the total adsorption energy may not be high enough to ensure irreversible adsorption. The latter can be achieved by using polymeric surfactants that contain specific groups that have high affinity to the surface and that are randomly attached to the polymer chain. This is exemplified by partially hydrolysed poly(vinyl acetate) (PVAc), technically referred to as poly(vinyl alcohol) (PVA). The polymer is prepared by partial hydrolysis of PVAc, leaving some residual vinyl acetate groups. Most commercially available PVA molecules contain 4–20 mol % acetate groups. These acetate groups, which are hydrophobic, give the molecule its amphipathic character. This blocky distribution of acetate groups on the poly(vinyl alcohol) backbone provides more effective anchoring of the polymeric surfactant chain on the particles. On a hydrophobic surface such as polystyrene, the polymer adsorbs with preferential attachment of the acetate groups on the surface, leaving the more hydrophilic vinyl alcohol segments dangling in the aqueous medium.

The most convenient polymeric surfactants for the stabilization of suspensions are those of the block and graft copolymer type [6]. As mentioned in Chapter 2, a block copolymer is a linear arrangement of blocks of variable monomer composition. The nomenclature for a diblock is poly-A-block-poly-B and for a triblock it is poly-A-block-poly-B-poly-A. One of the most widely used triblock polymeric surfactants for suspensions are the "Pluronics" or "Poloxamers" (BASF, Germany), which consist of two poly-A blocks of poly(ethylene oxide) (PEO) and one block of poly(propylene oxide) (PPO). Several chain lengths of PEO and PPO are available as indicated in the chemical structure below.

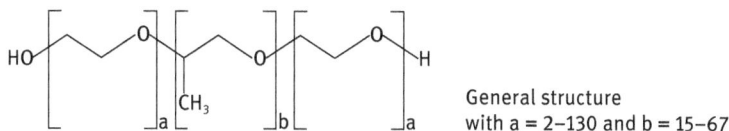

General structure
with a = 2–130 and b = 15–67

For Pluronics, the tradename is coded with a letter L (liquid), P (paste) and F (flake) that defines its physical form at room temperature. This is followed by two or three digits; the first one or two digits multiplied by 300 indicate the approximate molar mass of PPO and the last digit multiplied by 10 indicates the percentage of PEO. For example, Pluronic L61 indicates a liquid with PPO molar mass of 1800 and 10 % PEO. Pluronic F127 indicates a flake with PPO molar mass 3600 and 70 % PEO. The poloxamers (which are FDA approved) are commonly named with the letter P followed by three digits; the first two digits multiplied by 10 give the approximate molar mass of PPO and the last digit multiplied by 10 gives the percentage of PEO. For example, Poloxamer P407 has a molar mass of PPO of 400 and 70 % PEO. These polymeric triblocks can be applied as dispersants, whereby the assumption is made that the hydrophobic PPO chain resides at the hydrophobic surface, leaving the two PEO chains dangling in aqueous solution and hence providing steric repulsion. Although these triblock polymeric surfactants have been widely used in various applications in suspensions, some doubt has arisen on how effective these can be. It is generally accepted that the PPO chain is not sufficiently hydrophobic to provide a strong "anchor" to a hydrophobic surface. Several other di- and triblock copolymers have been synthesized, although these are of limited commercial availability. Typical examples are diblocks of polystyrene-block-polyvinyl alcohol, triblocks of poly(methyl methacrylate)-block poly(ethylene oxide)-block poly(methyl methacrylate), diblocks of polystyrene block-polyethylene oxide and triblocks of polyethylene oxide-block polystyrene-polyethylene oxide.

An alternative (and perhaps more efficient) polymeric surfactant is the amphipathic graft copolymer consisting of a polymeric backbone B (polystyrene or polymethyl methacrylate) and several A chains ("teeth") such as polyethylene oxide. This graft copolymer is sometimes referred to as a "comb" stabilizer. This copolymer is usually prepared by grafting a macromonomer such methoxy polyethylene oxide methacrylate with polymethyl methacrylate. BA$_n$ graft copolymer based on polymethyl methacrylate (PMMA) backbone (with some polymethacrylic acid) on which several PEO chains (with average molecular weight of 750) are grafted (e.g. TERSPERSE ® 2500 supplied by Huntsman, Belgium):

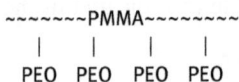

~~~~~~~PMMA~~~~~~~~
   | | | |
PEO PEO PEO PEO

Another example is styrene/maleic anhydride copolymer with grafted pendants chains (PEO/PPO) (e.g. TERSPERSE® 2612, Huntsman), schematically shown in Fig. 8.1.

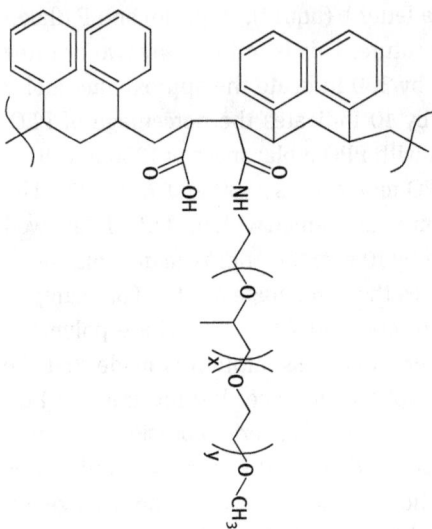

Fig. 8.1: Schematic structure of TERSPERSE 2612.

Graft copolymers based on polysaccharides have been developed for the stabilization of disperse systems. One of the most useful graft copolymers is that based on inulin obtained from chicory roots [7–9]. It is a linear polyfructose chain (with a degree of polymerization > 23) with a glucose end. The latter molecule is used to prepare a series of graft copolymers by random grafting of alkyl chains (using alkyl isocyanate) onto the inulin backbone. The first molecule of this series is INUTEC® SP1 (Beneo-Remy, Belgium) that is obtained by random grafting of C_{12} alkyl chains. It has an average molecular weight of ≈ 5000 Da and its structure is given in Fig. 8.2. The molecule is schematically illustrated in Fig. 8.3, which shows the hydrophilic polyfructose chain (backbone) and the randomly attached alkyl chains [7].

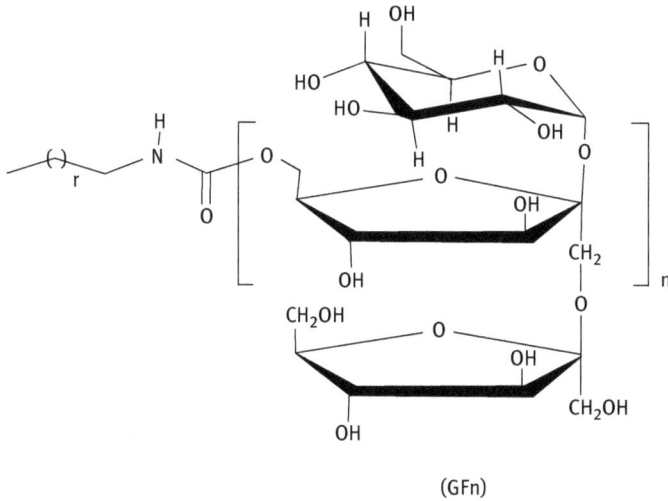

(GFn)

Fig. 8.2: Structure of INUTEC® SP1.

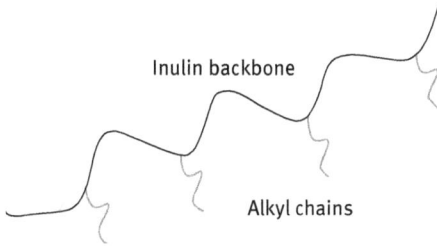

Fig. 8.3: Schematic representation of INUTEC® SP1 polymeric surfactant.

The main advantages of INUTEC® SP1 as a stabilizer for suspensions [9] are:
(i) Strong adsorption to the particle by multipoint attachment with several alkyl chains. This ensures lack of desorption and displacement of the molecule from the interface.
(ii) Strong hydration of the linear polyfructose chains both in water and in the presence of high electrolyte concentrations and high temperatures. This ensures effective steric stabilization.

8.3 Criteria for effective stabilization of suspensions

For effective stabilization of the suspension against flocculation one requires a dispersing agent which will be strongly adsorbed at the S/L interface and produce an effective repulsive barrier on close approach of the particles [2]. The above repulsive barrier is particularly important for concentrated suspensions (that contain more than 50 % by volume of solids).

It is essential to have strong "anchoring" (adsorption) to the particle surface: this requires multipoint attachment of the dispersant to the particle surface. The "anchor'" B chain should be insoluble in the medium and should have strong affinity to the surface. The adsorption strength is measured in terms of the segment/surface energy of adsorption χ^s. The total adsorption energy is given by the product of the number of attachment points, n, and χ^s. The total value of $n\chi^s$ should be sufficiently high for strong and irreversible adsorption. The χ^s of one segment of the B chain of the dispersant must be sufficiently high to prevent displacement of the polymeric surfactant by other surfactant molecules used as wetting agents. It is, therefore, essential to make sure that the χ^s per segment of the B chain is higher than that of the wetter or surfactant molecule.

The stabilizing chain A of the polymeric surfactant must provide an effective repulsive barrier to prevent flocculation by van der Waals attraction. Two main mechanisms of stabilization by polymeric surfactants can be considered: steric, as produced by nonionic polymeric surfactants of the A–B, A–B–A or BA_n graft copolymers; and electrosteric, as produced by mixtures of polymeric surfactants and ionic surfactants or polyelectrolytes [2].

The third and most important criterion for effective stabilization by polymeric surfactants is the solvency of the stabilizing chain by the molecules of the dispersion medium. The stabilizing chain A of the A–B, A–B–A or BA_n graft copolymer should be in a good solvent condition, i.e. very soluble in the medium and strongly solvated by its molecules [4]. As discussed in Chapter 5, solvation of the chain by the medium is determined by the chain/solvent (Flory–Huggins) interaction parameter, χ. In good solvent conditions, $\chi < 0.5$ and hence the mixing or osmotic interaction is positive (repulsive). χ should be maintained at < 0.5 under all conditions, e.g. low and high temperature, in the presence of electrolytes and other components of the formulation such as addition of anti-freeze (mostly propylene glycol).

It is essential to have an optimum adsorbed layer thickness to prevent the formation of a deep minimum in the energy-distance curve which may result in flocculation, particularly when using concentrated suspensions. The adsorbed layer thickness of the A chains, usually described by a hydrodynamic value δ_h (i.e. the thickness δ plus any contribution from the solvation shell), should be sufficiently large (> 5 nm) to prevent flocculation (although reversible) and an increase in the viscosity of the suspension.

8.4 Polymeric surfactants for stabilizing preformed latex dispersions

For this purpose polystyrene (PS) latexes were prepared using surfactant-free emulsion polymerization. Two latexes with z-average diameter of 427 and 867 (as measured using Photon Correlation Spectroscopy, PCS) that are reasonably monodisperse were prepared [10]. Two polymeric surfactants, namely Hypermer CG-6 and Atlox 4913 (UNIQEMA, UK) were used [10]. Both are graft ("comb") type, consisting of polymethylmethacrylate/polymethacrylic acid (PMMA/PMA) backbone with methoxy-capped polyethylene oxide (PEO) side chains ($M = 750$ Da). Hypermer CG-6 is the same graft copolymer as Atlox 4913 but it contains a higher proportion of methacrylic acid in the backbone. The average molecular weight of the polymer is ≈ 5000 Da. Figure 8.4 shows a typical adsorption isotherm of Atlox 4913 on the two latexes.

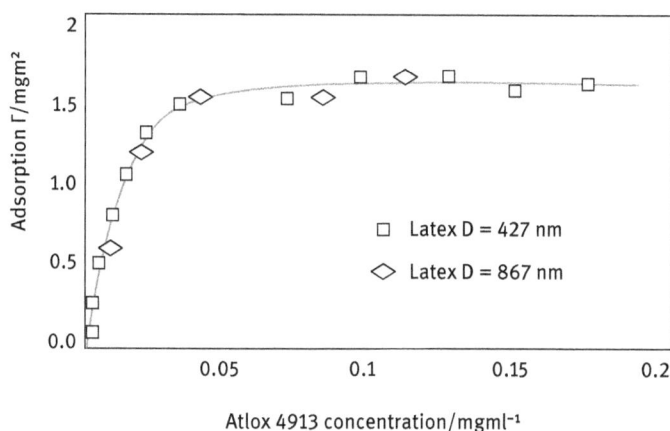

Fig. 8.4: Adsorption isotherms of Atlox 4913 on the two latexes at 25 °C.

Similar results were obtained for Hypermer CG-6, but the plateau adsorption was lower (1.2 mg m^{-2} compared with 1.5 mg m^{-2} for Atlox 4913). It is likely that the backbone of Hypermer CG-6, which contains more PMA, is more polar and hence less strongly adsorbed. The amount of adsorption was independent of particle size.

The influence of temperature on adsorption is shown in Fig. 8.5. The amount of adsorption increases with increasing temperature. This is due to the poorer solvency of the medium for the PEO chains. The PEO chains become less hydrated at higher temperature and the reduction of solubility of the polymer enhances adsorption.

The adsorbed layer thickness of the graft copolymer on the latexes was determined using rheological measurements. Steady state (shear stress σ – shear rate γ) measurements were carried out and the results were fitted to the Bingham equation to

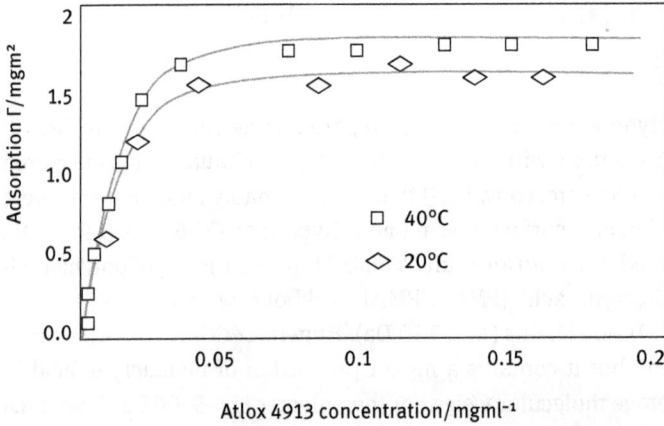

Fig. 8.5: Effect of temperature on adsorption of Atlox 4913 on PS.

obtain the yield value σ_β and the high shear viscosity η of the suspension [10]:

$$\sigma = \sigma_\beta + \eta\dot\gamma. \tag{8.2}$$

As an illustration, Fig. 8.6 shows a plot of σ_β versus volume fraction ϕ of the latex for Atlox 4913. Similar results were obtained for latexes stabilized using Hypermer CG-6.

At any given volume fraction, the smaller latex has higher σ_β when compared to the larger latex. This is due to the higher ratio of adsorbed layer thickness to particle radius, Δ/R, for the smaller latex. The effective volume fraction of the latex, ϕ_{eff}, is

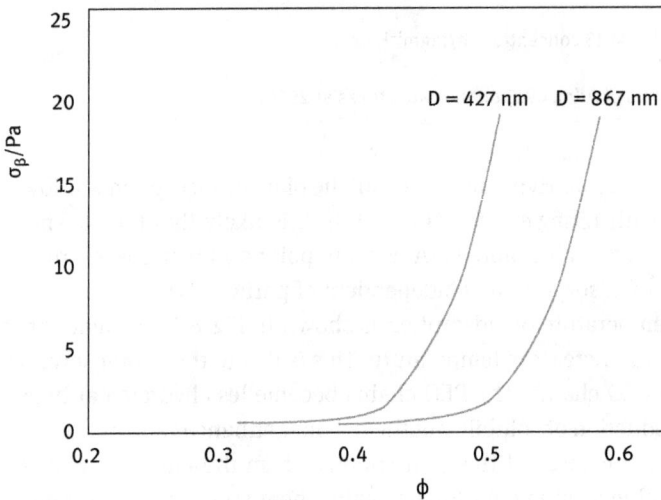

Fig. 8.6: Variation of yield stress with latex volume fraction for Atlox 4913.

related to the core volume fraction ϕ by the equation [10]

$$\phi_{\text{eff}} = \phi \left[1 + \frac{\Delta}{R} \right]^3 . \tag{8.3}$$

ϕ_{eff} can be calculated from the relative viscosity, η_r, using the Dougherty–Krieger equation [11, 12],

$$\eta_r = \left[1 - \left(\frac{\phi_{\text{eff}}}{\phi_p} \right) \right]^{-[\eta]\phi_p} , \tag{8.4}$$

where ϕ_p is the maximum packing fraction.

The maximum packing fraction, ϕ_p, can be calculated using the following empirical equation:

$$\frac{(\eta_r^{1/2} - 1)}{\phi} = \left(\frac{1}{\phi_p} \right) (\eta^{1/2} - 1) + 1.25. \tag{8.5}$$

The results showed a gradual decrease of adsorbed layer thickness Δ with increasing volume fraction ϕ. For the latex with diameter D of 867 nm and Atlox 4913, Δ decreased from 17.5 nm at $\phi = 0.36$ to 6.5 nm at $\phi = 0.57$. For Hypermer CG-6 with the same latex, Δ decreased from 11.8 nm at $\phi = 0.49$ to 6.5 nm at $\phi = 0.57$. The reduction of Δ with increasing ϕ may be due to overlap and/or compression of the adsorbed layers as the particles come close to each other at higher volume fraction of the latex.

The stability of the latexes was determined using viscoelastic measurements [13]. For this purpose, dynamic (oscillatory) measurements were used to obtain the storage modulus G^*, the elastic modulus G' and the viscous modulus G'' as a function of strain amplitude, γ_0, and frequency, ω (rad s^{-1}). The method relies on the application of a sinusoidal strain or stress and the resulting stress or strain is measured simultaneously. For a viscoelastic system, the strain and stress sine waves oscillate with the same frequency but out of phase. From the time shift Δt and ω, one can obtain the phase angle shift δ. This will be discussed below.

G' is measured as a function of electrolyte concentration and/or temperature to assess the latex's stability. As an illustration, Fig. 8.7 shows the variation of G' with temperature for latex stabilized with Atlox 4913 in the absence of any added electrolyte and in the presence of 0.1, 0.2 and 0.3 mol dm^{-3} Na$_2$SO$_4$. In the absence of electrolyte G' showed no change with temperature up to 65 °C.

In the presence of 0.1 mol dm^{-3} Na$_2$SO$_4$, G' remained constant up to 40 °C above which G' increased with a further increase in temperature. This temperature is denoted as the critical flocculation temperature (CFT). The CFT decreases with increasing electrolyte concentration reaching ≈ 30 °C in 0.2 and 0.3 mol dm^{-3} Na$_2$SO$_4$. This reduction in CFT with increasing electrolyte concentration is due to the reduction in solvency of the PEO chains with increasing electrolyte concentrations. The latex stabilized with Hypermer CG-6 gave relatively higher CFT values when compared to that stabilized using Atlox 4913.

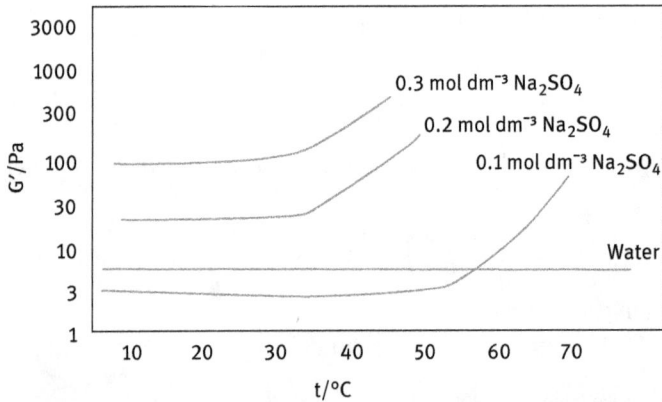

Fig. 8.7: Variation of G' with temperature in water and at various Na_2SO_4 concentrations.

8.5 Polymeric surfactants for stabilizing aqueous suspensions of hydrophobic organic particles

Results were obtained using two polymeric surfactants, namely PVA and Atlox 4913, and a suspension of ethirimol, an agrochemical fungicide [14, 15]. The adsorption isotherms for the two polymeric surfactants are shown in Fig. 8.8 and 8.9. The high affinity type isotherm is clearly demonstrated, indicating that in both cases adsorption was strong and irreversible.

However, the amount of adsorption per unit area (using the BET surface area of ethirimol of $0.22\,m^2\,g^{-1}$ obtained by Kr adsorption) was significantly higher than the values obtained on the model particles of polystyrene. This could be due to the errors involved in surface area determination of such coarse particles using BET gas adsorption.

These polymeric surfactants contain sufficient hydrophobic groups to ensure their adsorption and enough ethylene oxide units to provide an adequate energy barrier. This is particularly the case with the graft copolymer Atlox 4913 whereby the B chain (polymethyl methacrylate-methacrylic acid, the anchoring portion) adsorbs with sev-

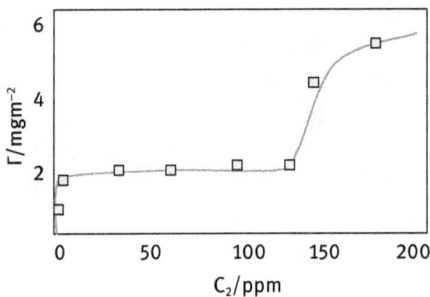

Fig. 8.8: Adsorption of PVA on ethirimol.

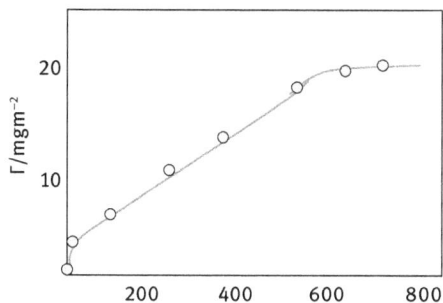

Fig. 8.9: Adsorption of the "comb" graft copolymer on ethirimol.

eral segments (small loops) on the hydrophobic particles leaving several strongly hydrated PEO chains (the "teeth") that provide effective steric repulsion. Using such a polymeric surfactant, a suspension of high volume fractions can be prepared.

8.6 Assessment of the long-term physical stability of suspensions

For the full assessment of the properties of suspension concentrates, three main types of investigations are needed:
(i)   Fundamental investigation of the system at a molecular level.
(ii) Investigations into the state of the suspension on standing.
(iii) Bulk properties of the suspension.

All the above investigations require a number of sophisticated techniques such as polymer adsorption and their conformation at the solid/liquid interface, measurement of the rate of flocculation and several rheological measurements. Apart from these practical methods which are feasible in most industrial laboratories, more fundamental information can be obtained using modern sophisticated techniques such as small angle X-ray and neutron scattering measurements, ultrasonic absorption techniques, etc. Several other modern techniques are also now available for investigating the state of the suspension: Freeze fracture and electron microscopy, atomic force microscopy, scanning tunnelling microscopy and confocal laser microscopy.

In all the above methods, care should be taken in sampling the suspension, which should cause as little disturbance as possible to the "structure" to be investigated. For example, when one investigates the flocculation of a concentrated suspension, dilution of the system for microscopic investigation may lead to a breakdown of the flocs and a false assessment is obtained. The same applies when one investigates the rheology of a concentrated suspension, since transfer of the system from its container to the rheometer may lead to a breakdown of the structure.

For the above reasons one must establish well-defined procedures for every technique and this requires a great deal of skill and experience. It is advisable in all cases to develop standard operation procedures for the above investigations.

8.6.1 Assessment of the structure of the solid/liquid interface

8.6.1.1 Polymer adsorption

A representative sample of the solid with known mass m and surface area A per gram is equilibrated with a surfactant or polymer concentration with concentration C_1. After equilibrium is reached (at a given constant temperature), the solid is removed by centrifugation and the equilibrium concentration C_2 is determined analytically [16, 17].

The amount of adsorption Γ in mol m^{-2} is given by

$$\Gamma = \frac{(C_1 - C_2)}{mA} = \frac{\Delta C}{mA}. \tag{8.6}$$

In most cases (particularly with surfactants) a plot of Γ versus C_2 gives a Langmuir type isotherm. The data can be fitted using the Langmuir equation,

$$\Gamma = \frac{\Gamma_\infty b C_2}{(1 + b C_2)}, \tag{8.7}$$

where b is a constant that is related to the free energy of adsorption

$$b \propto \left(-\frac{\Delta G_{ads}}{RT}\right). \tag{8.8}$$

Most polymers (particularly those with high molecular weight) give a high affinity isotherm, as discussed before.

8.6.1.2 Assessment of the state of the dispersion

Two general techniques may be applied for measuring the rate of flocculation of suspensions, both of which can only be applied for dilute systems. The first method is based on measuring the scattering of light by the particles. For monodisperse particles with a radius that is less than $\lambda/20$ (where λ is the wavelength of light), one can apply the Rayleigh equation, whereby the turbidity τ_0 is given by

$$\tau_0 = A' n_0 V_1^2, \tag{8.9}$$

where A' is an optical constant (which is related to the refractive index of the particle and medium and the wavelength of light) and n_0 is the number of particles, each with a volume V_1.

By combining the Rayleigh theory with the Smoluchowski–Fuchs theory of flocculation kinetics [18, 19], one can obtain the following expression for the variation of turbidity with time:

$$\tau = A' n_0 V_1^2 (1 + 2 n_0 k t), \tag{8.10}$$

where k is the rate constant of flocculation.

The second method for obtaining the rate constant of flocculation is by direct particle counting as a function of time. For this purpose optical microscopy or image analysis may be used, provided the particle size is within the resolution limit of the microscope. Alternatively, the particle number may be determined using electronic devices such as the Coulter counter or the flow ultramicroscope.

The rate constant of flocculation is determined by plotting $1/n$ versus t, where n is the number of particles after time t, i.e.

$$\left(\frac{1}{n}\right) = \left(\frac{1}{n_0}\right) + kt. \tag{8.11}$$

The rate constant k of slow flocculation is usually related to the rapid rate constant k_0 (the Smoluchowski rate) by the stability ratio W [19]:

$$W = \left(\frac{k}{k_0}\right). \tag{8.12}$$

One usually plots $\log W$ versus $\log C$ (where C is the electrolyte concentration) to obtain the critical coagulation concentration (ccc), which is the point at which $\log W = 0$.

Incipient flocculation for sterically stabilized suspensions is determined when the medium for the chains becomes a θ-solvent [5]. This occurs, for example, on heating an aqueous suspension stabilized with poly(ethylene oxide) (PEO) or poly(vinyl alcohol) chains. Above a certain temperature (the θ-temperature) that depends on electrolyte concentration, flocculation of the suspension occurs. The temperature at which this occurs is defined as the critical flocculation temperature (CFT).

This process of incipient flocculation can be followed by measuring the turbidity of the suspension as a function of temperature. Above the CFT, the turbidity of the suspension rises very sharply. For the above purpose, the cell in the spectrophotometer that is used to measure the turbidity is placed in a metal block that is connected to a temperature programming unit (which allows one to increase the temperature at a controlled rate).

8.6.2 Bulk properties of suspensions

For a "structured" suspension, obtained by "controlled flocculation" or addition of "thickeners" (such polysaccharides, clays or oxides), the "flocs" sediment at a rate depending on their size and porosity of the aggregated mass. After this initial sedimentation, compaction and rearrangement of the floc structure occurs, a phenomenon referred to as consolidation.

Normally in sediment volume measurements, one compares the initial volume V_0 (or height H_0) with the ultimately reached value V (or H). A colloidally stable suspension gives a "close-packed" structure with relatively small sediment volume (dilatant sediment referred to as clay). A weakly "flocculated" or "structured" suspension gives a more open sediment and hence a higher sediment volume. Thus by comparing the relative sediment volume V/V_0 or height H/H_0, one can distinguish between a clayed and a flocculated suspension [20].

A more quantitative method for assessing the bulk properties of suspensions is to use rheological measurements [13]. Three different rheological measurements may be applied:
(i) Steady state shear stress-shear rate measurements (using a controlled shear rate instrument).
(ii) Constant stress (creep) measurements (carried out using a constant stress instrument).
(iii) Dynamic (oscillatory) measurements (preferably carried out using a constant strain instrument).

The above rheological techniques can be used to assess sedimentation and flocculation of suspensions [13]. The rate of sedimentation decreases with increasing volume fraction of the disperse phase, ϕ, and ultimately it approaches zero at a critical volume fraction ϕ_p (the maximum packing fraction). However, at $\phi \approx \phi_p$, the viscosity of the system approaches ∞. Thus, for most practical emulsions, the system is prepared at ϕ values below ϕ_p and then "thickeners" are added to reduce sedimentation. These "thickeners" are usually high molecular weight polymers (such as xanthan gum, hydroxyethyl cellulose or associative thickeners), finely divided inert solids (such as silica or swelling clays) or a combination of the two.

In all cases, a "gel" network is produced in the continuous phase which is shear thinning (i.e. its viscosity decreases with increasing shear rate) and viscoelastic (i.e. it has viscous and elastic components of the modulus). If the viscosity of the elastic network, at shear stresses (or shear rates) comparable to those exerted by the particles, exceeds a certain value, then sedimentation is completely eliminated.

The shear stress, σ_p, exerted by a particle (force/area) can be simply calculated,

$$\sigma_p = \frac{(4/3)\pi R^3 \Delta \rho g}{4\pi R^2} = \frac{\Delta \rho R g}{3}. \tag{8.13}$$

For a 10 μm radius particle with density difference $\Delta \rho = 0.2$, σ_p is equal to

$$\sigma_p = \frac{0.2 \times 10^3 \times 10 \times 10^{-6} \times 9.8}{3} \approx 6 \times 10^{-3}. \tag{8.14}$$

For smaller particles smaller stresses are exerted.

Thus, to predict sedimentation, one has to measure the viscosity at very low stresses (or shear rates). These measurements can be carried out using a constant stress rheometer (Carrimed, Bohlin, Rheometrics or Physica). A constant stress σ (using for example a drag cup motor that can apply very small torques and using an air bearing system to reduce the frictional torque) is applied on the system (which may be placed in the gap between two concentric cylinders or a cone-plate geometry) and the deformation (strain γ or compliance $J = \gamma/\sigma = Pa^{-1}$) is followed as a function of time [13].

For a viscoelastic system, the compliance shows a rapid elastic response J_0 at $t \to 0$ (instantaneous compliance $J_0 = 1/G_0$, where G_0 is the instantaneous modulus

Creep is the sum of a constant value $J_e\sigma_0$ (elastic part) and a viscous contribution $\sigma_0 t/\eta_0$

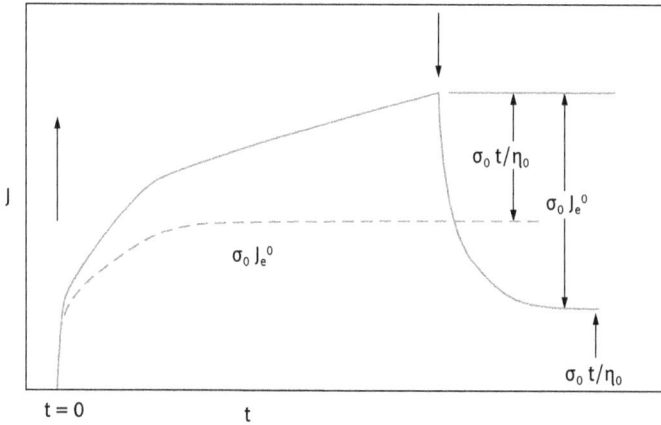

Fig. 8.10: Creep curve for a viscoelastic liquid.

that is a measure of the elastic (i.e. "solid-like") component). At $t > 0$, J increases slowly with time and this corresponds to the retarded response ("bonds" are broken and reformed but not at the same rate). After a certain time period (that depends on the system), the compliance shows a linear increase with time (i.e. the system reaches a steady state with constant shear rate). If the stress is removed after the steady state is reached, elastic recovery occurs and the strain changes sign. The above behaviour (usually referred to as "creep") is schematically represented in Fig. 8.10.

The slope of the linear part of the creep curve gives the value of the viscosity at the applied stress, η_σ:

$$\frac{J}{t} = \frac{Pa^{-1}}{s} = \frac{1}{Pa\,s} = \eta_\sigma. \tag{8.15}$$

The recovery curve will only give the elastic component, which if superimposed on the ascending part of the curve will give the viscous component.

Thus, one measures creep curves as a function of the applied stress (starting from a very small stress of the order of 0.01 Pa). This is illustrated in Fig. 8.11.

The viscosity η_σ (which is equal to the reciprocal of the slope of the straight portion of the creep curve) is plotted as a function of the applied stress. This is schematically shown in Fig. 8.12.

Below a critical stress, σ_{cr}, the viscosity reaches a limiting value, $\eta(0)$ namely the residual (or zero shear) viscosity. σ_{cr} may be denoted as the "true yield stress" of the emulsion, i.e. the stress above which the "structure" of the system is broken down. Above σ_{cr}, η_σ decreases rapidly with a further increase in shear stress (the shear thinning regime). It reaches another Newtonian value, $\eta(\infty)$, which is the high shear limiting viscosity.

Creep measurements (Constant stress) can be used to obtain the residual or zero shear viscosity

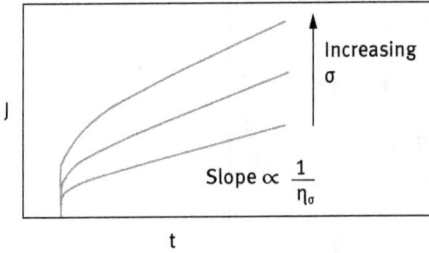

Fig. 8.11: Creep curves at increasing applied stress.

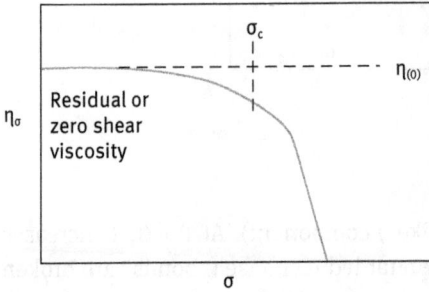

Critical stress is a useful parameter (related to yield stress) as denotes the stress at which stucture "breaks down"

Fig. 8.12: Variation of viscosity with applied stress.

$\eta(0)$ could be several orders of magnitudes (10^4–10^8) higher than $\eta(\infty)$. Usually one obtains good correlation between the rate of sedimentation, v, and the residual viscosity, $\eta(0)$. Above a certain value of $\eta(0)$, v becomes equal to 0. Clearly, to minimize sedimentation one has to increase $\eta(0)$; an acceptable level for the high shear viscosity $\eta(\infty)$ must be achieved, depending on the application. In some cases, a high $\eta(0)$ may be accompanied by a high $\eta(\infty)$ (which may not be acceptable for the application, for example if spontaneous dispersion on dilution is required). If this is the case, the formulation chemist should look for an alternative thickener.

Another problem encountered with many suspensions is that of "syneresis", i.e. the appearance of a clear liquid film at the top of the suspension. "Syneresis" occurs with most "flocculated" and/or "structured" (i.e. those containing a thickener in the continuous phase) suspensions. "Syneresis" may be predicted from measuring the yield value (using steady state measurements of shear stress as a function of shear rate) as a function of time or using oscillatory techniques (whereby the storage and loss modulus are measured as a function of strain amplitude and frequency of oscillation). The above techniques will be discussed in detail below.

It is sufficient to state in this section that when a network of the suspension particles (either alone or combined with the thickener) is produced, the gravity force will cause some contraction of the network (which behaves as a porous plug), thus causing some separation of the continuous phase which is entrapped between the droplets in the network.

As mentioned in Chapter 6, flocculation of suspensions is the result of long-range van der Waals attraction. Flocculation can be weak (and reversible) or strong, depending on the magnitude of the net attractive forces. Weak flocculation may result in reversible time dependency of the viscosity, i.e. on shearing the suspension at a given shear rate, the viscosity decreases and on standing the viscosity recovers to its original value. This phenomenon is referred to as thixotropy (sol $\leftrightarrow$ gel transformation).

Rheological techniques are most convenient to assess suspension flocculation without the need of any dilution (which in most cases results in a breakdown of the floc structure). In steady state measurements the suspension is carefully placed in the gap between concentric cylinder or cone-and-plate platens. For the concentric cylinder geometry, the gap width should be at least 10× larger than the largest particle size (a gap width that is greater than 1 mm is usually used). For the cone-and-plate geometry, a cone angle of 4° or smaller is usually used.

A controlled rate instrument is usually used for the above measurements; the inner (or outer) cylinder, the cone (or the plate) is rotated at various angular velocities (which allows one to obtain the shear rate, $\dot{\gamma}$) and the torque is measured on the other element (this allows one to obtain the stress σ).

For a Newtonian system (such as is the case of a dilute suspension, with a volume fraction ϕ less than 0.1), σ is related to $\dot{\gamma}$ by the equation

$$\sigma = \eta\dot{\gamma}, \tag{8.16}$$

where η is the Newtonian viscosity (that is independent of the applied shear rate).

For most practical suspensions (with $\phi > 0.1$ and containing thickeners to reduce sedimentation) a plot of σ versus γ is not linear (i.e. the viscosity depends on the applied shear rate). The most common flow curve is shown in Fig. 8.13 (usually described as a pseudoplastic or shear thinning system). In this case the viscosity decreases with increasing shear rate, reaching a Newtonian value above a critical shear rate.

Several models may be applied to analyse the results of Fig. 8.13.

(a) Power law model

$$\sigma = k\dot{\gamma}^n, \tag{8.17}$$

where k is the consistency index of the suspension and n is the power (shear thinning) index ($n < 1$); the lower the value of n, the more shear thinning the suspension is. This is usually the case with weakly flocculated suspensions or those to which a "thickener" has been added.

By fitting the results of Fig. 8.13 to equation (8.17) (this is usually in the software of the computer connected to the rheometer), one can obtain the viscosity of the

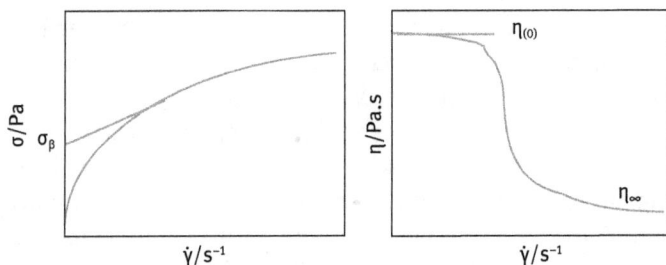

Fig. 8.13: Shear stress and viscosity versus shear rate for a pseudoplastic system.

suspension at a given shear rate,

$$\eta(\text{at a given shear rate}) = \frac{\sigma}{\gamma} = k\dot{\gamma}^{n-1}. \tag{8.18}$$

(b) Bingham model

$$\sigma = \sigma_\beta + \eta_{pl}\dot{\gamma}, \tag{8.19}$$

where σ_β is the extrapolated yield value (obtained by extrapolation of the shear stress-shear rate curve to $\gamma = 0$). Again this is provided in the software of the rheometer. η_{pl} is the slope of the linear portion of the $\sigma - \gamma$ curve (usually referred to as the plastic viscosity).

Both σ_β and η_{pl} may be related to the flocculation of the suspension. At any given volume fraction of the emulsion and at a given particle size distribution, the higher the value of σ_β and η_{pl}, the more flocculated the suspension is. Thus, if one stores a suspension at any given temperature and makes sure that the particle size distribution remains constant (i.e. no Ostwald ripening occurred), an increase in the above parameters indicates flocculation of the suspension on storage. Clearly, if Ostwald ripening occurs simultaneously, σ_β and η_{pl} may change in a complex manner with storage time. Ostwald ripening results in a shift of the particle size distribution to higher diameters; this has the effect of reducing σ_β and η_{pl}. If flocculation occurs simultaneously (having the effect of increasing these rheological parameters), the net effect may be an increase or decrease in the rheological parameters.

The above trend depends on the extent of flocculation relative to Ostwald ripening. Therefore, following σ_β and η_{pl} with storage time requires knowledge of Ostwald ripening. Only in the absence of this latter breakdown process can one use rheological measurements as a guide to assessing flocculation.

(c) Herschel–Buckley model [13]

In many cases, the shear stress-shear rate curve may not show a linear portion at high shear rates. In this case, the data may be fitted with a Hershel–Buckley model,

$$\sigma = \sigma_\beta + k\dot{\gamma}^n. \tag{8.20}$$

(d) Casson's model [13]

This is another semi-empirical model that may be used to fit the data of Fig. 8.13,

$$\sigma^{1/2} = \sigma_C^{1/2} + \eta_C^{1/2} \dot{\gamma}^{1/2}. \tag{8.21}$$

Note that σ_β is not equal to σ_C.

Equation (8.21) shows that a plot of $\sigma^{1/2}$ versus $\gamma^{1/2}$ gives a straight line from which σ_C and η_C can be evaluated.

In all the above analyses, the assumption was made that a steady state was reached. In other words, no time effects occurred during the duration of the flow experiment. Many suspensions (particularly those that are weakly flocculated or "structured" to reduce sedimentation) show time effects during flow. At any given shear rate, the viscosity of the suspension continues to decrease with increasing time of shear; on stopping the shear, the viscosity recovers to its initial value. This reversible decrease in viscosity is referred to as thixotropy.

The most common procedure for studying thixotropy is to apply a sequence of shear stress-shear rate regimes within controlled periods. If the flow curve is carried out within a very short time (say increasing the rate from 0 to say $500\,\mathrm{s}^{-1}$ in 30 s and then reducing it again from 500 to $0\,\mathrm{s}^{-1}$ within the same period), one finds that the descending curve is below the ascending one.

The above behaviour can be explained by considering the structure of the system. If, for example, the suspension is weakly flocculated, then on applying a shear force on the system, this flocculated structure is broken down (and this is the cause of the shear thinning behaviour). On reducing the shear rate back to zero the structure builds up only in part within the duration of the experiment (30 s).

The ascending and descending flow curves show hysteresis that is usually referred to as "thixotropic loop". If now the same experiment is repeated within a longer time experiment (say 120 s for the ascending and 120 s for the descending curves), the hysteresis decreases, i.e. the "thixotropic loop" becomes smaller.

The above study may be used to investigate the state of flocculation of a suspension. Weakly flocculated suspensions usually show thixotropy and the change of thixotropy with applied time may be used as an indication of the strength of this weak flocculation.

The above analysis is only qualitative and one cannot use the results in a quantitative manner. This is due to the possible breakdown of the structure on transferring the suspension to the rheometer and also during the uncontrolled shear experiment.

A very important point that must be considered during rheological measurements is the possibility of "slip" during the measurements. This is particularly the case with highly concentrated suspensions, where the flocculated system may form a "plug" in the gap of the platens leaving a thin liquid film at the walls of the concentric cylinder or cone-and-plate geometry. To reduce "slip", one should use roughened walls for the platens.

Strongly flocculated suspensions usually show much less thixotropy than weakly flocculated systems. Again one must be careful in drawing definite conclusions without other independent techniques (e.g. microscopy).

Constant stress (creep) measurements can also be applied to study the flocculation of suspensions. As mentioned above, a constant stress, σ, is applied on the system and the compliance, J (Pa^{-1}), is plotted as a function of time as was illustrated in Fig. 8.10 (creep curve). This above experiment is repeated several times increasing the stress from the smallest possible value (that can be applied by the instrument) in small increments. A set of creep curves are produced at various applied stresses as was illustrated in Fig. 8.12. From the slope of the linear portion of the creep curve (after the system reaches a steady state), the viscosity at each applied stress, η_σ, is calculated. As shown in Fig. 8.13, a plot of η_σ versus σ allows one to obtain the limiting (or zero shear) viscosity $\eta(0)$ and the critical stress σ_{cr} (which may be identified with the "true" yield stress of the system). The values of $\eta(0)$ and σ_{cr} may be used to assess the flocculation of the suspension on storage. If flocculation occurs on storage (without any Ostwald ripening), the values of $\eta(0)$ and σ_{cr} may show a gradual increase with increasing storage time.

As discussed in the previous section (on steady state measurements), the trend becomes complicated if Ostwald ripening occurs simultaneously (both have the effect of reducing $\eta(0)$ and σ_{cr}). Thus these measurements should be supplemented by particle size distribution measurements of the diluted suspension (making sure that no flocs are present after dilution) to assess the extent of Ostwald ripening. Another complication may arise from the nature of the flocculation. If it occurs in an irregular way (producing strong and tight flocs), $\eta(0)$ may increase, while σ_{cr} may show some decrease and this complicates the analysis of the results.

In spite of the above complications, constant stress measurements may provide valuable information on the state of the suspension on storage. Carrying out creep experiments and ensuring that a steady state is reached can be time consuming. One usually carries out a stress sweep experiment, whereby the stress is gradually increased (within a predetermined time period to ensure that one is not too far from reaching the steady state) and plots of η_σ versus σ are established.

The above experiments are carried out at various storage times (say every two weeks) and temperatures. From the change of $\eta(0)$ and σ_{cr} with storage time and temperature, one may obtain information on the degree and the rate of flocculation of the system.

One main problem in carrying out the above experiments is sample preparation. When a flocculated emulsion is removed from the container, care should be taken not to cause much disturbance to that structure (minimum shear should be applied on transferring the emulsion to the rheometer). It is also advisable to use separate containers for assessment of the flocculation. A relatively large sample is prepared and this is then transferred to a number of separate containers.

Dynamic (oscillatory) measurements are by far the most commonly used method to obtain information on the flocculation of a suspension. A strain is applied in a sinusoidal manner, with an amplitude γ_0 and a frequency ν (cycles/s or Hz) or ω (rad s^{-1}). In a viscoelastic system (such as is the case with a flocculated suspension), the stress oscillates with the same frequency, but out of phase from the strain. By measuring the time shift between strain and stress amplitudes (Δt) one can obtain the phase angle shift δ:

$$\delta = \Delta t \omega. \tag{8.22}$$

A schematic representation of the variation of strain and stress with ω is shown in Fig. 8.14. From the amplitudes of stress and strain and the phase angle shift, one can obtain the various viscoelastic parameters: the complex modulus G^*, the storage modulus (the elastic component of the complex modulus) G', the loss modulus (the viscous component of the complex modulus) G'', $\tan\delta$ and the dynamic viscosity η':

$$\text{Complex modulus} \quad |G^*| = \frac{\sigma_0}{\gamma_0} \tag{8.23}$$

$$\text{Storage modulus} \quad G' = |G^*|\cos\delta \tag{8.24}$$

$$\text{Loss modulus} \quad G'' = |G^*|\sin\delta \tag{8.25}$$

$$\tan\delta = \frac{G''}{G'}, \tag{8.26}$$

$$\text{Dynamic Viscosity} \quad \eta' = \frac{G''}{\omega}. \tag{8.27}$$

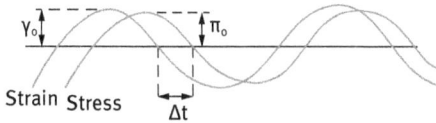

Fig. 8.14: Stress-strain relationship for a viscoelastic system.

G' is a measure of the energy stored in a cycle of oscillation. G'' is a measure of the energy dissipated as viscous flow in a cycle of oscillation. $\tan\delta$ is a measure of the relative magnitudes of the viscous and elastic components. Clearly, the smaller the value of $\tan\delta$, the more elastic the system is and vice versa.

η', the dynamic viscosity, shows a decrease with increasing frequency ω. η' reaches a limiting value as $\omega \to 0$. The value of η' in this limit is identical to the residual (or zero shear) viscosity $\eta(0)$. This is referred to as the Cox–Mertz rule.

In oscillatory measurements one carries out two sets of experiments.

8.6.2.1 Strain sweep measurements
In this case, the oscillation is fixed (say at 0.1 or 1 Hz) and the viscoelastic parameters are measured as a function of strain amplitude. This allows one to obtain the linear

viscoelastic region. In this region all moduli are independent of the applied strain amplitude and become only a function of time or frequency. This is illustrated in Fig. 8.15, which shows a schematic representation of the variation of G^*, G' and G'' with strain amplitude (at a fixed frequency).

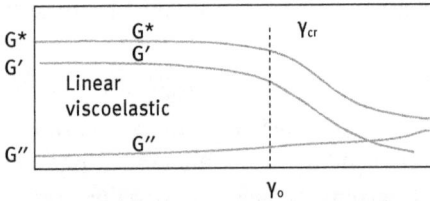

Fig. 8.15: Strain sweep results.

It can be seen from Fig. 8.15, that G^*, G' and G'' remain virtually constant up to a critical strain value, γ_{cr}. This region is the linear viscoelastic region. Above γ_{cr}, G^* and G' start to fall, whereas G'' starts to increase. This is the nonlinear region. The value of γ_{cr} may be identified with the minimum strain above which the "structure" of the suspension starts to break down (for example breakdown of flocs into smaller units and/or breakdown of a "structuring" agent).

From γ_{cr} and G', one can obtain the cohesive energy E_c $(J\,m^{-3})$ of the flocculated structure:

$$E_c = \int_0^{\gamma_{cr}} \sigma\,d\gamma = \int_0^{\gamma_{cr}} G'\gamma\,d\gamma = \frac{1}{2}G'\gamma_{cr}^2. \tag{8.28}$$

E_c may be used in a quantitative manner as a measure of the extent and strength of the flocculated structure in a suspension. The higher the value of E_c, the more flocculated the structure is. Clearly, E_c depends on the volume fraction of the suspension as well as the particle size distribution (which determines the number of contact points in a floc). Therefore, for quantitative comparisons between various systems, one has to make sure that the volume fraction of the disperse particles is the same and the suspensions have very similar particle size distribution.

E_c also depends on the strength of the flocculated structure, i.e. the energy of attraction between the particles. This depends on whether the flocculation is in the primary or secondary minimum. Flocculation in the primary minimum is associated with a large attractive energy and this leads to higher values of E_c when compared with the values obtained for secondary minimum flocculation. For a weakly flocculated suspension, such as is the case with secondary minimum flocculation of an electrostatically stabilized suspension, the deeper the secondary minimum, the higher the value of E_c (at any given volume fraction and particle size distribution of the suspension).

With a sterically stabilized suspension, weak flocculation can also occur when the thickness of the adsorbed layer decreases. Again, the value of E_c can be used as a measure of the flocculation; the higher the value of E_c, the stronger the flocculation. If incipient flocculation occurs (on reducing the solvency of the medium for the change to worse than θ-condition), a much deeper minimum is observed and this is accompanied by a much larger increase in E_c.

To apply the above analysis, one must have an independent method for assessing the nature of the flocculation. Rheology is a bulk property that can give information on the inter-particle interaction (whether repulsive or attractive) and to apply it in a quantitative manner one must know the nature of these interaction forces. However, rheology can be used in a qualitative manner to follow the change of the suspension on storage. Providing the system does not undergo any Ostwald ripening, the change of the moduli with time and in particular the change of the linear viscoelastic region may be used as an indication of flocculation. Strong flocculation is usually accompanied by a rapid increase in G' and this may be accompanied by a decrease in the critical strain above which the "structure" breaks down. This may be used as an indication of the formation of "irregular" flocs which become sensitive to the applied strain. The floc structure will entrap a large amount of the continuous phase and this leads to an apparent increase in the volume fraction of the suspension and hence an increase in G'.

8.6.2.2 Oscillatory sweep

In this case, the strain amplitude is kept constant in the linear viscoelastic region (one usually takes a point far from γ_{cr} but not too low, i.e. in the midpoint of the linear viscoelastic region) and measurements are carried out as a function of frequency. This is schematically represented in Fig. 8.16 for a viscoelastic liquid system.

Both G^* and G' increase with increasing frequency and ultimately above a certain frequency, they reach a limiting value and show little dependence on frequency. G'' is higher than G' in the low frequency regime; it also increases with increasing frequency and at a certain characteristic frequency, ω^* (that depends on the system), it

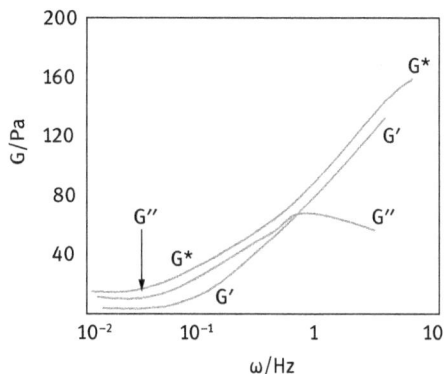

Fig. 8.16: Schematic representation of oscillatory measurements for a viscoelastic liquid.

becomes equal to G' (usually referred to as the crossover point), after which it reaches a maximum and then shows a reduction with a further increase in frequency.

In the low frequency regime, i.e. below ω^*, $G'' > G'$; this regime corresponds to longer times (remember that time is reciprocal of frequency) and under these conditions the response is more viscous than elastic. In the high frequency regime, i.e. above ω^*, $G' > G''$; this regime corresponds to short times and under these conditions the response is more elastic than viscous.

At sufficiently high frequency, G'' approaches zero and G' becomes nearly equal to G^*; this corresponds to very short time scales whereby the system behaves as a near elastic solid. Very little energy dissipation occurs at such high frequency.

The characteristic frequency, ω^*, can be used to calculate the relaxation time of the system t^*:

$$t^* = \frac{1}{\omega^*} \tag{8.29}$$

The relaxation time may be used as a guide for the state of the suspension. For a colloidally stable suspension (at a given particle size distribution), t^* increases with increasing volume fraction of the oil phase, ϕ. In other words, the crossover point shifts to a lower frequency with increasing ϕ. For a given suspension, t^* increases with increasing flocculation, provided that the particle size distribution remains the same (i.e. no Ostwald ripening).

The value of G' also increases with increasing flocculation, since aggregation of particles usually results in liquid entrapment and the effective volume fraction of the suspension shows an apparent increase. With flocculation, the net attraction between the droplets also increases and this results in an increase in G'. G' is determined by the number of contacts between the particles and the strength of each contact (which is determined by the attractive energy.

It should be mentioned that in practice one may not obtain the full curve, due to the frequency limit of the instrument and also measurements at low frequency are time consuming. Usually one obtains part of the frequency dependency of G' and G''. In most cases, one has a more elastic than viscous system. Most suspension systems used in practice are weakly flocculated and they also contain "thickeners" or "structuring" agents to reduce sedimentation and to acquire the right rheological characteristics for applications.

The exact values of G' and G'' required depend on the system and its applications. In most cases a compromise has to be made between acquiring the right rheological characteristics for the application and the optimum rheological parameters for long-term physical stability.

References

[1]   Tadros, Th. F., "Applied Surfactants", Wiley-VCH, Germany (2005).
[2]   Tadros, Th. F., "Dispersion of Powders in Liquids and Stabilisation of Suspensions", Wiley-VCH, Germany (2012).
[3]   Fleer, G. J., Cohen-Stuart, M. A., Scheutjens, J. M. H. M. Cosgrove, T. and Vincent, B., "Polymers at Interfaces", Chapman and Hall, London (1993).
[4]   Tadros, Th. F., Advances Colloid and Interface Sci., 147-148, 281 (2009).
[5]   Napper, D. H., "Polymeric Stabilisation of Colloidal Dispersions", Academic Press, London (1983).
[6]   Piirma, I., "Polymeric Surfactants", Surfactant Science Series, No. 42, NY, Marcel Dekker, (1992).
[7]   Stevens, C. V., Meriggi, A., Peristerpoulou, M., Christov, P. P., Booten, K., Levecke, B., Vandamme, A., Pittevils, N. and Tadros, Th. F., Biomacromolecules, 2, 1256 (2001).
[8]   Hirst, E. L., McGilvary, D. I. and Percival, E. G., J. Chem. Soc., 1297 (1950).
[9]   Tadros, Th. F., Vandamme, A., Levecke, B., Booten, K. and Stevens, C. V., Advances Colloid Interface Sci., 108–109, 207 (2004).
[10] Liang, W., Bognolo, G. and Tadros, Th. F., Langmuir, 11, 2899 (1995).
[11] Krieger, I. M., and Dougherty, T. J., Trans. Soc. Rheol, 3, 137 (1959).
[12] Krieger, I. M., Advances Colloid and Interface Sci., 3, 111 (1972).
[13] Tadros, Th. F., "Rheology of Dispersions", Wiley-VCH, Germany (2010).
[14] Tadros, Th. F., Adv. Chem. Ser., 9, 173 (1975).
[15] Heath, D. Knott, R. D. Knowles D. A. and Tadros, Th. F. ACS Symposium Ser., 254, 2 (1984).
[16] Garvey, M. J., Tadros, Th. F. and Vincent, B., J. Colloid Interface Sci., 49, 57 (1974).
[17] van den Boomgaard, Th., King, T. A., Tadros, Th. F., Tang, H. and Vincent, B., J. Colloid Interface Sci., 61, 68 (1978).
[18] von Smoluchowski, M. V., "Handbuch der Electricität und des Magnetismus", Vol. II, Barth, Leipzig (1914).
[19] Fuchs, N. Z., Phys., 89, 736 (1936)
[20] Tadros, Th. F., Advances Colloid and Interface Sci., 12, 141 (1980).

9 Polymeric surfactants in emulsion and dispersion polymerization

9.1 Introduction

Emulsion polymers (latexes) are commonly used in many industrial applications, e.g. as film formers in the coating industry. This is particularly the case with aqueous emulsion paints that are used for home decoration. These aqueous emulsion paints are applied at room temperature and the latexes coalesce on the substrate forming a thermoplastic film. Sometimes functional polymers are used for crosslinking in the coating system. The polymer particles are typically submicron (0.1–0.5 μm).

Generally speaking, there are two methods for the preparation of polymer dispersions, namely emulsion and dispersion polymerization. In emulsion polymerization, monomer is emulsified in a nonsolvent, commonly water, usually in the presence of an anionic or nonionic surfactant or nonionic polymeric surfactant. A water soluble initiator is added, and particles of polymer form and grow in the aqueous medium as the reservoir of the monomer in the emulsified droplets is gradually used up. In dispersion polymerization (which is usually applied for preparing nonaqueous polymer dispersion, commonly referred to as nonaqueous dispersion polymerization, NAD) monomer, initiator, stabilizer (usually a polymeric surfactant that is referred to as protective agent) and solvent initially form a homogeneous solution. The polymer particles precipitate when the solubility limit of the polymer is exceeded. The particles continue to grow until the monomer is consumed. In suspension polymerization the monomer is emulsified in the continuous phase using a polymeric surfactant. The initiator (which is oil soluble) is dissolved in the monomer droplets and the droplets are converted into insoluble particles, but no new particles are formed.

A description of both emulsion and dispersion polymerization is given below, with particular reference to the control of their particle size and colloid stability, which is greatly influenced by the emulsifier used. Particular emphasis will be given to the effect of the polymeric surfactants that have been more recently used for the preparation of emulsion polymers.

9.2 Emulsion polymerization

As mentioned above, in emulsion polymerization, the monomer, e.g. styrene or methyl methacrylate that is insoluble in the continuous phase, is emulsified using a surfactant that adsorbs at the monomer/water interface [1]. The surfactant micelles in bulk solution solubilize some of the monomer. A water soluble initiator such as potassium persulphate $K_2S_2O_8$ is added and this decomposes in the aqueous phase forming free radicals that interact with the monomers forming oligomeric chains. It has long been

DOI 10.1515/9783110487282-010

assumed that nucleation occurs in the "monomer swollen micelles". The reasoning behind this mechanism was the sharp increase in the rate of reaction above the critical micelle concentration and that the number of particles formed and their size depend to a large extent on the nature of the surfactant and its concentration (which determines the number of micelles formed). However, this mechanism was later disputed and it was suggested that the presence of micelles means that excess surfactant is available and molecules will readily diffuse to any interface.

The most accepted theory of emulsion polymerization is referred to as the coagulative nucleation theory [2, 3]. A two-step coagulative nucleation model has been proposed by Napper and co-workers [2, 3]. In this process the oligomers grow by propagation and this is followed by a termination process in the continuous phase. A random coil is produced which is insoluble in the medium and this produces a precursor oligomer at the θ-point. The precursor particles subsequently grow primarily by coagulation to form true latex particles. Some growth may also occur by further polymerization. The colloidal instability of the precursor particles may arise from their small size, and the slow rate of polymerization can be due to reduced swelling of the particles by the hydrophilic monomer [2, 3]. The role of surfactants in these processes is crucial since they determine the stabilizing efficiency, and the effectiveness of the surface active agent ultimately determines the number of particles formed. This was confirmed by using surface active agents of different nature. The effectiveness of any surface active agent in stabilizing the particles was the dominant factor and the number of micelles formed was relatively unimportant.

A typical emulsion polymerization formulation contains water, 50 % monomer blended for the required T_g, surfactant (and often "protective" colloid), initiator, pH buffer and fungicide. Hard monomers with a high T_g used in emulsion polymerization may be vinyl acetate, methyl methacrylate and styrene. Soft monomers with a low T_g include butyl acrylate, 2-ethylhexyl acrylate, vinyl versatate and maleate esters. Most suitable monomers are those with low, but not too low, water solubility. Other monomers such as acrylic acid, methacrylic acid, and adhesion promoting monomers may be included in the formulation. It is important that the latex particles coalesce as the diluent evaporates. The minimum film forming temperature (MFFT) of the paint is a characteristic of the paint system. It is closely related to the T_g of the polymer but the latter can be affected by materials present such as surfactant and the inhomogeneity of the polymer composition at the surface. High T_g polymers will not coalesce at room temperature and in this case a plasticizer ("coalescing agent") such as benzyl alcohol is incorporated in the formulation to reduce the T_g of the polymer, thus reducing the MFFT of the paint. Clearly, for any paint system one must determine the MFFT since, as mentioned above, the T_g of the polymer is greatly affected by the ingredients in the paint formulation. Several types of surfactants of the anionic, cationic, zwitterionic and nonionic types can be used in emulsion polymerization.

The role of surfactants is two-fold, firstly to provide a locus for the monomer to polymerize and secondly to stabilize the polymer particles as they form. In addition, surfactants aggregate to form micelles (above the critical micelle concentration) and these can solubilize the monomers. In most cases a mixture of anionic and nonionic surfactants is used for optimum preparation of polymer latexes. Cationic surfactants are seldom used, except for some specific applications where a positive charge is required on the surface of the polymer particles.

In addition to surfactants, most latex preparations require the addition of a polymeric surfactant (sometimes referred to as "protective colloid") such as partially hydrolyzed polyvinyl acetate (commercially referred to as polyvinyl alcohol, PVA), hydroxyethyl cellulose or a block copolymer of polyethylene oxide (PEO) and polypropylene oxide (PPO). These polymers can be supplied with various molecular weights or proportion of PEO and PPO. When used in emulsion polymerization they can be grafted by the growing chain of the polymer being formed. They assist in controlling the particle size of the latex, enhance the stability of the polymer dispersion and help control the rheology of the final paint.

A typical emulsion polymerization process involves two stages known as the seed stage and the feed stage. In the seed stage, an aqueous charge of water, surfactant, and colloid is raised to the reaction temperature (85–90 °C) and 5–10 % of the monomer mixture is added along with a proportion of the initiator (a water soluble persulphate). In this seed stage, the formulation contains monomer droplets stabilized by surfactant, a small amount of monomer in solution as well as surfactant monomers and micelles. Radicals are formed in solution from the breakdown of the initiator and these radicals polymerize the small amount of monomer in solution. These oligomeric chains will grow to some critical size, the length of which depends on the solubility of the monomer in water. The oligomers build up to a limiting concentration and this is followed by a precipitous formation of aggregates (seeds), a process similar to micelle formation, except in this case the aggregation process is irreversible (unlike surfactant micelles which are in dynamic equilibrium with monomers).

In the feed stage, the remaining monomer and initiator are fed together and the monomer droplets become emulsified by the surfactant remaining in solution (or by extra addition of surfactant). Polymerization proceeds as the monomer diffuses from the droplets, through the water phase, into the already forming growing particles. At the same time radicals enter the monomer-swollen particles causing both termination and re-initiation of polymerization. As the particles grow, the remaining surfactant from the water phase is adsorbed onto the surface of particles to stabilize the polymer particles. The stabilization mechanism involves both electrostatic and steric repulsion as was discussed in Chapter 5. The final stage of polymerization may include a further shot of initiator to complete the conversion.

According to the theory of Smith and Ewart [4] of the kinetics of emulsion polymerization, the rate of propagation R_p is related to the number of particles N formed in a reaction by the equation

$$-\frac{d[M]}{dt} = R_p k_p N n_{av}[M],$$ (9.1)

where [M] is the monomer concentration in the particles, k_p is the propagation rate constant and n_{av} is the average number of radicals per particle.

According to equation (9.1), the rate of polymerization and the number of particles are directly related to each other, i.e. an increase in the number of particles will increase the rate. This has been found for many polymerizations, although there are some exceptions. The number of particles is related to the surfactant concentration [S] by the equation [5]

$$N \approx [S]^{3/5}. \tag{9.2}$$

Using the coagulative nucleation model, Napper et al. [2, 3] found that the final particle number increases with increasing surfactant concentration with a monotonically diminishing exponent. The slope of $d(\log N_c)/d(\log t)$ varies from 0.4 to 1.2. At high surfactant concentration, the nucleation time will be long in duration since the new precursor particles will be readily stabilized. As a result, more latex particles are formed and eventually will outnumber the very small precursor particles at long times. The precursor/particle collisions will become more frequent and fewer latex particles are produced. The dN_c/dt will approach zero and at long times the number of latex particles remains constant. This shows the inadequacy of the Smith–Ewart theory which predicts a constant exponent (3/5) at all surfactant concentrations. For this reason, the coagulative nucleation mechanism has now been accepted as the most probable theory for emulsion polymerization. In all cases, the nature and concentration of surfactant used is very crucial and this is very important in the industrial preparation of latex systems.

Most reports on emulsion polymerization have been limited to commercially available surfactants, which in many cases are relatively simple molecules such as sodium dodecyl sulphate and simple nonionic surfactants. However, studies on the effect of surfactant structure on latex formation have revealed the importance of the structure of the molecule. Block and graft copolymers (polymeric surfactants) are expected to be better stabilizers when compared to simple surfactants. The use of these polymeric surfactants in emulsion polymerization and the stabilization of the resulting polymer particles is crucial. Most aqueous emulsion and dispersion polymerization reported in the literature are based on a few commercial products with a broad molecular weight distribution and varying block composition. The results obtained from these studies could not establish what effect the structural features of the block copolymer has on their stabilizing ability and effectiveness in polymerization. Fortunately, model block copolymers with well-defined structures could be synthesized and their role in emulsion polymerization has been investigated out using model polymers and model latexes.

A series of well-defined A–B block copolymers of polystyrene-block-polyethylene oxide (PS–PEO) were synthesized [6] and used for emulsion polymerization of styrene. These molecules are "ideal", since the polystyrene block is compatible with the polystyrene formed and thus it forms the best anchor chain. The PEO chain (the stabilizing chain) is strongly hydrated with water molecules and it extends into the aqueous phase forming the steric layer necessary for stabilization (see Chapter 5). However, the PEO chain can become dehydrated at high temperature (due to the breaking of hydrogen bonds), thus reducing the effective steric stabilization. Thus emulsion polymerization should be carried out at temperatures well below the θ-temperature of PEO (see Chapter 5).

Five block copolymers were synthesized [6] with various molecular weights of the PS and PEO blocks. The molecular weight of the polystyrene block and the resulting PS–PEO polymer was determined using gel permeation chromatography. The mole percent of ethylene oxide and the percent of PEO in the block were determined using H^1 NMR spectroscopy. The molecular weight of the blocks varied from M_n = 1000–7000 for PS and M_w = 3000–9000 for PEO. These five block copolymers were used for emulsion polymerization of styrene at 50 °C (well below the θ-temperature of PEO). The results indicated that for efficient anchoring the PS blocks need not be more than 10 monomer units. The PEO block should have a $M_w \geq 3000$. However, the ratio of the two blocks is very important; for example if the wt % of PEO is ≤ 3000, the molecule becomes insoluble in water (not sufficiently hydrophilic) and no polymerization could occur when using this block copolymer. In addition, the 50 % PEO block could produce a latex, but it was unstable and coagulated at 35 % conversion. It became clear from these studies that the % PEO in the block copolymer plays an important role and this should exceed 75 %. However, the overall molecular weight of the block copolymer is also very important. For example if one uses a PS block with M_n = 7000, the PEO molecular weight have to be 21 000, which is too high and may result in bridging flocculation, unless one prepares a very dilute latex.

Another systematic study of the effect of block copolymer on emulsion polymerization was carried out using blocks of poly(methylmethacrylate)-block-polyethylene oxide (PMMA–PEO) for the preparation of PMMA latexes. The ratio and molecular weight of PMMA to PEO in the block copolymer was varied. Ten different PMMA–PEO blocks were synthesized [6] with M_n for PMMA varying between 400 and 2500. The M_w of PEO was varied between 750 and 5000. The recipe for MMA polymerization consisted of 100 monomer, 800 g water, 20 g PMMA–PEO block copolymer and 0.5 g potassium persulphate. The polymerization was carried out at 45 °C, which is well below the θ-temperature of PEO. The rate of polymerization, R_p, was calculated by using latex samples drawn from the reaction mixture at various time intervals (the amount of latex was determined gravimetrically). The particle size of each latex sample was determined by dynamic light scattering (Photon Correlation Spectroscopy, PCS). The number of particles, N, in each case was calculated from the weight of the latex and the z-average diameter. The results obtained were used to study the effect of the anchoring group PMMA, molecular weight, the effect of PEO molecular weight and the effect of the total molecular weight of the block copolymer. The results are summarized in Tab. 9.1 and 9.2.

Tab. 9.1: Effect of PMMA and PEO molecular weight in the diblock.

M_n PMMA	M_w PEO	wt % PEO	$R_p \times 10^4$ mol/l s	D / nm	$N \times 10^{-13}$ cm^{-3}
400	750	65	1.3	213	1.7
400	2000	83	1.5	103	14.7
400	5000	93	2.4	116	10.3
900	750	46	Unstable latex	—	—
800	2000	71	3.4	92	20.6
800	5000	86	3.2	106	13.5
1300	2000	61	2.4	116	10.3
1200	5000	81	4.6	99	16.6
1900	5000	72	3.4	110	11.4
2500	5000	67	2.2	322	0.4

Tab. 9.2: Effect of total molecular weight of the PMMA–PEO diblock.

M_w	wt % PEO	$R_p \times 10^4$ mol/l s	D / nm	$N \times 10^{-13}$ cm^{-3}
1150	65	1.3	213	1.7
2400	83	1.5	103	14.7
2800	71	3.4	92	20.6
3300	61	2.4	99	16.6
6200	81	4.6	99	16.6
6900	72	3.4	110	11.4
7500	67	2.2	322	0.4

The results of the systematic study (Tab. 9.1 and 9.2) of varying the PMMA and PEO block molecular weight, the % PEO in the chain as well as the overall molecular weight clearly show the effect of these factors on the resulting latex. For example, when using a block copolymer with 400 molecular weight of PMMA and 750 molecular weight of PEO (i.e. containing 65 wt % PEO) the resulting latex has fewer particles when compared with the other surfactants. The most dramatic effect was obtained when the PMMA molecular weight was increased to 900 while keeping the PEO molecular weight (750) the same. This block copolymer contains only 46 wt % PEO and it became insoluble in water due to the lack of hydrophilicity. The latex produced was unstable and it collapsed at the early stage of polymerization. The PEO molecular weight of 750 is insufficient to provide effective steric stabilization (see Chapter 5). By increasing the molecular weight of PEO to 2000 or 5000 while keeping the PMMA molecular weight at 400 or 800, a stable latex was produced with a small particle diameter and large number of particles. The best results were obtained by keeping the molecular weight of PMMA at 800 and that of PEO at 2000. This block copolymer gave the highest conversion rate, the smallest particle diameter and the largest number of particles (see Tab. 9.2). It is interesting to note that by increasing the PEO molecular weight to 5000 while keeping the PMMA molecular weight at 800, the rate of conversion decreased, the average diameter increased and the number of particles decreased when compared with the results obtained using 2000 molecular weight for PEO. It seems that when the PEO molecular weight is increased the hydrophilicity of the molecule increased (86 wt % PEO) and this reduced the efficiency of the copolymer. It seems that by increasing hydrophilicity of the block copolymer and its overall molecular weight, the rate of adsorption of the polymer to the latex particles and its overall adsorption strength may have decreased. The effect of the overall molecular weight of the block copolymer and its overall hydrophilicity have a big effect on the latex production (see Tab. 9.2). Increasing the overall molecular weight of the block copolymer above 6200 resulted in a reduction in the rate of conversion, an increase in the particle diameter and a reduction in the number of latex particles. The worst results were obtained with an overall molecular weight of 7500 while reducing the PEO wt %. In this case particles with 322 nm diameter were obtained and the number of latex particles was significantly reduced.

The importance of the affinity of the anchor chain (PMMA) to the latex particles was investigated by using different monomers [6]. For example, when using styrene as the monomer the resulting latex was unstable and it showed the presence of co-agulum. This can be attributed to the lack of chemical compatibility of the anchor chain (PMMA) and the polymer to be stabilized, namely polystyrene. This clearly indicates that block copolymers of PMMA–PEO are not suitable for emulsion polymerization of styrene. However, when using vinyl acetate monomer, whereby the resulting poly(vinyl acetate) latex should have strong affinity to the PMMA anchor, no latex was produced when the reaction was carried out at 45 °C. It was speculated that the water solubility of the vinyl acetate monomer resulted in the formation of oligomeric chain radicals which could exist in solution without nucleation. Polymerization at 60 °C, which did nucleate particles, was found to be controlled by chain transfer of the vinyl acetate radical with the surfactant, resulting in broad molecular weight distributions.

Emulsion polymerization of MMA using triblock copolymers was carried out using PMMA-block-PEO-PMMA with the same PMMA molecular weight (800 or 900) while varying the PEO molecular weight from 3400 to 14 000 in order to vary the loop size. Although the rate of polymerization was not affected by the loop size, the particles with the smallest diameter were obtained with the 10 000 molecular weight PEO. Comparing the results obtained using the triblock copolymer with those obtained using diblock copolymer (while keeping the PMMA block molecular weight the same) showed the same rate of polymerization. However, the average particle diameter was smaller and the total number of particles larger when using the diblock copolymer. This clearly shows the higher efficacy of the diblock copolymer when compared with the triblock copolymer.

The first systematic study of the effect of graft copolymers on emulsion polymerization was carried out by Piirma and Lenzotti [7] who synthesized well characterized graft copolymers with different backbone and side chain lengths. Several grafts of poly(p-methylstyrene)-graft-polyethylene oxide, $(PMSt)-(PEO)_n$, were synthesized

and used in styrene emulsion polymerization. Three different PMSt chain lengths (with molecular weight of 750, 2000 and 5000) and three different PEO chain lengths were prepared. In this way the structure of the amphipathic graft copolymer could be changed in three different ways:
(i)   three different PEO graft chain lengths;
(ii) three different backbone chain lengths with the same wt % PEO;
(iii) four different wt % PEO grafts.

Piirma and Lenzotti [7] first investigated the graft copolymer concentration required to produce the highest conversion rate, the smallest particle size and the largest number of latex particles. The monomer-to-water ratio was kept at 0.15 to avoid overcrowding of the resulting particles. They found that a concentration of 18 g/100 g monomer (2.7 % aqueous phase) was necessary to obtain the above results, after which a further increase in graft copolymer concentration did not significantly increase the rate of polymerization or increase the number of particles used. Using the graft copolymer concentration of 2.7 % aqueous phase, the results showed an increase in the number of particles with increasing conversion reaching a steady value at about 35 % conversion. Obviously, before that conversion, new particles are still being stabilized from the oligomeric precursor particles, after which all precursor particles are assimilated by the existing particles. The small size of the latex produced, namely 30–40 nm, clearly indicates the efficiency with which this graft copolymer stabilizes the dispersion.

Three different backbone chain lengths of M_n = 1140, 4270 and 24 000 were used while the weight percent of PEO (82 %) was kept the same, which is equivalent to 3, 10 and 55 PEO chains per backbone respectively. The results showed that the rate of polymerization, particle diameter and number of particles was similar for the three cases. Since the graft copolymer concentration was the same in each case, it can be concluded that one molecule of the highest molecular graft is just as effective as 18 molecules of the lowest molecular weight graft in stabilizing the particles.

Four graft copolymers were synthesized with a PMSt backbone with M_w = 4540 while increasing the wt % of PEO: 68, 73, 82 and 92 wt % (corresponding to 4.8, 6, 10 and 36 grafts per chain). The results showed a sharp decrease (by more than one order of magnitude) in the number of particles when the wt % of PEO was increased from 82 to 94 %. The reason for this reduction in the number of particles is the increased hydrophilicity of the graft copolymer which could result in desorption of the molecule from the surface of the particle. In addition, a graft with 36 side chains does not leave enough space for anchoring by the backbone.

The effect of PEO side chain length on emulsion polymerization using graft copolymers was systematically studied by keeping the backbone molecular weight the same (1380) while gradually increasing the PEO molecular weight of the side chains from 750 to 5000. For example by increasing M_w of PEO from 750 to 2000 while keeping the wt % of PEO roughly the same (84 and 82 wt % respectively) the number of side chains in the graft decreases from 10 to 3. The results showed a decrease in the rate of polymerization as the number of side chains in the graft increases. This is followed by a sharp reduction in the number of particles produced. This clearly shows the importance of spacing of the side chains to ensure anchoring of the graft copolymer to the particle surface, which is stronger with the graft containing a smaller number of side chains. If the number of side chains for the PEO with M_w of 2000 is increased from 3 to 9 (93 wt % of PEO), the rate of polymerization and number of particles decrease. Using a PEO chain with M_w of 5000 (92 wt % PEO) and 3 chains per graft gives the same result as the PEO 2000 with 3 side chains. Any increase in the number of side chains in the graft results in a reduction in the rate of polymerization and the number of latex particles produced. This clearly shows the importance of spacing of the side chains of the graft copolymer.

Similar results were obtained using a graft copolymer of poly(methyl methacrylate-co-2-hydroxypropyl methacrylate)-graft-polyethylene oxide, PMMA(PEO)$_n$, for emulsion polymerization of methyl methacrylate. As with PMSt(PEO)$_n$ graft, the backbone molecular weight had little effect on the rate of polymerization or the number of particles used. The molecular weight of the PEO side chains was varied at constant M_w of the backbone (10 000). Three PEO grafts with M_w of 750, 2000 and 5000 were used. Although the rate of polymerization was similar for the three graft copolymers, the number of particles was significantly lower with the graft containing PEO 750. This shows that this short PEO chain is not sufficient to stabilize the particles. The overall content of PEO in the graft also has a large effect. Using the same backbone chain length while changing the wt % of PEO 200, it was found that the molecule containing 67 wt % PEO is not sufficient to stabilize the particles when compared with a graft containing 82 wt % PEO. This shows that a high concentration of PEO in the adsorbed layer is required for effective steric stabilization (see Chapter 5).

The chemical nature of the monomer also plays an important role. For example, stable latexes could be produced using PMSt(PEO)$_n$ graft but not with PMMA(PEO)$_n$ graft. More recently, a novel graft copolymer of hydrophobically modified inulin (INUTEC® SP1) has been used in emulsion polymerization of styrene, methyl methacrylate, butyl acrylate and several other monomers [8]. All lattices were prepared by emulsion polymerization using potassium persulphate as initiator. The z-average particle size was determined by photon correlation spectroscopy (PCS) and electron micrographs were also taken.

Emulsion polymerization of styrene or methylmethacrylate showed an optimum weight ratio of (INUTEC)/monomer of 0.0033 for PS and 0.001 for PMMA particles. The (initiator)/(monomer) ratio was kept constant at 0.00125. The monomer conversion was higher than 85 % in all cases. Latex dispersions of PS reaching 50 % and of PMMA reaching 40 % could be obtained using such a low concentration of INUTEC® SP1. Figure 9.1 shows the variation of particle diameter with monomer concentration.

The stability of the latexes was determined by determining the critical coagulation concentration (ccc) using CaCl$_2$. The ccc was low (0.0175–0.05 mol dm^{-3}) but this was higher than that for the latex prepared without surfactant. Post addition of

INUTEC® SP1 resulted in a large increase in the ccc as illustrated in Fig. 9.2 which shows log W–log C curves (where W is the ratio between the fast flocculation rate constant to the slow flocculation rate constant, referred to as the stability ratio) at various additions of INUTEC® SP1.

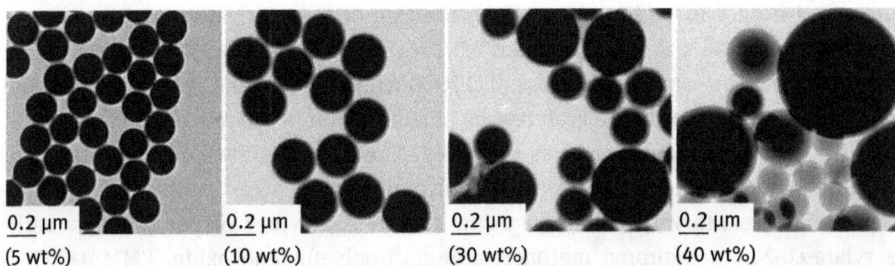

(a) PS latexes

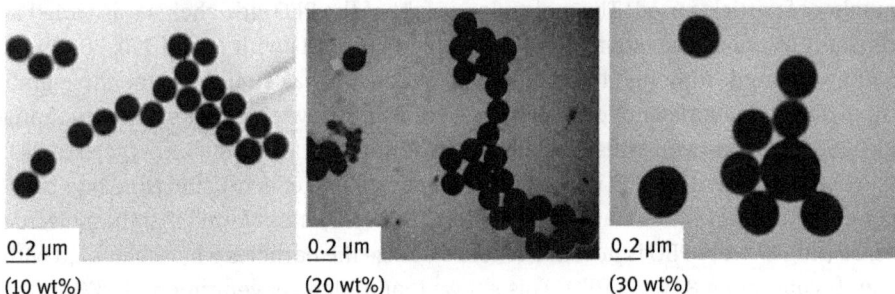

(b) PMMA Latexes

Fig. 9.1: Electron micrographs of the latexes.

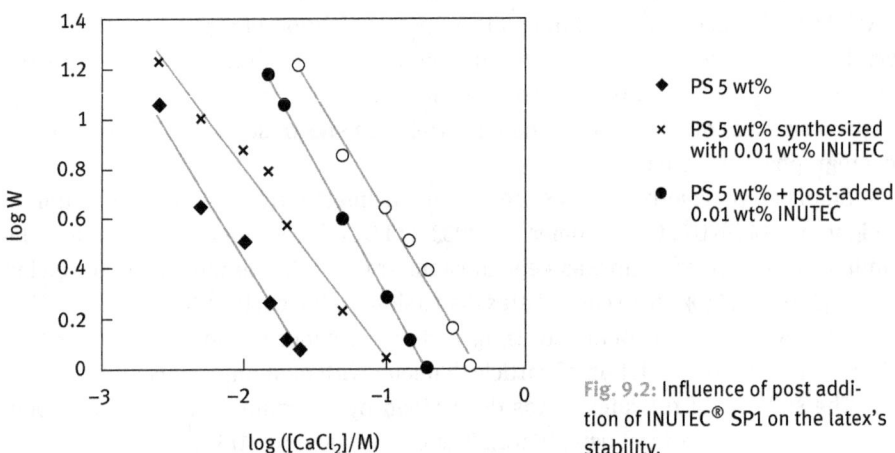

Fig. 9.2: Influence of post addition of INUTEC® SP1 on the latex's stability.

As with the emulsions, the high stability of the latex when using INUTEC® SP1 is due to the strong adsorption of the polymeric surfactant on the latex particles and formation of strongly hydrated loops and tails of polyfructose that provide effective steric stabilization. Evidence for the strong repulsion produced when using INUTEC® SP1 was obtained from atomic force microscopy investigations [9] in which the force between hydrophobic glass spheres and hydrophobic glass plate, both containing an adsorbed layer of INUTEC® SP1, was measured as a function of distance of separation both in water and in the presence of various Na_2SO_4 concentrations. The results are shown in Fig. 9.3 and 9.4.

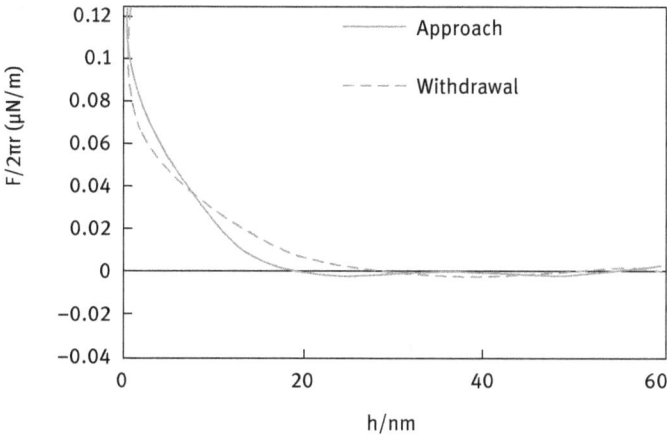

Fig. 9.3: Force-distance curves between hydrophobized glass surfaces containing adsorbed INUTEC® SP1 in water.

9.3 Dispersion polymerization

This method is usually applied for the preparation of nonaqueous latex dispersions and hence it is referred to as NAD. The method has also been adapted to prepare aqueous latex dispersions by using an alcohol-water mixture. In the NAD process the monomer, normally an acrylic, is dissolved in a nonaqueous solvent, normally an aliphatic hydrocarbon, and an oil soluble initiator and a stabilizer (to protect the resulting particles from flocculation, sometimes referred to as "protective colloid") are added to the reaction mixture. The most successful stabilizers used in NAD are block and graft copolymers. These block and graft copolymers are assembled in a variety of ways to provide the molecule with an "anchor chain" and a stabilizing chain. The anchor chain should be sufficiently insoluble in the medium and have a strong affinity to the polymer particles produced. In contrast, the stabilizing chain should be soluble in the medium and strongly solvated by its molecules to provide effective steric

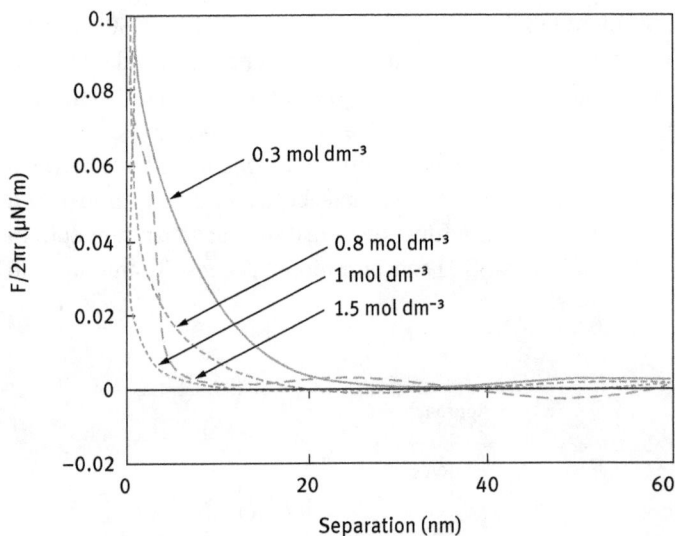

Fig. 9.4: Force-distance curves for hydrophobized glass surfaces containing adsorbed INUTEC® SP1 at various Na_2SO_4 concentrations.

stabilization (see Chapter 5). The length of the anchor and stabilizing chains has to be carefully adjusted to ensure strong adsorption (by multipoint attachment of the anchor chain to the particle surface) and a sufficiently "thick" layer of the stabilizing chain that prevents close approach of the particles to a distance where the van der Waals attraction becomes strong. The criteria for effective steric stabilization by block and graft copolymers are discussed in detail in Chapter 5. Several configurations of block and graft copolymers are possible, as illustrated in Fig. 9.5.

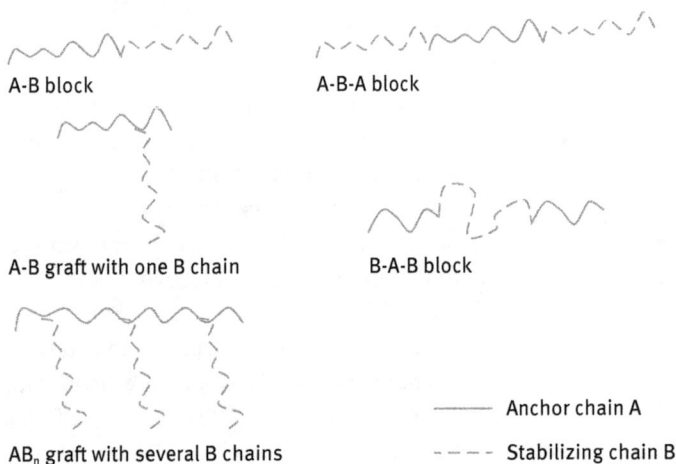

Fig. 9.5: Configurations of block and graft copolymers.

Typical preformed graft stabilizers based on poly(12-hydroxy stearic acid) (PHS) are simple to prepare and effective in NAD polymerization. Commercial 12-hydroxystearic acid contains 8–15 % palmitic and stearic acids which limit the molecular weight during polymerization to an average of 1500–2000. This oligomer may be converted to a "macromonomer" by reacting the carboxylic group with glycidyl methacrylate. The macromonomer is then copolymerized with an equal weight of methyl methacrylate (MMA) or similar monomer to give a "comb" graft copolymer with an average molecular weight of 10 000–20 000. The graft copolymer contains on average 5–10 PHS chains pendent from a polymeric anchor backbone of PMMA. This graft copolymer can stabilize latex particles of various monomers. The major limitation of the monomer composition is that the polymer produced should be insoluble in the medium used.

Several other examples of block and graft copolymers that are used in dispersion polymerization are given in Tab. 9.3 which also shows the continuous phase and disperse polymer that can be used with these polymers.

Tab. 9.3: Block and graft copolymers used in emulsion polymerization.

Polymeric Surfactant	Continuous phase	Disperse polymer
Polystyrene-block-poly(dimethyl siloxane)	Hexane	Polystyrene
Polystyrene-block-poly(methacrylic acid)	Ethanol	Polystyrene
Polybutadiene-graft-poly(methacrylic acid)	Ethanol	Polystyrene
Poly(2-ethylhexyl acrylate)-graft-poly(vinyl acetate)	Aliphatic hydrocarbon	Poly(methyl methacrylate)
Polystyrene-block-poly(t-butylstyrene)	Aliphatic hydrocarbon	Polystyrene

Two main criteria must be considered in the process of dispersion polymerization:
(i)   the insolubility of the formed polymer in the continuous phase;
(ii) the solubility of the monomer and initiator in the continuous phase.

Initially, dispersion polymerization starts as a homogeneous system but after sufficient polymerization, the insolubility of the resulting polymer in the medium forces them to precipitate. Initially, polymer nuclei are produced which then grow to polymer particles. The latter are stabilized against aggregation by the block or graft copolymer that is added to the continuous phase before the process of polymerization starts. It is essential to choose the right block or graft copolymer which should have a strong anchor chain A and good stabilizing chain B as schematically represented in Fig. 9.5. Dispersion polymerization may be considered a heterogeneous process which may include emulsion, suspension, precipitation and dispersion polymerization. In dispersion and precipitation polymerization, the initiator must be soluble in the continuous phase, whereas in emulsion and suspension polymerization the initiator is chosen to be soluble in the disperse phase of the monomer. A comparison of the rate of polymerization of methylmethacrylate at 80 °C for the three systems was given by Barrett and Thomas [10] and is illustrated in Fig. 9.6. The rate of dispersion polymerization is much faster than precipitation or solution polymerization. The enhancement of the rate in precipitation polymerization over solution polymerization has been attributed to the hindered termination of the growing polymer radicals.

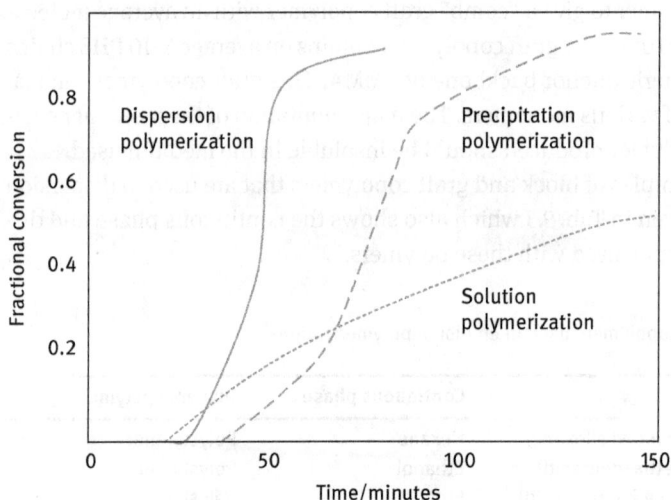

Fig. 9.6: Comparison of rates of polymerization.

Several mechanisms have been proposed to explain the mechanism of dispersion polymerization; however, no single mechanism can explain all happenings in emulsion polymerization. Barrett and Thomas [10] suggested that particles are formed in dispersion polymerization by two main steps:
(i) Initiation of monomer in the continuous phase and subsequent growth of the polymer chains until the latter become insoluble. This process clearly depends on the nature of the polymer and medium.
(ii) The growing oligomeric chains associate with each other forming aggregates which below a certain size are unstable and they become stabilized by the block or graft copolymer added.

As mentioned before, this aggregative nucleation theory cannot explain all happenings in dispersion polymerization. An alternative mechanism based on Napper's theory [2, 3] for aqueous emulsion polymerization can be adapted to the process of dispersion polymerization. This theory includes coagulation of the nuclei formed and not just association of the oligomeric species. The precursor particles (nuclei), being unstable, can undergo one of the following events to become colloidally stable:

(i)   homocoagulation, i.e. collision with other precursor particles;
(ii) growth by propagation, adsorption of stabilizer;
(iii) swelling with monomer.

The nucleation-terminating events are diffusional capture of oligomers and hetero-coagulation.

The number of particles formed in the final latex does not depend on particle nucleation alone, since other steps are involved which determine how many precursor particles created are involved in the formation of a colloidally stable particle. This clearly depends on the effectiveness of the block or graft copolymer used in stabilizing the particles (see below).

In most cases, increasing polymeric surfactant concentration (at any given monomer amount) results in the production of a larger number of particles with smaller size. This is to be expected since the larger number of particles with smaller size (i.e. larger total surface area of the disperse particles) requires more polymeric surfactant for their formation. The molecular weight of the polymeric surfactant can also influence the number of particles formed. For example, Dawkins and Taylor [11] found that in dispersion polymerization of styrene in hexane, increasing the molecular weight of the block copolymer of polydimethyl siloxane-block-polystyrene resulted in the formation of smaller particles which was attributed to the more effective steric stabilization by the higher molecular weight block.

A systematic study of the effect of monomer solubility and concentration in the continuous phase was carried out by Antl and co-workers [12]. Dispersion polymerization of methyl methacrylate in hexane mixed with a high boiling point aliphatic hydrocarbon was investigated using poly(12-hydoxystyearic acid)-glycidyl methacrylate block copolymer. They found that the methyl methacrylate concentration had a drastic effect on the size of the particles produced. When the monomer concentration was kept below 8.5 %, very small particles (80 nm) were produced and these remained very stable. However, between 8.5 and 35 % monomer the latex produced was initially stable but flocculated during polymerization. An increase in monomer concentration from 35 to 50 % resulted in the formation of a stable latex but the particle size increased sharply from 180 nm to 2.6 μm as the monomer concentration increased. The authors suggested that the final particle size and stability of the latex are strongly affected by increased monomer concentration in the continuous phase. The presence of monomer in the continuous phase increases the solvency of the medium for the polymer formed. In a good solvent for the polymer, the growing chain is capable of reaching higher molecular weight before it is forced to phase separate and precipitate.

NAD polymerization is carried in two steps:
(i)   Seed stage: the diluent, portion of the monomer, portion of dispersant and initiator (azo or peroxy type) are heated to form an initial low-concentration fine dispersion.

(ii) Growth stage: the remaining monomer together with more dispersant and initiator are then fed over the course of several hours to complete the growth of the particles.

A small amount of transfer agent is usually added to control the molecular weight. Excellent control of particle size is achieved by proper choice of the designed dispersant and correct distribution of dispersant between the seed and growth stages. NAD acrylic polymers are applied in automotive thermosetting polymers and hydroxy monomers may be included in the monomer blend used.

Two main factors must be considered when considering the long-term stability of a nonaqueous polymer dispersion. The first and very important factor is the nature of the "anchor chain" A. As mentioned above, this should have a strong affinity to the produced latex and in most cases it can be designed to be "chemically" attached to the polymer surface. Once this criterion is satisfied, the second important factor in determining the stability is the solvency of the medium for the stabilizing chain B. As will be discussed in detail, the solvency of the medium is characterized by the Flory–Huggins interaction parameter χ. Three main conditions can be identified: $\chi < 0.5$ (good solvent for the stabilizing chain); $\chi > 0.5$ (poor solvent for the stabilizing chain); and $\chi = 0.5$ (referred to as the θ-solvent). Clearly, to maintain stability of the latex dispersion, the solvent must be better than a θ-solvent. The solvency of the medium for the B chain is affected by addition of a nonsolvent and/or temperature changes. It is, therefore, essential to determine the critical volume fraction (CFV) of a nonsolvent, above which flocculation (sometimes referred to as incipient flocculation) occurs. One should also determine the critical flocculation temperature at any given solvent composition, below which flocculation occurs. The correlation between CFV or CFT and the flocculation of the nonaqueous polymer dispersion has been demonstrated by Napper [13] who investigated the flocculation of poly(methyl methacrylate) dispersions stabilized by poly(12-hydroxy stearic acid) or poly(n-lauryl methacrylate-co-glycidyl methacrylate) in hexane by adding a nonsolvent such as ethanol or propanol and cooling the dispersion. The dispersions remained stable until the addition of ethanol transformed the medium to a θ-solvent for the stabilizing chains in solution. However, flocculation did occur under conditions of slightly better than θ-solvent for the chains. The same was found for the CFT which was 5–15 K above the θ-temperature. This difference was accounted for by the polydispersity of the polymer chains. The θ-condition is usually determined by cloud point measurements and the least soluble component will precipitate first, giving values that are lower than the CFV or higher than the CFT.

The process of dispersion polymerization has been applied in many cases using completely polar solvents such as alcohol or alcohol-water mixtures [14]. The results obtained showed completely different behaviour when compared with dispersion polymerization in nonpolar media. For example, results obtained by Lok and Ober [14] using styrene as monomer and hydroxypropyl cellulose as stabilizer showed a linear increase of particle diameter with increasing weight percent of the monomer. There was no region in monomer concentration where instability occurred (as has been observed for the dispersion polymerization of methyl methacrylate in aliphatic hydrocarbons). Replacing water in the continuous phase with 2-methoxyethanol, Lok and Ober [14] were able to grow large, monodisperse particles up to 15 μm in diameter. They concluded from these results that the polarity of the medium is the controlling factor in the formation of particles and their final size. The authors suggested a mechanism in which the polymeric surfactant molecule grafts to the polystyrene chain, forming a physically anchored stabilizer (nuclei). These nuclei grow to form the polymer particles. Paine [15] carried out dispersion polymerization of styrene by systematically increasing the alcohol chain length from methanol to octadecanol and using hydroxypropyl cellulose as stabilizer. The results showed an increase in particle diameter with increasing number of carbon atoms in the alcohol, reaching a maximum when hexanol was used as the medium, after which there was a sharp decrease in the particle diameter with any further increase in the number of carbon atoms in the alcohol. Paine explained his results in terms of the solubility parameter of the dispersion medium. The largest particles are produced when the solubility parameter of the medium is closest to those of styrene and hydroxypropyl cellulose.

References

[1]   Blakely, D. C., "Emulsion Polymerization", Elsevier Applied Science, London (1975).
[2]   Barrett, K. E. J. (ed.), "Dispersion Polymerization in Organic Media", John Wiley & Sons Ltd, Chichester (1975).
[3]   Smith, W. V. and Ewart, R. H., J. Chem. Phys., 16, 592 (1948).
[4]   Litchi, G., Gilbert, R. G. and Napper, D. H., J. Polym. Sci., 21, 269 (1983).
[5]   Feeney, P. J., Napper, D. H. and Gilbert, R. G., Macromolecules, 17, 2520 (1984); 20, 2922 (1987).
[6]   Piirma, I., "Polymeric Surfactants", Surfactant Science Series, No. 42, Marcel Dekker, NY, (1992).
[7]   Piirma, I. and Lenzotti, J. R., Br. Polymer J., 21, 45 (1959).
[8]   Nestor, J., Esquena, J., Solans, C., Levecke, B., Booten, K. and Tadros, Th. F., Langmuir, 21, 4837 (2005).
[9]   Nestor, J., Esquena, J., Solans, C., Luckham, P. F., Levecke, B. and Tadros, Th. F., J. Colloid Interface Sci., 311, 430 (2007).
[10] Barrett, K. E. J. and Thomas, H. R., J. Polym. Sci., Part A1, 7, 2627 (1969).
[11] Dawkins, J. V. and Taylor, G., Polymer, 20, 171 (1987).
[12] Antl, I., Goodwin, J. W., Hill, R. D., Ottewill, R. H., Owen, S. M., Papworth, S. and Waters, J. A., Colloids Surf., 1, 67 (1986).
[13] Napper, D. H., "Polymeric Stabilisation of Colloidal Dispersions", Academic Press, London (1983).
[14] Lok, K. P. and Ober, C. K., Can. J. Chem., 63, 209 (1985).
[15] Paine, A. J., J. Polymer Sci., Part A, 28, 2485 (1990).

10 Polymeric surfactants in pharmacy

10.1 Introduction

Polymeric surfactants are applied in most pharmaceutical disperse systems such as suspensions and emulsions. These disperse systems cover a wide range of sizes ranging from particles or droplets ≤ 1 μm (colloidal dispersions) to 50–75 μm. The formulation and stabilization of these disperse systems in pharmacy require in many cases the use of homopolymers, block and graft copolymers to provide effective steric stabilization as described in detail in Chapter 5. One of the most commonly used homopolymers for pharmaceutical suspensions and nanosuspensions is poly(vinylpyrrolidone) (PVP) which in many cases is used in combination with an anionic surfactant such as sodium dodecyl sulphate (SDS) that acts as a wetting agent and gives a synergistic mixture with PVP to provide electrosteric repulsion. This will be discussed in some detail below. Further commonly used A–B–A block copolymers for stabilization of suspensions and emulsions are the poloxamers, where A is polyethylene oxide (PEO) and B is poly(propylene oxide) (PPO). This block copolymer adsorbs on hydrophobic drug particles or emulsion droplets with PPO forming the "anchor" chain and PEO forming the "stabilizing" chain. The mechanism of the stabilizing action of poloxamers is described below.

One of the most important applications of polymeric surfactants in pharmacy is the formation of biodegradable nanoparticles that can be used for targeted delivery of drugs. A very useful polymeric surfactant is an A–B block of poly(lactic)-polyglycolic acid (PLA–PEO) that assembles by aggregation to form nanoparticles of the polymer as will be discussed below.

10.2 PVP/SDS system for stabilizing nanosuspensions of drugs

Nanosuspensions have wide applications in drug delivery systems of poorly insoluble compounds, whereby reducing the particle size to nanoscale dimensions enhances the drug bioavailability. These systems must be stabilized against aggregation, Ostwald ripening (crystal growth) and sintering.

At present, the small molecular entities produced by the current state of pharmaceutical discovery show an increasing trend to be highly water insoluble [1, 2]. This low water solubility is a challenge to achieve adequate bioavailability [3] for oral administration. It also limits the types of formulation suitable for parenteral administration [4]. In recent years, nanocrystalline suspensions have been applied for drug delivery of highly water insoluble ingredients (APIs) [5]. By reducing the particle size of the API, the rate of the dissolution, dC/dt, which is directly proportional to the surface

DOI 10.1515/9783110487282-011

specific area A, is increased as described by the Noyes and Whitney equation [6]:

$$\frac{dC}{dt} = KA(C_s - C),$$ (10.1)

where K is a constant and C_s is the saturation solubility.

The solubility of the API can be significantly enhanced. This is due to the increase in solubility of the active ingredient on reduction of particle radius as given by the Kelvin equation [7]

$$S(r) = S(\infty) \exp\left(\frac{2\gamma V_m}{rRT}\right),$$ (10.2)

where S(r) is the solubility of a particle with radius r and S(∞) is the solubility of a particle with infinite radius (the bulk solubility), γ is the S/L interfacial tension, R is the gas constant and T is the absolute temperature.

Equation (10.2) shows a significant increase in solubility of the particle with the reduction of particle radius, particularly when the latter becomes significantly smaller than 1 μm. This is illustrated in Fig. 10.1.

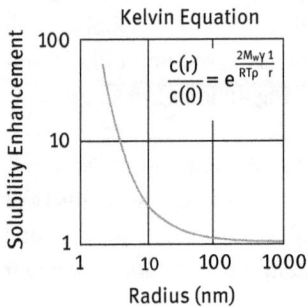

Fig. 10.1: Solubility enhancement with decreasing particle radius.

It can be seen from Fig. 10.1 that the solubility of nanodispersion particles increase very rapidly with decreasing radius, particularly when r < 100 nm. This means that a particle with a radius of say 4 nm will have about 10 times solubility enhancement compared say with a particle with 10 nm radius, which has a solubility enhancement of only 2 times.

Significant increases in solubility are typically observed when r is less than 200 nm. In addition, the injectable dose can be increased for parenteral administration since nanoparticle formulations are essentially made of pure drug and typically use a small amount of excipients. In contrast, standard formulations using solvents such as polysorbate limit the dose due to poor tolerability of the excipient.

For the preparation of nanocrystalline suspensions, the particle size reduction by a top-down process is the most commonly used method due to the possibility to control particle size by proper choice of wetting/dispersing agent, as well as by control of milling conditions. The wetting agent is essential to prevent aggregates and agglomerates of particles in the formulation. A wet-milling process is applied to reduce the

particle size of the active pharmaceutical ingredient (API). The nanocrystalline particles in the formulation must be stabilized against flocculation and crystal growth [8]
and this is best achieved by the use of polymeric surfactants.

Recently, Nakach et al. [9] investigated the methods that can be applied to select the appropriate wetting/dispersing agent in a top-down process. Wetting can be
assessed by using the sinking time test method, as well as by measuring contact angle using direct observation of a sessile drop of liquid on a powder compact, or by
measuring the rate of penetration of surfactant solution through a powder plug [10].
The ability of the dispersant to reduce or eliminate flocculation of the nanodispersion can be assessed by measuring the average particle size as a function of time
after milling. Flocculation results in an increase in the average particle size on storage since the size of floc produced is larger than the size of the single particle. Even
in the absence of flocculation, the average particle size may increase with time as a
result of Ostwald ripening [10]. The driving force of Ostwald ripening is the higher
solubility of the smaller particles when compared to larger ones [7]. This results in
a shift of the particle size distribution to larger values when the nanosuspension is
stored, particularly at higher temperature. When using ionic dispersant, the efficiency
of electrostatic repulsion can be assessed from a knowledge of the ionic concentration and ion valency, as well as measurement of the zeta potential of the particles
[10]. It is well known that electrostatic repulsion increases with decreasing electrolyte
concentration, decreasing ion valency and increasing zeta potential [10]. Nonionic
dispersants reduce flocculation through steric repulsion [11] as discussed in Chapter 5.
These agents, mostly polymeric surfactants, form adsorbed layers with thickness (δ)
which is strongly hydrated in water. When two particles, each with an adsorbed layer
of thickness δ, approach each other at a surface-to-surface distance h that is smaller
than 2δ, strong repulsion occurs as a result of two phenomena:
(i) unfavourable mixing of the stabilizing chains when these are in good solvent;
(ii) reduction of configurational entropy on considerable overlap of the stabilizing
     chains [11].

To apply the above principles, a model hydrophobic highly insoluble API that was provided by Sanofi (Paris) was micronized by jet milling before use. The physicochemical
properties of the API are given in Tab. 10.1.

Several dispersing/wetting agents were used for the investigation ranging from
cellulose derivatives, polyvinyl pyrrolidone, phospholipids, poloxamers (A–B–A
block copolymers of polyethylene oxide A and polypropylene oxide B), polyethylene glycol and derivatives. For the screening of dispersant/wetting agents, low shear
milling was applied using 20 % (w/w) of API, 3 % (w/w) of dispersant/wetting agents,
and 77 % (w/w) of water for injection (WFI). An aliquot of 10 ml suspension and 20 ml
of zirconium oxide beads (700 μm diameter supplied by Netzsch (Germany)) were
introduced into a 30 ml vial. The vial was agitated in an orbital roller mill for 5 days at
0.03 m/s and at room temperature.

Tab. 10.1: Physicochemical properties of the API.

Average particle diameter	5 μm
Specific surface area ($m^2\,g^{-1}$)**	1.5
Molecular weight (g/mol)	497.4
Water solubility (μg/ml)	0.2
pK_a	no pK_a
log P*	6.9
Density (g/ml)	1.42
Melting point (°C)	156.7

* P is the partition coefficient between octanol and water.
** Measurement done using Blaine method [12].

For the assessment of process ability using high shear milling, a suspension containing 20 % (w/w) of API, 3 % dispersant/wetting agent and 77 % (w/w) of WFI was prepared. An aliquot of 50 ml suspension and 50 ml of Polymill® crosslinked polystyrene beads milling media (500 μm diameter) supplied by Alkermes Inc. (Waltham, MA, USA) were introduced into a NanoMill® 01 milling system (annular mill purchased from Alkermes Inc., Waltham, MA, USA, with a stator of 80 mm diameter and rotor of 73 mm). The mill was operated during 1 h at 20 °C and 3 m/s.

For the optimization of the dispersant/wetting agent content a suspension was prepared using 20 % (w/w) of API, the dispersant/wetting agents concentration was varied from 0.3 to 3 % (w/w) and WFI was varied accordingly from 79.3 to 77 % (w/w). An aliquot of 50 ml suspension and 50 ml of Polymill® crosslinked polystyrene beads milling media (500 μm diameter supplied by Alkermes Inc., Waltham, MA, USA) were introduced into a NanoMill® 01 milling system (specifications as above). The mill was operated at 20 °C and 3 m/s. The milling operation was performed during 105–240 min. The resulting nanosuspension was characterized by using several techniques briefly described below.

Particle size measurement was performed using two methods:

(i) Dynamic light scattering, referred to as photon correlation spectroscopy (PCS), using Coulter N4+ equipment (supplied by Beckman Coulter, France). The method is based on measuring the intensity fluctuation of scattered light as the particles undergo Brownian diffusion. From the intensity fluctuation, the diffusion coefficient, D, can be calculated. The particle radius, r, is estimated from the diffusion coefficient, D, using the Stokes–Einstein equation [13]. The measurements were carried out using a scattering angle of 90°. The refractive index was fixed at 1.332 and the temperature at 20 °C. The suspension was diluted from 20 % (w/w) to 0.1 % (w/w) with distilled water. 10 μl of diluted suspension was added to 1 ml distilled water.

(ii) Laser diffraction using Malvern Mastersizer 2000. This method is based on measuring the angle of light diffracted by particles, which depends on the particle's radius, using Fraunhofer diffraction theory. This method can measure particle sizes down to 1 μm. For smaller particles, forward light scattering is measured by applying the Mie theory of light scattering. By combining results obtained with light diffraction and forward light scattering, particle size distributions in the range 0.02 to 10 μm can be obtained.

Scanning electron microscopy (SEM) evaluations were carried out using JOEL JSM-6300F field emission SEM. The suspensions were diluted 10 000 times using WFI. Then 1 ml of the obtained suspension was filtered through Millipore filter Isopore 0.1 μm. The filter was then rinsed 3 times with 1 ml of WFI for each rinse. The filter was then bonded to an aluminium pad using conductive adhesive on both sides. The filter was then metalized with gold using metallizer Xenosput XE200 EDWARDS. The gold deposit was approximately 1.5 to 2 nm thickness. Nanoparticles were observed at 15 kV and the observation was done at several magnifications (1000×, 5000×, 10 000×, 20 000×), for an overview and a detailed view.

The short-term stability was assessed by measuring particle size right after milling, and then after 7 and 15 days' storage at ambient temperature. For the selected formulations, the stability was assessed for a period of 8 weeks at ambient temperature. Zeta potential measurement were carried out using a ZetaSizer Nano ZS from Malvern, UK, which applies the M3-PALS technique, a combination of laser Doppler velocimetry (LDV) and phase analysis light scattering (PALS). The equipment uses an He-Ne laser (red light of 633 nm wavelength), which first splits into two, providing an incident and reference beam.

Zeta potential, ζ, is calculated from the electrophoretic mobility, μ, using the Smoluchowski equation [14], that is valid when $\kappa r \gg 1$ (where κ^{-1} is the Debye length and r is the particle radius). For the case of small particles and low electrolyte concentration, the Huckel equation [14] is applied to calculate zeta potential.

Steady state, shear stress vs. shear rate curves, were carried out using a HAAKE VT550 (Germany) rheometer. A concentric cylinder device was used for this measurement. The measurement was carried out at 20 °C. The shear rate was gradually increased from 0 to 1500 s^{-1} (up curve) over a period of 2 min and decreased from 1500 to 0 s^{-1} (down curve) over another period of 2 min. The test samples were 25 ml of unmilled suspension, which contains 20 % API (w/w), 3 % (w/w) stabilizer, and 77 % (w/w) of WFI. Those samples were homogenized by using an Ultra-Turrax for 10 min at 6000 rpm. When the system is Newtonian, the shear stress increases linearly with the applied shear rate and the slope of the line gives the viscosity of the suspension. In this case the up and down curves coincide with each other. When the system is non-Newtonian, the viscosity of the suspension decreases with the applied shear rate. When it is a thixotropic system, the down curve is below the up curve, showing hysteresis. The latter could be assessed by measuring the area under the loop. In summary, Newtonian, non-Newtonian, as well as thixotropy fluid, can be distinguished from the shear stress and shear rate curves.

The surface tension, γ, of the selected dispersant/wetting agent was measured by using a KRUSS K12 tensiometer (Germany). In these measurements, the Wilhelmy plate method was applied under quasi-equilibrium conditions. Therefore, the force required to detach the plate from the interface was accurately determined. From the γ versus log C curves, where C is the total surfactant concentration, the critical micelle concentration (cmc) was determined.

Wetting was assessed by measuring the rate of penetration of surfactant solution through a powder plug. The result shows a linear relationship between the rate of penetration and time. From the slope of the line a wettability factor can be calculated using the following equation:

$$H^2 = \frac{\gamma}{2\eta} CR \cos \theta t, \qquad (10.3)$$

where H is the height of liquid penetrated within the powder plug, θ is the contact angle, γ is the surface tension of the liquid, η is the liquid viscosity, R is the mean radius of the capillary within the powder plug, C is a tortuosity factor, and t is the time. Since all powder plugs are prepared at the same compression pressure, the parameter C can be assumed to be a constant.

To calculate H^2, the mass of the liquid penetrated within the powder plug was measured using a microbalance. The relationship of the mass (m) and the height of the liquid penetrated within the powder plug can be expressed by the equation

$$m = HS\varepsilon\rho, \qquad (10.4)$$

where m is the mass of the liquid penetrated within the powder plug, H is the height of the liquid penetrated within the powder plug, ρ is the volumetric mass of the liquid, S is the surface of the powder plug, and ε is the fraction of the dead volume of the powder.

Combining equations (10.3) and (10.4), the following equation is obtained:

$$m^2 = \frac{\gamma\rho^2}{\eta} \frac{s^2}{2} CR\varepsilon^2 \cos \theta t. \qquad (10.5)$$

From a plot of m^2 versus time (linear curve), the slope $(d(m^2)/dt)$ can be determined and the wettability factor can be calculated from a knowledge of the surface tension (γ) and the viscosity (η) of the liquid.

The wettability factor can be expressed by the equation

$$\frac{d(m^2)}{dt} \left(\frac{\eta}{\gamma} \right) = K = \frac{s^2\rho^2 CR \cos \theta}{2}. \qquad (10.6)$$

The adsorption isotherm of PVP was measured at room temperature. Known amounts of API were equilibrated at room temperature with various concentrations of PVP dispersant. Then, the bottles containing the various dispersions were rotated from several hours to up to 15 hours until equilibrium was achieved. Then the particles were removed from the dispersant solution by centrifugation. The dispersant concentration in the supernatant was analytically determined using UV spectrometry by Carry 50 at the wavelength of 200 nm. To obtain the amount of adsorption per unit area of the powder (Γ), the specific surface area of the powder (A) in m^2/g was determined using the gas flow method (Blaine).

Two criteria were used to select the optimum stabilizer. The first criterion is that the API particle diameter has to be in the range of 100–500 nm after milling. The second criterion is that the formulation should be free of flocculation after at least 2 weeks of storage at room temperature. The selection was performed with the following step-by-step approach:

(i) Measuring the viscosity as a function of shear rate as well as thixotropy. The samples that gave viscosity greater than 15 mPa s at shear rate of $1000\,s^{-1}$ were rejected. This criterion is essential to ensure faster milling kinetics as well as manufacturability at industrial scale.

(ii) Milling ability using the high shear mill, namely NanoMill® 01 milling system. This step is essential to ensure preparation of nanosuspension at industrial scale using high speed milling. All samples that gave particle size greater than 500 nm or showed instability due to flocculation or Ostwald ripening were rejected.

The combination of SDS/PVP appeared superior to other tested agents. Therefore, they were selected to be further evaluated as follows:

(i) Assessment of wettability, to ensure that the combined system of SDS/PVP gives better wettability than that of the wetting agent SDS alone. The synergistic effect obtained using the mixture was confirmed by measuring the surface tension as well as the critical micelle concentration (cmc), and that of the wetting agent alone. Furthermore, the optimum concentration of SDS/PVP system was used to confirm the milling ability using the high shear mill.

(ii) Measurement of the adsorption isotherm to ensure the strong adsorption of the dispersant (PVP) on the particles' surface.

(iii) Measurement of zeta potential to ensure the electrostatic stabilization. An absolute value greater than 20 mV is usually required for electrostatic repulsion to offer the overall stability, because electrostatic repulsion is proportional to the square of zeta potential.

(iv) Measurement of the long-term physical stability of the selected formulation. This was assessed by measuring the particle size distribution as a function of time over a period of 8 weeks at room temperature.

After roller milling, all samples were inspected for API suspendability; HPC, cyclodextrin, PEG, Montanov 68 and sodium polyacrylate showed obvious flocculation and the appearance of a "dry" sample. Therefore, they are not included for further evaluation. The remaining samples were assessed by measuring the particle size at time 0, after 7 and 14 days. Suspensions with a particle size greater than 500 nm and/or showing flocculation after 7 days were discontinued for further evaluation. These discarded samples were: HPMC, Poloxamer188, Poloxamer407, PVP-SDS (50/50 % w/w), PVP-SDS (30/70 % w/w) and SDS.

The rheological behaviour of the suspension was used to assess the stability of the resulting suspension. Figures 10.2 and 10.3 show typical flow curves for unmilled suspensions prepared using Solutol HS15 (hydroxystearate) and Phosal 50 PG (phospholipid). The suspension prepared using Solutol shows Newtonian behaviour with a low viscosity of 4.8 mPa s. In contrast, the suspension using Phosal 50 PG gives non-Newtonian behaviour with clear thixotropy, indicating flocculation of the suspension.

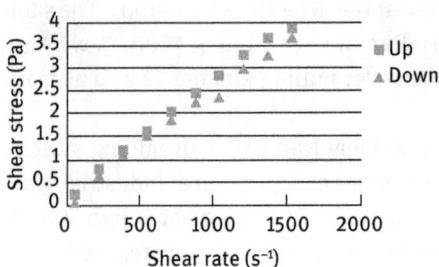

Fig. 10.2: Shear stress-shear rate curves for unmilled suspensions using Solutol HS15.

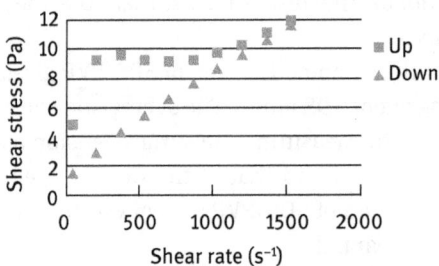

Fig. 10.3: Shear stress-shear rate curves for unmilled suspensions using Phosal 50 PG.

Suspensions with high viscosity greater than 15 mPa s at shear rate of $1000\,s^{-1}$ were excluded from further evaluation.

After high shear milling, the suspensions were assessed by measuring the particles size at time 0, after 7 and 14 days. The results are shown in Fig. 10.4. Two systems, PVP/SDS at the ratio of 60/40 or Vitamin E TPGS offered the best stabilization of the nanocrystalline formulations. Confirmation of these results was obtained by SEM measurement as illustrated in Fig. 10.5 for suspensions prepared using PVP-SDS and Montanox (ethoxylated sorbitan ester). These SEM pictures show large differences between an unstable formulation based on Montanox (needle-shaped particles) and a stable formulation based on PVP-SDS (small but irregular shaped particles). When using Montanox, the suspension shows Ostwald ripening and formation of needle-shaped crystals. This may be due to the specific adsorption of the Montanox molecules

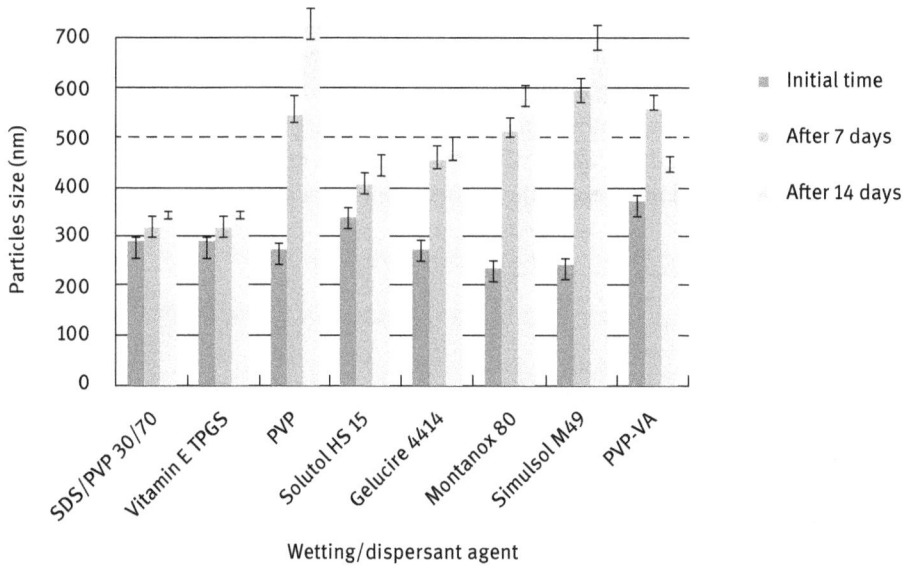

Fig. 10.4: Particle size results after high shear milling using different dispersants.

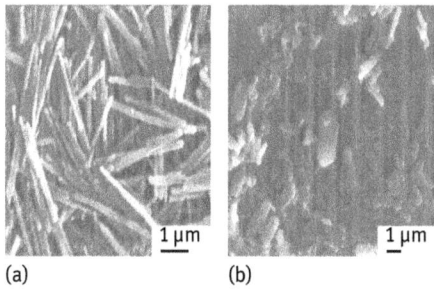

Fig. 10.5: SEM pictures of milled particles using PVP/SDS (a) and Montanox (b).

on certain crystal faces allowing growth to occur on the other faces and hence the formation of needles.

The selection of the final formulation should be based on the following criteria: wettability evaluation, adsorption isotherm measurement, and evaluation of stress tests (heating, freezing-thawing stability, centrifugation, ionic strength, dilution in biorelevant media). PVP-SDS was selected as a model to exemplify the methodology of wetting/dispersant agent selection.

Wettability was assessed by measuring the surface tension as a function of surfactant concentration. Figure 10.6 shows the γ–log C curve for a typical SDS-PVP mixture (80/20 % w/w). This graph shows a typical behaviour with γ decreasing with increasing log C until the critical micelle concentration (cmc) is reached, after which γ shows only a small decrease with increasing log C. A plot of cmc versus percent of PVP (Fig. 10.7) in the binary mixture shows a minimum at 50 to 70 % of PVP above which the cmc increases. This result implies a maximum of surface activity between 50 to 70 % in the binary mixture. To obtain maximum wetting of the API particles, a PVP-SDS mixture containing 60 % PVP in the minimum region of the cmc was chosen. Under this condition, maximum reduction in surface energy can be expected for the powder-liquid interface, which will offer enhanced crack propagation (Rehbinder effect), and enhanced breakage of the particles upon wet milling.

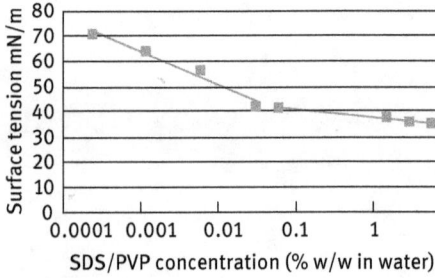

Fig. 10.6: γ-log C curves for SDS/PVP mixtures (80/20 % wt/wt).

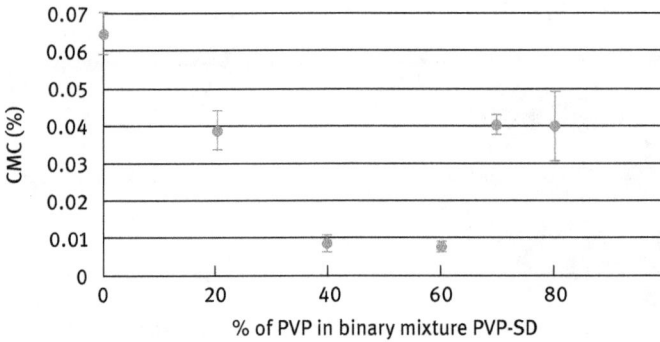

Fig. 10.7: Variation of cmc with percent PVP in the binary mixture of PVP/SDS.

Figure 10.8 shows a plot of wettability factor K versus PVP-SDS concentration. For comparison, the results obtained using SDS alone are shown in the same graph. It can be seen from Fig. 10.8 that K increases with increasing surfactant concentration, reaching a plateau at a certain surfactant concentration. For the PVP-SDS system, this plateau is reached at 1.2 %, consisting of 0.72 % PVP and 0.48 % (w/w). Using the same concentration of SDS alone (0.48 % w/w), the K value is much lower than that obtained with the combined system. This clearly demonstrates the synergistic effect obtained when a polymer surfactant mixture is used. The latter is a much more effective wetting system when compared with the individual components.

The milling ability was investigated using a kinetic experiment where the reduction in particle size or the equivalent increase in surface specific area was measured as a function of milling time. A typical result is shown in Fig. 10.9 at 1.2 % w/w PVP-SDS

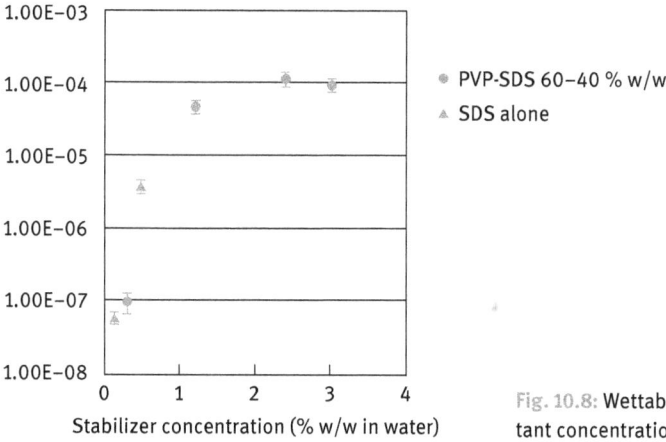

Fig. 10.8: Wettability factor versus surfactant concentration.

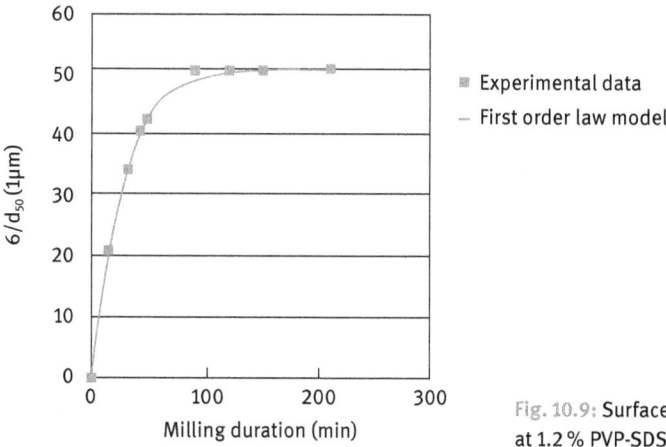

Fig. 10.9: Surface area versus milling time at 1.2 % PVP-SDS stabilizer.

stabilizer. The results obtained show an exponential increase in the surface area (or decrease in particles size) reaching a plateau at a certain milling time (Fig. 10.9). The results follow first order kinetics that can be represented by the equation:

$$\frac{6}{d_{50}} = \left(\frac{6}{d_{50}}\right)_{\infty} (1 - e^{\frac{-t}{\tau}}), \tag{10.7}$$

where, $6/d_{50}$ is the implicit specific surface area, d_{50} is the particles diameter at time t, $(d_{50})_{\infty}$ is the plateau value, and τ is the duration to reach 63 % of the maximum surface are. Values for $(6/d_{50})_{\infty}$ and τ were obtained at various stabilizer concentrations and the results are shown in Fig. 10.10. The results show an initial increase in $(6/d_{50})_{\infty}$ and τ with increasing stabilizer concentration reaching a plateau value at 1.2 %. These results are consistent with those obtained using wettability evaluation. It is clear that a minimum of 1.2 % stabilizer concentration is required to obtain the smallest particle size. Below this stabilizer concentration, there is not enough power

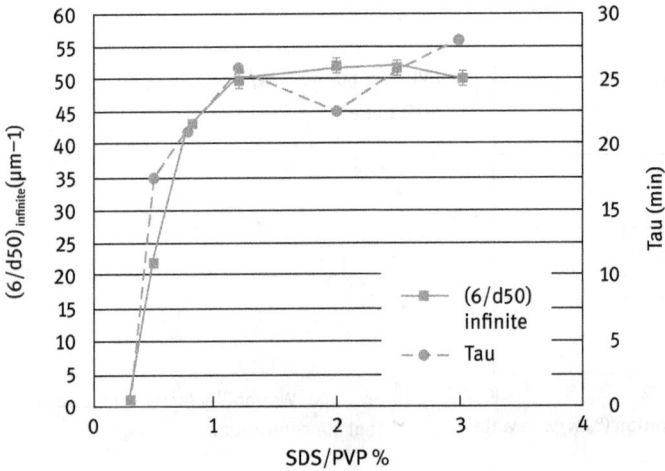

Fig. 10.10: Variation of surface area and duration τ to reach 63 % of maximum surface area with SDS/PVP %.

to completely saturate the particles with surfactant molecules and this may result in rejoining of the small particles after their formation during the milling process.

Figure 10.11 shows the adsorption isotherm of PVP alone on the API powder surface. The results show the high affinity type isotherm, as indicated by the complete adsorption of the first added PVP molecules. The results obtained at high PVP concentration show a great deal of scatter, which is likely due to the possible error of the UV method for determining the remaining PVP concentration. At high PVP concentration, one measures the difference between two large quantities. Any uncertainty in the estimated concentration using the UV method can produce a large error in the amount adsorbed. It is therefore difficult to ascertain an exact plateau value of the isotherm which appears to be between 0.6 and 0.9 mg m^{-2}. Assuming a plateau value

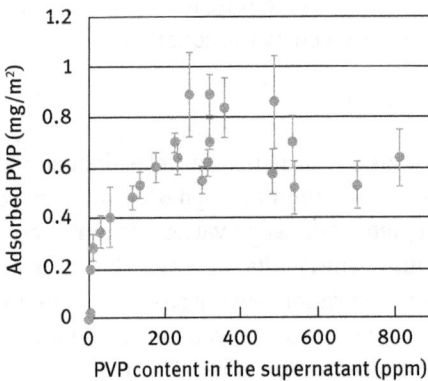

Fig. 10.11: Adsorption isotherm of PVP on API.

of 0.7 mg m^{-2}, the concentration of PVP required to completely saturate the particles can be roughly estimated. From Fig. 10.9 the smallest particles diameter obtained is about 120 nm. This gives a surface specific area of 42.8 m^2 g^{-1}. For a 20 % suspension the total surface area was calculated as 704 m^2 by using the equation,

$$\text{surface area} = \frac{6 \times 20}{\rho d_{50}}. \tag{10.8}$$

The total surface area coverage requires 493 mg or 0.493 % of PVP which corresponds to 0.82 % of PVP-SDS 60/40 % w/w. The results were in good agreement with the values obtained in milling ability and wettability tests.

Figure 10.12 shows the variation of zeta potential for the PVP-SDS system as a function of SDS concentration. In the absence of SDS, PVP alone gave a low negative zeta potential of –20 mV, which is insufficient to give electrostatic stabilization. In this case, the main stability arises from steric repulsion due to the adsorbed loops and tails of PVP molecules. Upon addition of SDS (30/70 SDS-PVP), the zeta potential increases sharply to –50 mV, which contributes to stability through electrostatic repulsion. With further increase of SDS concentration to 40/60 SDS-PVP, the zeta potential increases further to –54 mV and remains almost constant with any further increase of SDS concentration. Thus when using a mixture of SDS and PVP, the stabilizing mechanism is a combination of electrostatic repulsion, which shows an energy maximum at intermediate separation distance, and steric repulsion that occurs at shorter distances of separation that is comparable to twice the adsorbed layer thickness. This combined stabilization mechanism is referred to as electrosteric.

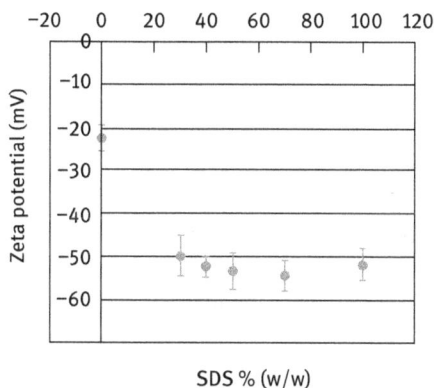

Fig. 10.12: Zeta potential as a function of SDS concentration.

Using the optimum PVP-SDS ratio of 60/40 at a concentration of 1.2 %, long-term stability results were obtained by following the particle size as a function of time at room temperature for the 20 % w/w API nanosuspension. Figure 10.13 shows the variation of d_{10}, d_{50} and d_{90} with storage duration over a period of 57 days. It can be seen from this figure that no change in particle size is observed during this period. This further

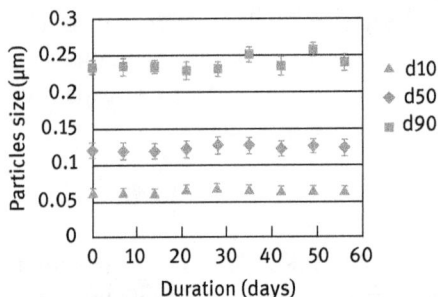

Fig. 10.13: Variation of particle size with storage time.

confirmed the high colloidal stability of the nanosuspension that was prepared using the method described above.

Using a colloidal and interfacial fundamental approach [15–18], an optimum wetting/dispersant agent can be selected for preparation of nanosuspensions with a d_{50} lower than 150 nm. These nanosuspensions can be prepared using a simple milling procedure, namely a roller mill combined with particle size measurement. This procedure is exemplified using a model hydrophobic drug (API) and nanosuspensions could be prepared using a dispersing-wetting agent of PVP-SDS mixture. The results clearly showed an optimum ratio of 60/40 PVP-SDS and a minimum total concentration of 1.2%. This composition gave the maximum wettability, the best milling results and the maximum stability. This approach can help the formulator to select the best wetting/dispersant system for any API. A step forward would be to introduce additional stress tests to assess the formulation's robustness such as thermal stability, freeze-thaw stability and effect of other ingredients in the formulation such as electrolytes and nonelectrolytes.

10.3 Poloxamers for stabilizing suspensions and emulsions in pharmacy

One of the most widely used triblock polymeric surfactants are the "Poloxamers" (BASF, Germany), which consist of two poly-A blocks of poly(ethylene oxide) (PEO) and one block of poly(propylene oxide) (PPO). Several chain lengths of PEO and PPO are available as indicated in the chemical structure below.

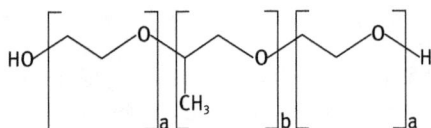

General structure
with a = 2–130 and b = 15–67

Tab. 10.2: Poloxamer grades and their composition.

Poloxamer	EO units (a)	PPO units (b)	Average molar mass
L124	10–15	18–23	2 090–2 360
F188	75–85	25–40	7 680–9 510
F237	60–80	35–40	6 840–8 830
F338	137–146	42–47	12 700–17 400
F407	95–105	54–60	9 840–14 600

The poloxamers (which are FDA approved) are commonly named with the letter L (liquid), P (paste) or F (flake) followed by three digits; the first two digits multiplied by 10 gives the approximate molar mas of PPO and the last digit multiplied by 10 gives the percentage of PEO. For example, Poloxamer F407 has a molar mass of PPO of 400 and 70 % PEO. A summary of the most commonly used poloxamers and their composition is given in Tab. 10.2.

These polymeric triblocks can be applied as emulsifiers or dispersants, whereby the assumption is made that the hydrophobic PPO chain resides at the hydrophobic surface, leaving the two PEO chains dangling in aqueous solution and hence providing steric repulsion. The conformation of the block copolymer at the oil/water (O/W) interface is shown in Fig. 10.14.

Although these triblock polymeric surfactants have been widely used in various applications in emulsions and suspensions, some doubt has arisen on how effective these can be. It is generally accepted that the PPO chain is not sufficiently hydrophobic to provide a strong "anchor" to a hydrophobic surface or to an oil droplet. Indeed, the reason for the surface activity of the PEO–PPO–PEO triblock copolymers at the O/W interface may stem from a process of "rejection" anchoring of the PPO chain since it is not soluble both in oil and water [15–18].

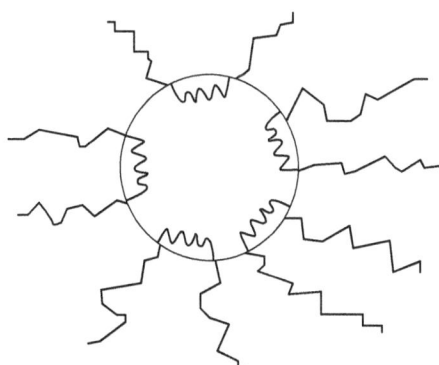

∿∿ Poly (propylene oxide) PPO
∿∿ PEO

Fig. 10.14: Conformation of poloxamer at the O/W interface.

10.4 Polymeric surfactants for the preparation of nanoparticles for drug delivery

The concept of delivering a drug to its pharmaceutical site of action in a controlled manner has attracted much attention in the pharmaceutical industries in recent years. A great deal of research is being carried out since the site delivery of a drug can be controlled at a rate and concentration that optimizes the therapeutic activity, while minimizing the adverse toxic effects [19, 20].

The use of biodegradable colloidal nanoparticles offer a number of advantages over more conventional dosage forms [20, 21]. Due to their small size (20–200 nm) they are suitable for intravenous administration, they can be applied as prolonged circulating drug depots and for targeting specific organs or sites. Several other advantages of nanoparticles can be listed: protection of drugs against metabolism or recognition by the immune system, reduction of toxic effects especially for chemotherapeutic drugs and improved patient compliance by avoiding repetitive administration [15, 22].

Various biodegradable colloidal drug carriers have been developed, of which polymeric nanoparticles are the most widely systems.

The most widely used polymeric nanoparticles are those of the A–B and A–B–A block copolymer type. Block copolymers of the B-A and B-A–B types are known to form micelles that can be used as drug carriers [15, 22]. These block copolymers consist of a hydrophobic B block that is insoluble in water and one or two A blocks which are very soluble in water and strongly hydrated by its molecules. In aqueous media the block copolymer will form a core of hydrophobic chains and a shell of hydrophilic chains. These self-assembled structures are referred to as micelles and they are schematically illustrated in Fig. 10.15. The core-shell structure is ideal for drug delivery where the water insoluble drug is incorporated in the core and the hydrophilic shell provides effective steric stabilization, thus minimizing adsorption of the blood plasma components and preventing adhesion to phagocytic cells.

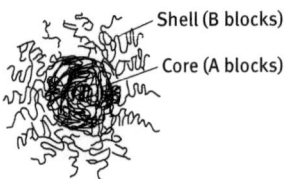

Fig. 10.15: Core-shell structure of block copolymers.

Polymeric nanoparticles, with the drug entrapped within the polymer matrix, have some advantages in terms of their stability both in storage and in vivo applications [22]. The choice of a polymer is restricted by its biodegradability. Both model non-biodegradable, biodegradable and these nanoparticles have been used for studies on their use of these as drug carriers.

Regardless of the type of nanoparticle, these colloidal systems are recognized as foreign bodies after administration to the systemic circulation. They can be quickly removed by the phagocytic cells (macrophages) of the reticuloendothelial system (RES), in particular by the Kupffer cells of the liver. The main approach is to design nanoparticles that avoid RES recognition. This can be achieved by controlling the size and surface properties of the nanoparticle. If the nanoparticles remain in circulation for a prolonged period of time, and avoid liver deposition, there is the possibility of redirecting the particles to other organs/tissues. Long-term circulating nanoparticles can potentially be actively directed to a particular site by the use of targeting moieties such as antibodies or sugar residues that can be specifically recognized by cell surface receptors [22].

As mentioned above, following intravenous (i.v.) injection, the colloidal nanoparticles are recognized as foreign bodies and they may be removed from the circulating blood by the phagocyte cells of the RES. Within 5 minutes after i.v. injection, $\approx 60-90\%$ of the nanoparticles can be phagocytosed by the macrophage of the liver and spleen. Site specific delivery to other organs must avoid this process from taking place. The design of any nanoparticle system with long-circulation requires understanding the mechanism of phagocytosis. The clearance of nanoparticles is mediated by adsorption of blood components to the surface of the particles, a process referred to as opsonization that is described below.

The adsorption of proteins (a component of blood) at the surface of the nanoparticles can result in the surface becoming hydrophobic and this may lead to enhanced phagocytosis. The hydrophobic segments of a protein molecule may adsorb on a hydrophobic surface. While on the surface, the protein may be denatured due to the loss of configurational liability. There may be also a gain in configurational entropy on going from a globular to a more extended state. However, the process of protein adsorption is quite complex due to the presence of more than one type in the blood plasma. The process of protein adsorption is summarized below.

Opsinons refer to proteins that enhance phagocytosis, whereas dyopsinons are molecules that suppress phagocytosis. This depends on the hydrophobic/hydrophilic nature of the protein. Opsinons are immunoglobulin molecules that adsorb on the particle surface, thus making them more "palatable" to macrophages. Dyopsinons are immunoglobulin molecules that render the surface of the particles more hydrophilic, thus suppressing phagocytosis. The interaction of the blood components with the nanoparticles is a complex process, although its control is the key to avoiding phagocytosis [22].

When considering nanoparticles as drug delivery systems one must consider three main characteristics: (i) Particle size, which determine the deposition of colloidal nanoparticles containing the drug following intravenous administration. (ii) Surface charge, which determines the interaction between the nanoparticles and the macrophages. (iii) Surface hydrophobicity, which determines the interaction of the serum components with the nanoparticle surface. This determines the degree of opsonization. A description of each of the above characteristics is given below [22].

(i)   Influence of particle size: Particles > 7 μm are larger than the blood capillaries (≈ 6 μm) and they become entrapped in the capillary beds of the lungs. Thus, aggregated or flocculated particles tend to accumulate in the lung with fatal consequences. Most of the particles that pass the lung capillary bed become accumulated by the RES of the spleen, liver and bone marrow. The degree of splenic uptake increases with increasing particle size. The splenic removal of particles and liposomes > 200 nm is due to a nonphagocytic process whereby the splenic architecture acts as a sieve or filter bed. As the particle size is reduced below 200 nm, the extent of splenic uptake decreases and the majority of particles are mostly cleared by the liver. Colloidal particles not cleared by the RES can exit the blood circulation via the sinusoidal fenestrations of the liver and bone marrow provided they are smaller than 150 nm.

(ii) Influence of surface charge: The surface charge determines the electrostatic repulsion between the colloidal nanoparticle and the blood components or a cell surface. However, the range of electrostatic repulsion decreases with increasing ionic strength. The blood has an ionic strength of ≈ 0.15 mol dm^{-3} and hence the range of electrostatic repulsion is less than 1 nm. This means the surface charge only influences the protein-protein or particle-macrophage interactions at very short distances. Thus the effect of surface charge on phagocytosis is not due to its effect on electrostatic repulsion, but due to its influence on hydrophobicity of the particles that can determine protein adsorption.

(iii) Influence of surface hydrophobicity: As mentioned above, the hydrophobic sites on a nanoparticle determine the adsorption of the serum components. Increasing surface hydrophobicity increases protein adsorption thus increasing the degree of opsonization. It has been shown by in vitro studies that the increased adsorption of proteins on a hydrophobic surface leads to enhanced uptake by phagocytic cells. As will be shown later, the surface modification of nanoparticles by adsorbed or grafted polymers can affect their surface hydrophobicity and hence their ability to be captured by the phagocytic cells.

The surface of polystyrene latex particles is relatively hydrophobic and can be easily modified by adsorbed nonionic polymers. Poloxamers and poloxamines are composed of a central poly(propylene oxide) (PPO) block and terminal poly(ethylene oxide) (PEO) chains. The general structure of poloxamers and poloxamines is given in Fig. 10.16. As mentioned above, the hydrophobic central PPO chain anchors the copolymer to the surface of the particle, whereas the hydrophilic PEO blocks provide the required hydrophilic steric barrier. In general, the thickness of the adsorbed layer increases with increasing length of the PEO chains [15, 22].

The coating of polystyrene particles with poloxamers and poloxamines dramatically reduces their sequestration by the liver. The thickness of the PEO layer is cru-

(a)
$$HO(CH_2CH_2O)_a(CHCH_2O)_b(CH_2CH_2O)_aH$$
$$|$$
$$CH_3$$

(b)
$$HO(CH_2CH_2O)_a(CHCH_2O)_b \quad \quad \quad CH_3 \quad \quad \quad (OCH_2CH)_b(CH_2CH_2O)_aOH$$

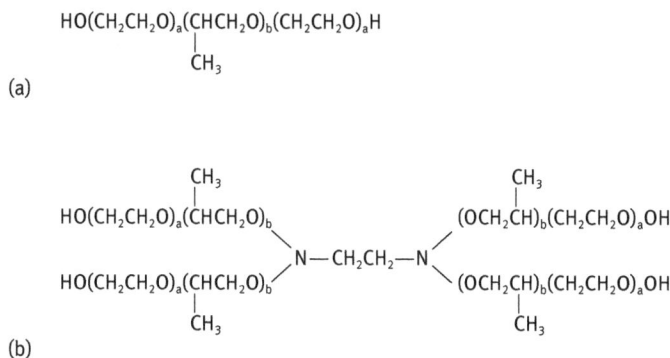

Fig. 10.16: General structure of poloxamers (a) and poloxamines (b).

cial to altering the biological fate of the nanoparticles. Poloxamers with short PEO chains, such as Poloxamer 108 (M_w of the PEO is 1800 Da) do not provide an effective steric barrier against in vitro phagocytosis. In contrast, Poloxamer 338 (M_w of the PEO is 5600 Da) is sufficient to suppress in vitro phagocytosis in the presence of serum and dramatically reduce the liver/spleen uptake of 60 nm polystyrene nanoparticles from 90 % to 45 % following i.v. injection. Coating with Poloxamine 908 (M_w of the PEO is 5200 Da) had a more pronounced effect decreasing the amount cleared to less than 25 %.

The presence of a hydrated PEO layer alone does not necessarily prolong the circulatory half-life of all drug carriers. The particle size plays a major role. Particles > 200 nm in diameter with coated Poloxamine 908 enhanced spleen uptake and decreased blood levels following i.v. administration to rats. Polystyrene particles with 60 and 150 nm diameters and coated with Poloxamer 407 were redirected to the sinusoidal endothelial cells of rabbit bone marrow following i.v. administration. In contrast, Poloxamer 407 particles with diameter 250 nm were mostly sequestered by the liver and spleen and only a small portion reached the bone marrow [15, 22].

Polystyrene particles with chemically grafted PEO chains (M_w of the PEO is 2000 Da) were prepared with different surface densities of PEO. In vitro cell interaction studies demonstrated that particle uptake by nonparenchymal rat liver cells (primary Kupffer cells) decreased with increasing PEO surface density until an optimum density is reached. In vivo studies showed that only particles with very low PEO surface density result in considerable liver deposition. However, the results showed that the liver avoidance and blood circulation was not improved above that obtained with Poloxamine 908, even though the surface density of the grafted PEO particles was higher than that of Poloxamine 908.

As mentioned above, studies using polystyrene nanoparticles as model drug carriers have demonstrated that optimizing the particle size and modifying the surface using a hydrophilic PEO layer (as a steric barrier) can result in an increase in circu-

Tab. 10.3: Biodegradable polymers for drug carriers.

Poly(lactic acid)/Poly(lactic-co-glycolic acid) – PLA/PLGA
Poly(anhydrides)
Poly(caprolactone)
Poly(ortho esters)
Poly(β-maleic acid-co-benzyl malate)
Poly(alkylcyanoacrylate)

lation lifetime and to some extent selective targeting may be achieved. For practical applications in drug targeting, polymeric nanoparticles that are constructed from biodegradable and biocompatible materials must be constructed [15, 22]. These polymeric nanoparticles can act as drug carriers by incorporation of the active substance in the core of the nanoparticle. Natural materials such as albumin and gelatine are poorly characterized and in some cases can produce an adverse immune response. This led to the use of synthetic, chemically well-defined biodegradable polymers which do not cause any adverse immune response. A list of these biodegradable polymers is given in Tab. 10.3.

The most widely used biodegradable polymers are the aliphatic polyesters based on lactic and glycolic acid which have the following structures.

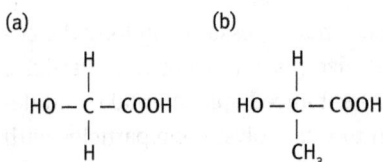

(a)

$$\begin{array}{c} H \\ | \\ HO - C - COOH \\ | \\ H \end{array}$$

(b)

$$\begin{array}{c} H \\ | \\ HO - C - COOH \\ | \\ CH_3 \end{array}$$

Poly(lactic acid) (PLA) and poly(lactic acid-co-glycolic acid) (PLGA) have been used in the production of a wide range of drug carrier nanoparticles. PLA and PLGA degrade by bulk hydrolysis of the ester linkages. The polymers degrade to lactic and glycolic acids which are eliminated in the body, primarily as carbon dioxide and urine.

The preparation of biodegradable nanoparticles with a diameter less than 200 nm (to avoid splenic uptake) remains a technical challenge. Particle formation by in situ emulsion polymerization (that is commonly used for the preparation of polystyrene latex) is not applicable to biodegradable polymers such as polyesters. Instead, the biodegradable polymer is directly synthesized by chemical polymerization methods. The polymer is dissolved in a water immiscible solvent such as dichloroethane which is then emulsified into water using a convenient emulsifier such as poly(vinyl alcohol) (PVA). Nanoemulsions can be produced by sonication or homogenization and the organic solvent is then removed by evaporation. Using this procedure, nanoparticles of PLA and PLGA with a diameter ≈ 250 nm could be produced. Unfortunately, the emulsifier could not be completely removed from the particle surface and hence this procedure was abandoned.

To overcome the above problem nanoparticles were prepared using a surfactant-free method. In this case the polymer is dissolved in a water miscible solvent such as acetone. The acetone solution is carefully added to water while stirring [22]. The polymer precipitates out as nanoparticles which are stabilized against flocculation by electrostatic repulsion (resulting from the presence of COOH groups on the particle surface). Using this above procedure surfactant-free nanoparticles with diameter < 150 nm could be prepared. Later, the procedure was modified by incorporation of poloxamers or poloxamines in the aqueous phase. These block copolymers are essential for surface modification of the nanoparticle as is discussed below.

Following the encouraging in vivo results using polystyrene latex with surface modification using poloxamer and poloxamine, investigations were carried out using surfactant-free PLGA, $\approx$ 140 nm diameter, which was surface modified using the following block copolymers: water soluble poly(lactic)-poly(ethylene) glycol (PLA-PEG); poloxamers and poloxamines. The results showed that both PLA-PEG 2 : 5 (M_w of PLA is 2000 Da and M_w of PEO is 5000 Da) Poloxamine 908 form an adsorbed layer of 10 nm. The coated PLGA nanoparticles were effectively sterically stabilized against electrolyte induced flocculation and in vivo studies demonstrated a prolonged systemic circulation and reduced liver/spleen accumulation when compared to the uncoated particles. The main drawback of the polymer adsorption approach is the possibility of desorption in vivo by the blood components. Chemical attachment of the PEG chain to the biodegradable carrier would certainly be beneficial [15, 22].

The best approach is to use block copolymer assemblies as colloidal drug carriers [23–31]. Block copolymers of the B–A and B–A–B types are known to form micelles that can be used as drug carriers. These block copolymers consist of a hydrophobic B block that is insoluble in water and one or two A blocks which are very soluble in water and strongly hydrated by its molecules. In aqueous media the block copolymer will form a core of hydrophobic chains and a shell of the hydrophilic chains. These self-assembled structures are referred to as micelles and they were schematically illustrated in Fig. 10.15. The core-shell structure is ideal for drug delivery when the water insoluble drug is incorporated in the core and the hydrophilic shell provides effective steric stabilization, thus minimizing adsorption of the blood plasma components and preventing adhesion to phagocytic cells [15, 22].

The critical micelle concentration (cmc) of block copolymers is much lower than that obtained with surfactants. Typically, the cmc is of the order of 10^{-5} g ml^{-1} or less. The aggregation number N (number of copolymer molecules forming a micelle) is typically several tens or even hundreds. This results in assemblies of the order of 10–30 nm which are ideal as drug carriers. The thermodynamic tendency for micellization to occur is significantly higher for block copolymers when compared with low molecular weight surfactants.

The inherent core-shell structure of aqueous block copolymer micelles enhances their potential as a colloidal drug carrier. As mentioned before, the hydrophobic core can be used to solubilize water insoluble substances such as hydrophobic drug molecules. The core acts as a reservoir for the drug which also can be protected against in vivo degradation. Drugs may be incorporated by covalent or noncosolvent binding, such as hydrophobic interaction. The hydrophilic shell minimizes the adsorption of blood plasma components. It also prevents the adhesion of phagocytic cells and influences the parakinetics and biodistribution of micelles. The stabilizing chains (PEG) are chemically grafted to the core surface, thus eliminating the possibility of desorption or displacement by serum components. The size of the block copolymer micelles is advantageous for drug delivery [15, 22].

The water solubility of PLA-PEG and PLGA-PEG copolymers depends on the molecular weight of the hydrophobic (PLGA-PEG) and hydrophilic (PEG) blocks. Water soluble PLA-PEG copolymers with relatively low molecular weight PLA blocks self-disperse in water to form block copolymer micelles. For example, water soluble PLA-PEG 2:5 (M_W of PLA is 2000 Da and M_W of PEO is 5000 Da) form spherical micelles ≈ 25 nm in diameter. These micelles solubilize model and anticancer drugs by micellar incorporation. However, in vivo, the systemic lifetimes produced were relatively short and the clearance rate was significantly faster when the micelles are administered at low concentration. This suggests micellar dissociation at concentrations below the cmc.

By increasing the PLA/PLGA core molecular weight, the block copolymer becomes insoluble in water and hence it cannot self-disperse to form micelles. In this case the block copolymer is dissolved in a water immiscible solvent such as dichloromethane and the solution is emulsified into water using an emulsifier such as PVA. The solvent is removed by evaporation resulting in the formation of self-assembled nanoparticles with a core-shell structure. Using this procedure, nanoparticles of PLGA-PEG copolymers (M_W of PLGA block of 45 000 Da and M_W of PEO of 5000, 12 000 or 20 000 Da) can be obtained. High drug loading (up to 45 % by nanoparticle weight) and entrapment efficiencies (more than 95 % of the initial drug used) can be achieved.

The PLGA-PEG nanoparticles show prolonged blood circulation times and reduced liver deposition when compared to the uncoated PLGA nanoparticles. The adsorption of plasma proteins onto the surfaces of the PEG coated particles is substantially reduced, in comparison with the uncoated PLGA nanoparticles. The qualitative composition of the adsorbed plasma protein is also altered by the presence of the PEG layer. Substantially reduced adsorption of opsinon proteins such as fibrinogen, immunoglobulin G and some apolipoproteins is achieved. These results clearly show the importance of the presence of the hydrophilic PEG chain on the surface of the nanoparticles, which prevents opsonization [23–32].

The particle size and surface properties are strongly dependent on the emulsification conditions and the choice of the emulsifier. By using a water miscible solvent such as acetone, the nanoparticles can be directly precipitated and the solvent is removed by evaporation. Using this procedure, one can produce a series of PLA-PEG nanoparticles. The blood circulation of the nanoparticles (e.g. PLA-PEG 30 : 2) is considerably increased when compared with albumin coated PLA nanoparticles. The albumin molecules are rapidly displaced by the protein in the plasma leading to phagocytosis by Kupffer cells in the liver. The PLA-PEG nanoparticles show a low deposition of proteins on the particle surface.

Functionality is introduced in the core-forming A block in the form of polymers such as poly(L-lysine) or poly(aspartic acid). Both these polymers are biodegradable but not hydrophobic. Hydrophobicity is imparted by covalent or ionic attachment of the drug molecule. In this way potent anticancer drugs can be coupled to the aspartic acid residues of poly(aspartic acid)-poly(ethylene glycol) (P(Asp)-PEG) copolymer. In aqueous media the block copolymer-drug conjugate form micelles, but some of the drug may become physically entrapped in the core of the micelle. These P(asp)-PEG micelles ($\approx$ 40 nm diameter) remain in the vascular system for prolonged periods, with 68 % of the injected dose remaining 4 hours after intravenous administration. These systems offer a promising route for drug delivery.

The mechanism of action of the hydrophilic PEG chains can be explained in terms of steric interaction that is well known in the theory of steric stabilization (described in detail in Chapter 5). Before considering the steric interaction one must know the polymer configuration at the particle/solution interface. The hydrophilic PEG chains can adopt a random coil (mushroom) or an extended (brush) configuration [33]. This depends on the graft density of the PEG chains as will be discussed below. The conformation of the PEG chains on the nanoparticle surface determines the magnitude of steric interaction. This configuration determines the interaction of the plasma proteins with the nanoparticles.

The hydrophilic PEG B chains (buoy blocks) can be regarded as chains terminally attached or grafted to the micellar core (A blocks). If the distance between the grafting points, D, is much greater than the radius of gyration, R_G, the chains will assume a "mushroom" type conformation, as is illustrated in Fig. 10.17 (a). The extension of the mushroom from the surface will be of the order of $2R_G$ and the volume fraction of the polymer exhibits a maximum away from the surface, as illustrated in Fig. 10.18 (a). If the graft density reaches a point whereby D < R_G, the chains stretch in solution forming a "brush". A constant segment density throughout the brush with all chains ending a distance Δ (the layer thickness) from the surface and the volume fraction of the polymer shows a step function, as illustrated in Fig. 10.18 (b).

The thickness of the block "brush", Δ, for a grafted chains of N bonds of length ℓ is given by

$$\frac{\Delta}{\ell} \approx N \left(\frac{\ell}{D}\right)^{2/3}. \tag{10.9}$$

This means that for terminally-attached chains at high graft density (brush), Δ depends linearly on N. This is in contrast to polymer chains in free solutions, where $R_G \approx N^{3/5}$ or $R_G \approx N^{1/2}$. In the case of micellar structures, the distance between grafting points, D, is determined by the aggregation number. Unless high aggregation

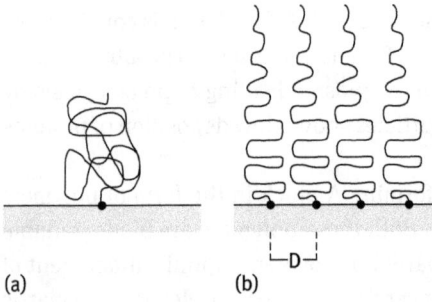

Fig. 10.17: Schematic representation of the conformation of terminally attached PEG chains.

(a)　　　　　(b)

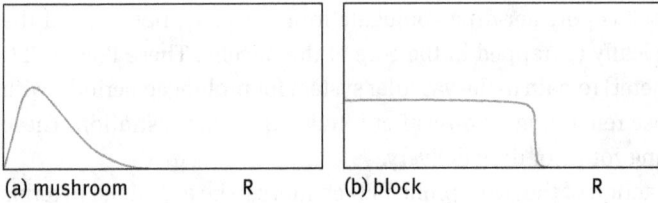

(a) mushroom　　　R　　(b) block　　　R

Fig. 10.18: Volume fraction profile for (a) mushroom (b) brush.

numbers and hence grafting densities can be achieved, a weaker dependence of Δ on chain length is expected.

For a brush on a flat surface, the attached chain is confined to a cylindrical volume of radius $D/2$ and height Δ. If the individual chains of the brush are attached to a spherical core (as is the case with nanoparticles), then the volume accessible to each chain increases and the polymer chains have an increased freedom to move laterally resulting in a smaller thickness Δ. This is schematically illustrated in Fig. 10.19, which shows the difference between particles with high surface curvature (Fig. 10.19 (a)) and those with a surface with low surface curvature (Fig. 10.19 (b)). The curvature effect

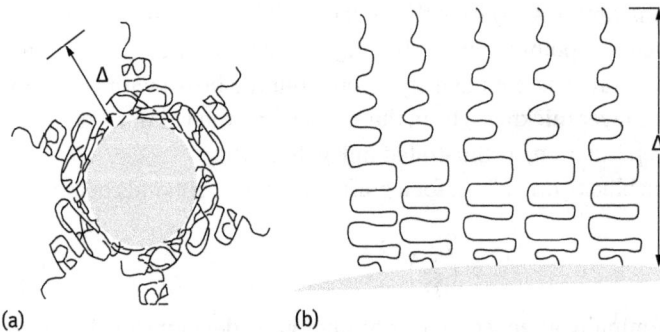

(a)　　　　　　　　(b)

Fig. 10.19: Effect of surface curvature on the adsorbed layer thickness Δ: (a) high surface curvature; (b) low surface curvature.

was illustrated for PEO and poloxamer block copolymers using polystyrene latex particles with different sizes. An increase in the layer thickness with increasing particle radius was observed [15, 22].

Most studies with model nonbiodegradable and biodegradable systems showed that the presence of a hydrated PEG steric barrier significantly increased the blood circulation of the nanoparticles following intravenous administration. The hydrophilic PEG layer minimizes the interactions with phagocytic cells and prevents the adsorption of opsinons. Hydrophilicity is necessary but not sufficient for achieving these two effects. This was demonstrated using dextran (which is considerably hydrophilic) coated liposomes which showed shorter circulation times when compared with their PEG counterparts. This clearly showed that chain flexibility is the second prerequisite for inhibiting phagocytic clearance [15, 22].

PEG chains only have a weak tendency to interact hydrophobically with the surrounding proteins. As the protein approaches the stabilizing PEG chains, the configurational entropy of both molecules is reduced. The more mobile the stabilizing PEG chains, the greater the loss in entropy and the more effective the repulsion from the surface. At sufficiently high surface density, the flexible PEG chains form an impermeable barrier, preventing the interaction of the opsinons with the particle surface. This repulsion is referred to as elastic interaction, G_{el}. On the approach of a second surface to a distance h smaller than the adsorbed layer thickness Δ, a reduction in configurational entropy of the chain occurs [34–36]. The mechanism of elastic interaction was described in detail in Chapter 5.

Four different types of interaction between a protein molecule and hydrophobic substrate can be considered [37]:
(i)   Hydrophobic attraction between the protein and substrate.
(ii) Steric repulsion (osmotic and elastic effects).
(iii) Van der Waals attraction between the protein and substrate.
(iv) Van der Waals attraction between the protein and PEG chains.

These interactions are schematically represented in Fig. 10.20. The interaction of plasma proteins with the PEG steric layer is dependent on the conformation of the chains, which is determined by the surface curvature as discussed above.

There is ample evidence to suggest the high surface coverage of long brush-like PEG chains is necessary for preventing serum protein adsorption. However, the precise surface characteristics required for successful PES avoidance are not well established and more research is still required. In vitro phagocytosis of poloxamer coated polystyrene (PS) nanoparticles (60 and 250 nm in diameter) decreases with increasing PEG molecular weight and hence its thickness. However, increasing the PEG molecular weight above 2000 Da did not improve the ability of the coated nanoparticles to avoid phagocytosis. Similar results were obtained in vivo for both coated PS particles and liposomes of phosphatidylamine-PEG. However, results using PLGA-PEG nano-

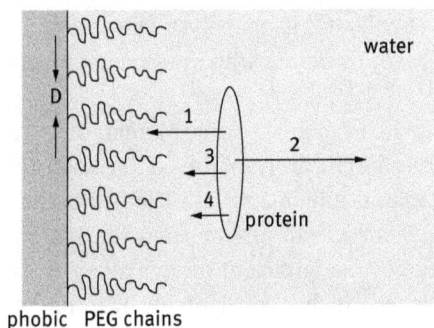

Fig. 10.20: Schematic representation of the various interactions between the PEG layer and a protein molecule.

particles showed an increase in performance when the PEG chain molecular weight was increased from 5000 to 20 000 Da.

The colloid stability of the PLA-PEG nanoparticles can be assessed by the addition of an electrolyte such as Na_2SO_4 which is known to reduce the solvency of the medium for the PEO chains [15, 22]. This reduction in solvency results in an increase in the Flory–Huggins parameter χ (from its value in water of < 0.5) and when χ reaches 0.5 (the θ-condition) flocculation occurs. In this way one can determine the critical flocculation point, CFPT. The CFPT can be easily determined by following the turbidity of the nanoparticle dispersion as a function of Na_2SO_4 concentration. At the CFPT a sharp increase in turbidity is observed. The reversibility of flocculation can be assessed by diluting the flocculated dispersion with water and observing if the flocs can be redispersed by gentle shaking. The effect of the presence of serum protein on nanodispersion stability can also be studied by firstly coating the nanoparticles with the protein and then determining the CFPT using Na_2SO_4.

The hydrodynamic diameter of the nanoparticles and polydispersity index is determined using PCS. The PEG molar mass is fixed at 5000 Da while gradually increasing the PLA molar mass. The nanoparticle composition is expressed as a ratio of PLA:PEG; for example PLA-PEG 2 : 5 refers to a nanoparticle with PLA molar mass of 2000 and PEG molar mass of 5000. For comparison, the hydrodynamic diameter of PLA is also determined at a given molar mass. The zeta potential of the nanoparticles in 1 mM HEPES buffer (adjusted to pH 7.4 by addition of HCl) is also determined. A summary of the results is given in Tab. 10.4.

The results of Tab. 10.4 show that the nanoparticles have a relatively low polydispersity index. They suggest that the particle size distribution is monomodal as is illustrated in Fig. 10.21. This is confirmed by absence of subpopulations in Fig. 10.21. Particles prepared from PLA ($M_w = 35\,000$) are significantly larger than those prepared with a near equivalent PLA block (30 : 5). The effect of increasing the polymer concentration in the acetone solution on the particle size is shown in Fig. 10.22 for PLA-PEG copolymers and in Fig. 10.23 for PLA homopolymer. The results of Fig. 10.22 show that up to PLA-PEG 30 : 5, the hydrodynamic diameter is independent of polymer concentration. In contrast, the results of Fig. 10.23 for the PLA homopolymer show a

Tab. 10.4: Hydrodynamic diameter, polydispersity index and zeta potential of PLA and PLA-PEG nanoparticles.

Polymer	Hydrodynamic diameter, D_{hyd} / nm mean ± SD	Polydispersity index mean ± SD	Zeta potential 1 mM HEPES / mV
PLA (M_w 35 kDa)	124.6 ± 2.5	0.11 ± 0.03	−49.6 ± 0.7
PLA-PEG 2 : 5	26.0 ± 1.6	0.19 ± 0.01	—
PLA-PEG 3 : 5	28.2 ± 0.6	0.14 ± 0.01	—
PLA-PEG 6 : 5	41.1 ± 1.8	0.10 ± 0.04	—
PLA-PEG 15 : 5	50.6 ± 2.0	0.06 ± 0.01	−6.5 ± 0.7
PLA-PEG 30 : 5	63.8 ± 1.8	0.08 ± 0.02	−6.4 ± 1.5
PLA-PEG 45 : 5	80.7 ± 4.8	0.10 ± 0.01	−6.1 ± 0.4
PLA-PEG 75 : 5	118.7 ± 4.9	0.10 ± 0.01	−14.2 ± 0.6
PLA-PEG 110 : 5	156.6 ± 5.0	0.13 ± 0.02	−28.0 ± 0.4

significant increase in particle diameter with increasing polymer concentration. It appears that the PEG block moderates the association of the PLA-PEG copolymer.

The results shown in Tab. 10.4 clearly indicate that the PLA nanoparticles have a high negative zeta potential of 49.6 mV. This is probably due to the presence of ionic carboxylic groups on the nanoparticle surface. At pH 7.4 (which is above the pK_a of COOH groups) and low ionic strength, a high negative surface charge is produced. This negative charge provides electrostatic stabilization of the nanoparticles.

The zeta potential of the PLA-PEG nanoparticles is significantly reduced (to ≈ −6 mV) up to a PLA : PEG ratio of 45 : 5. This reduction is due to the presence of the PEG layer that cause a significant shift in shear plane and hence reduction of ζ. However, when the PLA : PEG ratio is increased above 45 : 5, ζ starts to increase, since the PEG layer thickness become smaller relative to the core of PLA.

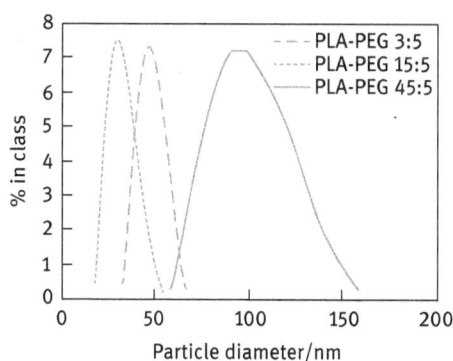

Fig. 10.21: Particle size distribution of PLA-PEG nanoparticles (intensity weighted CONTIN analysis).

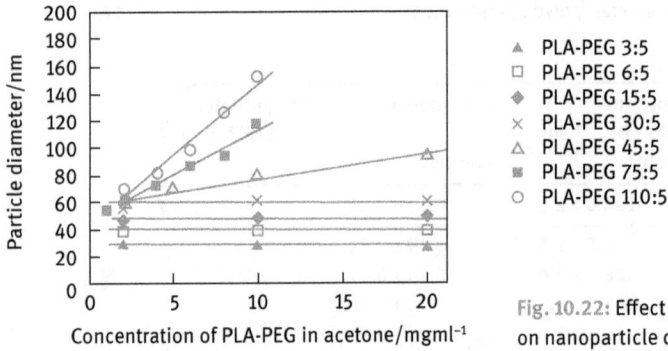

Fig. 10.22: Effect of PLA-PEG concentration on nanoparticle diameter.

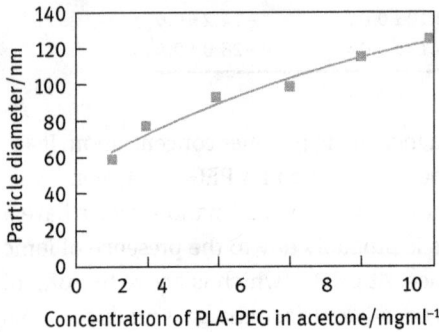

Fig. 10.23: Effect of PLA concentration on nanoparticle diameter.

Figure 10.24 shows the variation in turbidity with Na_2SO_4 concentration. Above a critical Na_2SO_4 concentration the turbidity shows a rapid increase with any further increase in electrolyte concentration. This critical concentration is defined as the critical flocculation point (CFPT). The CFPT decreases with increasing PLA block in the nanoparticle. The results are as expected for a sterically stabilized dispersion and show that at the CFPT, the medium becomes θ-solvent for the chains at which the Flory–Huggins interaction parameter χ become 0.5, and this is the onset of incipient

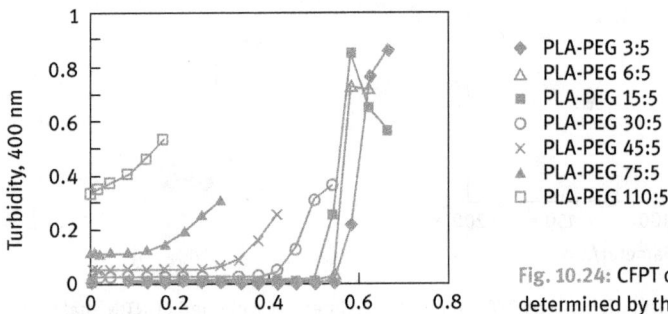

Fig. 10.24: CFPT of PLA-PEG nanoparticles determined by the turbidity method.

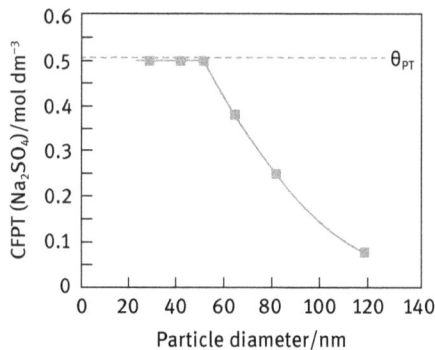

Fig. 10.25: CFPT of PLA-PEG with increasing amounts of PLA (3 : 5 to 75 : 5) as a function of particle diameter.

flocculation. Above the CFPT, χ become greater than 0.5, i.e. the medium becomes worse than a θ-solvent for the chains.

The effect of the PLA core on the CFPT is illustrated in Fig. 10.25, which shows the variation of CFPT with particle diameter.

The results of Fig. 10.25 clearly show that nanoparticles with PLA blocks of $M_w <$ 15 000 give a CFPT close to the θ-point of the PEG chain. When the PLA block's M_w is > 15 000, the nanoparticles give a CFPT below the θ-point of the PEG chain. The CFPT decreases with increasing M_w of the PLA block. This discrepancy between the CFPT and θ-point of the PEG chain may be due to the decrease in the surface coverage of the nanoparticles by the PEG chains when the PLA block M_w exceeds a certain value. This reduction in surface coverage leads to lateral movement of the PEG chains, which results in a smaller layer thickness. This smaller PEG thickness can result in a deep attractive minimum, causing flocculation under conditions of better than the θ-point of the PEG chain [15, 22].

Increasing the concentration of the PLA-PEG copolymer used during the solvent/precipitation preparation method results in an increase in nanoparticle size. This is illustrated in Fig. 10.26 for PLA-PEG 45 : 5 which shows the variation of CFPT with particle diameter. The results show a linear increase in the CFPT with increasing particle diameter approaching the θ-point for the PEG chain when the diameter reaches 94 nm (obtained when the concentration of the PLA-PEG reaches 20 mg ml^{-1}). This increase in the stability of the nanoparticle dispersion with increasing particle diameter may be due to the increased surface coverage of the particles with PEG chains as the particle size is increased.

The aggregation number of the PLA-PEG copolymer in aqueous solution determines the properties of the micellar-like structures of the nanoparticles. The aggregation number is the number of copolymer units per micelle and this emphasizes the self-assembly of PLA-PEG of the nanoparticle [15, 22]. The process is irreversible with no dynamic equilibrium between the self-assembled structure and unimers. Thus, the process is different from that of surfactant micelles where a dynamic equilibrium exists between the monomer and the micelle. The micellar

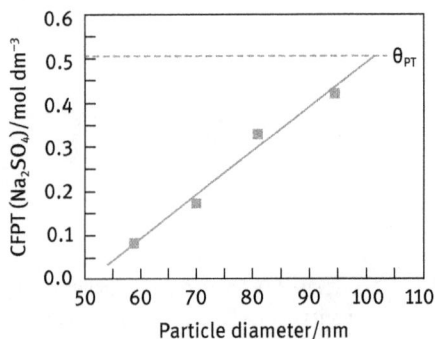

Fig. 10.26: CFPT of PLA-PEG 45 : 5 nanoparticles as a function of particle diameter.

Tab. 10.5: Molar mass and aggregation numbers of PLA-PEG micellar-like nanoparticles.

PLA-PEG[a]	N_{PLA}	R_{hyd}[b] (nm)	$\overline{M}_{w,mic}$ (Da)	N_{agg}	S_t/N_{agg} (nm^2)
2 : 5	28	13.0	2.01×10^5	29	73
3 : 5	42	14.1	3.13×10^5	39	64
4 : 5	56	17.5	7.02×10^5	78	49
6 : 5	83	20.6	1.99×10^6	180	30
9 : 5	125	23.4	2.85×10^6	203	34
15 : 5	208	25.3	5.57×10^6	278	29
30 : 5	417	31.9	1.31×10^7	375	34
45 : 5 (2 mg ml^{-1})	625	30.3	1.19×10^7	238	49
45 : 5 (10 mg ml^{-1})	625	40.4	3.37×10^7	674	30
45 : 5 (20 mg ml^{-1})	625	48.0	5.67×10^7	1134	26

a All prepared using 10 mg ml^{-1} solutions of PLA-PEG in acetone unless otherwise stated.
b Results from PCS measurements.

aggregation number (Tab. 10.5) scales with the increase in the number of monomeric units in the core-forming block N_A:

$$N_{agg} \approx N_A^{\beta}. \tag{10.10}$$

The value of the exponent β depends on the composition of the micelle-forming copolymer. In the large core limit ("crew-cut" micelles, $N_A \gg N_B$) mean density models may be used. The volume fraction of B segments in the corona, ϕ_B, is assumed to be independent of the distance from the core and N_{agg} is predicted to be proportional to N_A ($\beta = 1$). If $N_B \gg N_A$ (star model which assumes a concentration profile for ϕ_B) $N_{agg} \approx N_A^{4/5}$.

Figure 10.27 shows the variation in the aggregation number, N_{agg}, and hydrodynamic radius, R_{hyd}, with the number of monomeric units of PLA, N_{PLA}. Figure 10.28 shows log-log plots of N_{agg} versus N_{PLA}. For comparison, results for the N_{agg} of PLA : PEG with a lower molecular weight PEG, namely 1800, are shown in the same figure. For copolymers with relatively low PLA to PEG weight ratio (2 : 5 to 6 : 5) there is a sharp increase in the micellar aggregation number as the molar mass of PLA is increased. This trend can be rationalized in terms of the thermodynamics of micelle formation as discussed below.

In a selective solvent, where the block of the copolymer B is in a good solvent (PEO, $\chi_{BS} < 0.5$) whereas the other segments of A are in a worse than θ-solvent (PLA, $\chi_{AS} > 0.5$) there will be strong attraction between the A segments. These attractive hydrophobic interactions (enthalpic) must overcome the repulsive (entropic) forces between B chains in the corona. PLA-PEG copolymers with very low PLA to PEG ratios (e.g. 400 : 1800) do not form micelles in aqueous media. The PLA-PEG 2 : 5 copolymer

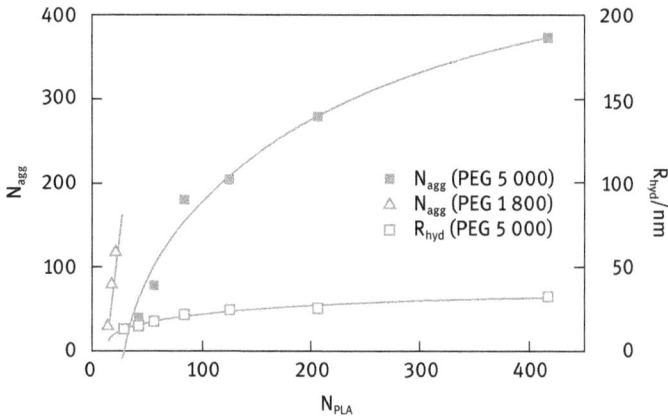

Fig. 10.27: Variation in the aggregation number and hydrodynamic radius of PLA-PEG micellar-like nanoparticles with number of monomeric units of PLA.

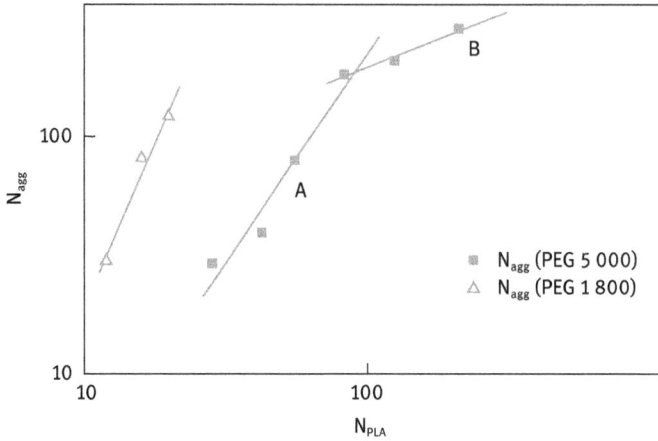

Fig. 10.28: log-log plots of N_{agg} versus N_{PLA}.

has the shortest PLA block and although the block copolymer is water soluble with the least number of unfavourable interactions between the lactic acid units and the aqueous solvent, it still spontaneously forms micellar structures. Since the interactions between the low molecular weight PLA chains within the core of the micelle are weak, the 2:5 PLA:PEG copolymers form loosely packed micellar assemblies with a lot of free space (solvent) within the corona region. This can be demonstrated by calculating the surface area per copolymer unit (S_t/N_{agg}) at the outer surface of the micelle.

A schematic picture of the PLA:PEG micelle is shown in Fig. 10.29. The external surface area of the 2:5 micelle ($4\pi R_{hyd}^2$) is 2124 nm^2 and its aggregation number is 29 (see Tab. 10.5). The area per PEG block in the micelle is $(2124/29) = 73$ nm^2. This area may be compared with the cross-sectional area of a PEG chain of molar mass of 5 kDa in a good solvent. The radius of gyration R_g of PEG is related to its molar mass by [15, 22]

$$R_g = 0.0215 M_w^{0.583}. \tag{10.11}$$

This gives an R_g of 3.1 nm, which in free solution occupies a sphere with maximum cross-sectional area (πR_g^2) of 30 nm^2. This clearly shows the high area per block copolymer (73 nm^2) in the 2:5 PLA:PEG micelle. The loosely packed nature of the micellar-like nanoparticles can be confirmed by ^{1}H NMR studies on nanoparticles in D$_2$O.

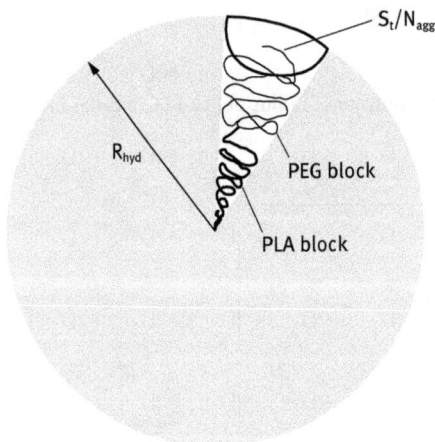

Fig. 10.29: Schematic representation of PLA-PEG micellar-like nanoparticles.

Increasing the length of the hydrophobic PLA block from 2 to 6 kDa increases the number of unfavourable interactions between the lactic acid units of the PLA chains and the aqueous media. This makes the copolymer water insoluble and it is forced to assemble into nanoparticles by precipitation into water from a water miscible solvent (e.g. acetone). The number of attractive hydrophobic interactions between the lactic acid units of the associating PLA chains increases with increasing length of the chain.

This results in an increased packing density of the PLA-PEG subunits and a sharp increase in the micellar aggregation number as illustrated in Fig. 10.30. The surface area per copolymer unit at the outer surface of the micelle (S_t/N_{agg}) falls rapidly as the molar mass of the PLA block is increased from 2 to 6 kDa (Tab. 10.5). (S_t/N_{agg}) appears to tend towards a value that is consistent with the maximum cross-sectional area of a PEG 5 kDa in solution ($\approx 30 \, nm^2$). The increasing aggregation number and decreasing surface curvature results in a decrease in the conical volume available to each of the coronal PEG chain [15, 22].

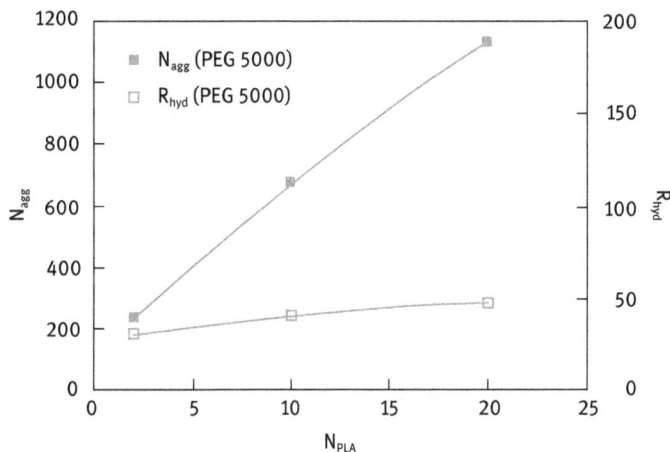

Fig. 10.30: Variation in aggregation number and hydrodynamic radius of PLA-PEG 45 : 5 nanoparticles with block copolymer concentration in the organic phase.

The log-log plots of N_{agg}-NPLA shown in Fig. 10.28 give the following scaling relationship [15, 22]:

$$N_{agg} \approx N_{PLA}^{1.74}. \tag{10.12}$$

The scaling exponent (1.74) is larger than that predicted theoretically for both "crew-cut" and "star" models. This points to a third class of block copolymer micelles, characterized by blocks of different chemical composition or polarity (strongly aggregated). This class predicts an N_A^2 dependency of the micellar aggregation number:

$$4\pi R_C^2 = N_{agg} D^2. \tag{10.13}$$

If the micelle core is strongly segregated, then the area and volume of the core are given by

$$\frac{4}{3}\pi R_C^3 = \frac{N_{agg} N_A}{\rho_{bulk}}, \tag{10.14}$$

$$N_{agg} = \frac{36\pi N_A^2}{D^6 \rho_{bulk}}, \tag{10.15}$$

where R_c is the core radius, D^2 is the interfacial area per chain and ρ_{bulk} is the mass density of the core.

The particle size of nanoparticles produced from the PLA-PEG 45 : 5 % copolymer depends on the concentration of the polymer dissolved in acetone [15, 22]. Figure 10.30 shows the variation in the aggregation number and hydrodynamic radius with copolymer concentration. Increasing the concentration of the acetonic copolymer solution increases the local concentration of PLA-PEG units available to aggregate at any particle formation site, following precipitation into water. A significant increase in the aggregation number from 238 to 674 occurs when the copolymer concentration is increased from 2 to 10 mg ml^{-1}. At 2 mg ml^{-1} small nanoparticles are produced with a high surface area (S_t/N_{agg}) per PEG block at the outer boundary of the micelle of 49 nm^2 that is similar to that of the smaller PLA-PEG 2 : 5 to 4 : 5 copolymers. These smaller PLA-PEG 45 : 5 nanoparticles have a loosely packed structure with a lot of solvent in the corona region. The particle radius of these 45 : 5 nanoparticles (30.3 nm) is considerably greater than that of those produced by the smaller 2 : 5 PLA-PEG block copolymer (13 nm). The largest PLA-PEG 45 : 5 nanoparticles have a low (S_t/N_{agg}) (26 nm^2) comparable to that of the PLA-PEG 6 : 5 micellar-like nanoparticles (30 nm^2). This implies that the largest PLA-PEG 45 : 5 nanoparticles have a high PEG surface coverage which gives them high colloid stability.

The main objective of using PLA-PEG nanoparticles for drug delivery is to have long-circulating particulate carriers and minimize opsonization by means of high coverage of brush-like PEG chains. On the basis of this rationale, copolymers with an intermediate PLA to PEG ratio (e.g. PLA-PEG 15 : 5) would appear to form assemblies with optimal protein-resistant surface properties [15, 22].

To test the above hypothesis, PLA and PLA-PEG nanoparticles were radiolabelled for in vivo studies by incorporation of the hydrophobic gamma-emitter, [111]IN-oxine (8-hydroxy quinoline). In vitro studies showed that [111]IN-oxine is released from PLA and PLA-PEG nanoparticles on incubation with rat serum [15, 22].

For each PLA/PLA-PEG system, a group of three male Wistar rats (150 ± 10 g) was injected intravenously via the lateral tail vein with 0.3 ml (equivalent to 1 mg of solid material) of the nanoparticle dispersion. A group of control rats was injected with 10 kBq of unincorporated (free) [111]IN-oxine. Blood samples of 20 µl were taken from the contralateral tail vein at various time intervals after administration (5, 15, 30, 60, 120 and 180 minutes). The animals were sacrificed after three hours by intravenous injection of phenobarbitone solution, and the liver, spleen lungs and kidneys removed. The organ- and blood-associated activity was counted using a gamma counter. The carcass-associated radioactivity was determined using a well counter. A total blood volume of 75 % of the body weight was assumed. The results for the blood- and organ-associated activity are expressed as a percentage of the injected dose and are mean values for the three rats ± standard deviation. The data for the lung and kidney are not presented, since the radioactivity associated with these organs was negligible (less than 1 % of the administered dose).

The results for the blood circulation and organ distribution showed some interesting trends [15, 22]. The PLA nanoparticles (uncoated with PEG) were rapidly cleared from the blood circulation, with only 13 % of the injected dose still circulating after 5 minutes. After 3 hours, 70 % of the i.v. administered nanoparticles had been removed by the liver. This is attributed to rapid opsonization of the particle surface and subsequent phagocytosis by the Kupffer cells of the liver. The smallest of the PLA-PEG nanoparticles studied in vivo (PLA-PEG 6 : 5, ≈ 40 nm in diameter) were cleared from the circulation, with a high percentage of the radioactivity (≈ 70 %) having accumulated in the liver three hours after the i.v. injection. However, the blood clearance rate was significantly slower than found for the PLA nanoparticles. An increase in the length of the PLA block produced larger particles which exhibited prolonged circulation times and a reduced liver uptake. For example, in the case of PLA-PEG 110 : 5 nanoparticles (≈ 160 nm), 43 % of the injected dose still remained in the systemic circulation after 3 hours, whilst only 23 % of the injected dose accumulated in the liver. Despite avoiding recognition by the Kupffer cells of the liver, 11 % of the injected dose of the PLA-PEG 110 : 5 nanoparticles was found to accumulate in the spleen.

The prolonged circulation times and reduced deposition of the larger PLA-PEG nanoparticles are surprising in view of the low PEG surface coverage of these systems, which are actually stabilized by the presence of adsorbed serum components. It appears that that such low PEG coverage is sufficient for restricting the adsorption of the high molecular weight opsinons. The layer thickness of terminally attached PEG chains with a molecular weight of 5 kDa is approximately 6.2 nm, which may adequately prohibit the adhesion of phagocytic cells.

It is surprising that the smaller micellar-like assemblies prepared from PLA-PEG copolymers with a low molecular weight PLA block were fairly rapidly cleared from the circulation and accumulated in the liver. These nanoparticles are the most colloidally stable of the PLA-PEG assemblies studied and hence the notion that effective steric stabilization is the most crucial effect for achieving blood circulation longevity is now questionable. It appears likely that the short circulation lifetime of the small PLA-PEG micelle-like nanoparticles is partly due to their ability to penetrate deep into the interstitial space of the liver [15, 22].

It seems from the above discussion that the circulatory lifetime of PLA-PEG nanoparticles in vivo does not correlate with their colloid stability in vitro. It seems that the particle size of the PLA assembly is crucial in determining its biological fate. The presence of even a low surface coverage of hydrated PEG chains is sufficient to enable relatively large (> 100 nm) PLA-PEG particles to remain in systemic circulation. Regardless of the characteristics of the PEG layer, small nanoparticles (≈ 40 nm) are cleared by the liver to a higher degree, with their small size possibly permitting access to all cell types [15, 22].

References

[1] Lee, E. M., "Nanocrystals: Resolving pharmaceutical formulation issues associated with poorly water-soluble compounds", in J. J. Marty (ed.), "Particles", Marcel Dekker, Orlando (2002).

[2] Sharma, D., Soni, M., Kumar, S. and Gupta, G., Res. J. Pharm. Technol., 2, 220 (2009).

[3] Kipp, J., Int. J. Pharm. 284, 109 (2004).

[4] Wong, J., Brugger, A., Khare, A., Chaubal, M., Papadopoulos, P., Rabinow, B., Kipp, J. and Ning, J., Advanced Drug Delivery Reviews, 60, 939 (2008).

[5] Shegokar, R. and Müller, R. H., International Journal of Pharmaceutics, 399, 129 (2010).

[6] Noyes, A. A. and Whitney, W. R., "The rate of solution of solid substances in their own solutions", Journal of the American Chemical Society, 19, 930 (1897).

[7] Thompson, W. (Lord Kelvin), Phil. Mag., 42, 448 (1871).

[8] Tadros, Th. F., "Formulation of Disperse Systems", Wiley-VCH, Germany (2014).

[9] Nakach, M., Authelin, J-R., Tadros, Th. F., Galet, L. and Chamayou, A., International J. Pharm., 476, 277 (2014).

[10] Tadros, Th. F., "Dispersion of Powders in Liquids and Stabilisation of Suspensions", Wiley-VCH, Germany (2012).

[11] Napper, D. H., "Polymeric Stabilisation of Colloidal Dispersions", Academic Press, London (1983).

[12] Kaye, B. H., Powder Technol., 1 (1967).

[13] Pecora, R., "Dynamic Light Scattering: Applications of Photon Correlation Spectroscopy", Springer, Germany (1985).

[14] Hunter, R. J., "Zeta Potential in Colloid Science: Principles and Application", Academic Press, London (1988).

[15] Tadros, Th. F., "Nanodispersions", De Gruyter, Germany (2016).

[16] Tadros, Th. F., "Interfacial Phenomena and Colloid Stability", De Gruyter, Germany (2015).

[17] Tadros, Th. F. (ed.), "Encyclopedia of Colloid and Interface Science", Springer, Germany (2013).

[18] Tadros, Th. F., "Applied Surfactants", Wiley-VCH, Germany (2005).

[19] Mills, S. N. and Davis, S. S., "Controlled Drug Delivery", in "Polymers in Controlled Drug Delivery", L. Illum and S. S. Davis (eds.), IOP Publishing, Bristol (1987), pp. 1–14.

[20] Krueter, J., "Colloidal Drug Delivery Systems", Marcel Dekker, New York (1994).

[21] Muller, R. H., "Colloidal Carriers for Controlled Drug Delivery: Modification, Characterisation and in Vivo Distribution", Wiss. Verl-Ges., Stuttgart, Germany (1990).

[22] Riley, T., Ph. D. Thesis, Nottingham University (1999).

[23] Mills, S. N. and Davis, S. S., in "Polymers in Controlled Drug Delivery", L. Illum and S. S. Davis (eds.), IOP Publishing, Bristol UK (1987) pp. 1–14.

[24] Kreuter, J, "Colloid Drug Delivery Systems", Marcel Dekker, New York (1994).

[25] Muller, R. H., "Colloidal Carriers for Controlled Drug Delivery: Modification, Characterization and in Vivo Distribution", Wiss. Verl-Ges, Stuttgart, Germany (1990).

[26] Kreuter, J., Nanoparticle based drug delivery systems, J. Control. Rel., 16, 169–176 (1991).

[27] Illum, L. and Davis S. S., "The organ uptake of intravenously administered colloidal particles can be altered using a non-ionic surfactant", (poloxamer 338), FEBS Lett., 167, 72–82 (1984).

[28] Muir, I. S., Moghimi, S. M., Illum, L., Davis, S. S. and Davies, M. C., "The effect of block copolymer on the uptake of model polystyrene microspheres by Kupffer cells – in vitro and in vivo studies", Biochem. Soc. Trans., 19, 329S (1991).

[29] Illum, L., Davis, S. S., Muller, R. H., Mak, E. and West, P., "The organ distribution and circulation life-time of intravenously injected colloidal carriers stabilized with a block copolymer poloxamine 908", Life Sci., 40, 367–374 (1987).

[30] Chasin, M., and Langer, R., (eds.), "Biodegradable Polymers as Drug Delivery Systems", Marcel Dekker, NY (1990).

[31] Stolnik, S., Dunn, S. E., Davies, M. C., Coombes, A. G. A., Taylor, D. C., Irving, M. P., Purkiss, S. C., Tadros, Th. F., Davis, S. S. and Illum, L., "Surface modification of Poly(lactide-co-glycolide) Nanospheres by Biodegradable Poly(lactide)-poly(ethylene glycol) Copolymers", Pharm. Res., 11, 1800–1808 (1994).

[32] Kwon, G. S. and Kataoka, K., Block copolymer micelles as long circulating drug vehicles, Advan. Drug Del. Rev., 16, 295–309 (1995).

[33] de Gennes, P. G., "Scaling Concepts in Polymer Physics", Cornell University Press, Ithaca, London (1979).

[34] Napper, D. H., "Polymeric Stabilization of Colloidal Dispersions", Academic Press, London (1983).

[35] Fleer, G. J., Cohen Stuart, M. A., Scheutjens, J. M. H. M., Cosgrove T. and Vincent, B., "Polymers at Interfaces", Chapman and Hall, London (1993).

[36] Jeon, S. I., Lee, J. H., Andrade, J. D. and de Gennes, P. G., "Protein surface interaction in the presence of polyethylene oxide. I. Simplified theory", J. Colloid Interface Sci., 142, 149–158 (1991).

[37] Jeon, S. I. and Andrade, J. D., "Protein surface interaction in the presence of polyethylene oxide", J. Colloid Interface Sci., 142, 159–166 (1991).

11 Polymeric surfactants in cosmetics and personal care products

11.1 Introduction

Polymeric surfactants play a major role in many cosmetic and personal care products of which the following are worth mentioning: stabilizing oil/water (O/W) emulsions in many formulations such as hand creams, and lotions; stabilizing W/O emulsions, which are sometimes applied in hand creams to enhance hydration; stabilizing nanoemulsions against Ostwald ripening; stabilizing multiple emulsions of the W/O/W type that are used in hand creams; stabilizing nonaqueous suspensions of titanium dioxide that are used in sunscreen formulations, stabilizing liposomes and vesicles that are applied in many formulations for enhancing the penetration of active ingredients such as anti-wrinkle agents. These polymeric surfactants of the A–B, A–B–A block and BA_n (or AB_n) types (where B is the "anchor" chain and A is the stabilizing chain(s)) provide effective stabilization of emulsions, nanoemulsions and multiple emulsions as described in detail in Chapter 5. A summary of the above applications of polymeric surfactants in cosmetics and personal care products is given below to emphasize the stabilizing role of these molecules.

11.2 Stabilization of O/W and W/O emulsions using polymeric surfactants

O/W emulsions are frequently used in lotions and hand creams where the absence of flocculation, Ostwald ripening and coalescence are required [1]. Flocculation refers to aggregation of the droplets (without any change in primary droplet size) into larger units. It is the result of van der Waals attraction which is universal in all disperse systems [1]. The main force of attraction arises from the London dispersion force that results from charge fluctuations of the atoms or molecules in the disperse droplets. Van der Waals attraction increases with decreasing separation distance between the droplets, and at small separation distances the attraction becomes very strong resulting in droplet aggregation or flocculation. The latter occurs when there is not sufficient repulsion to keep the droplets apart to distances where the van der Waals attraction is weak. Flocculation may be "strong" or "weak", depending on the magnitude of the attractive energy involved. In cases where the net attractive forces are relatively weak, an equilibrium degree of flocculation may be achieved (so-called weak flocculation), associated with the reversible nature of the aggregation process. The exact nature of the equilibrium state depends on the characteristics of the system. One can envisage the build-up of aggregate size distribution and an equilibrium may be established between single droplets and large aggregates. With a strongly flocculated system, one

DOI 10.1515/9783110487282-012

refers to a system in which all the droplets are present in aggregates due to the strong van der Waals attraction between the droplets.

Two main rules can be applied for reducing (eliminating) flocculation depending on the stabilization mechanism:

(i) With charge stabilized emulsions, e.g. using ionic surfactants, the most important criterion is to make the energy barrier in the energy-distance curve, G_{max}, as high as possible. This is achieved by three main conditions: high surface or zeta potential, low electrolyte concentration and low valency of ions.

(ii) Sterically stabilized emulsions obtained by using polymeric surfactants are more effective in reducing or eliminating flocculation. Four main criteria are necessary:

  (a) Complete coverage of the droplets by the stabilizing chains.

  (b) Firm attachment (strong anchoring) of the chains to the droplets. This requires the chains to be insoluble in the medium and soluble in the oil. However, this is incompatible with stabilization, which requires a chain that is soluble in the medium and strongly solvated by its molecules. These conflicting requirements are solved by the use of A–B, A–B–A block or BA_n graft copolymers (B is the "anchor" chain and A is the stabilizing chain(s)). Examples for the B chains for O/W emulsions are polystyrene, polymethylmethacrylate, polypropylene oxide and alkyl polypropylene oxide. For the A chain(s), polyethylene oxide (PEO) or polyvinyl alcohol are good examples.

  (c) Thick adsorbed layers; the adsorbed layer thickness should be in the region of 5–10 nm. This means that the molecular weight of the stabilizing chains could be in the region of 1000–5000.

  (d) The stabilizing chain should be maintained in good solvent conditions (the Flory–Huggins interaction parameter $\chi < 0.5$) under all conditions of temperature changes on storage.

Flocculation of a lotion or hand cream results in increased viscosity of the system and this affects the consistency ("skin feel") of the resulting formulation. In most cases, flocculation results in the formation of a "stringy" product with unacceptable "skin feel" for the customer [2].

Ostwald ripening (disproportionation) results from the finite solubility of the liquid phases. Liquids referred to as being immiscible often have mutual solubilities which are not negligible. With emulsions, which are usually polydisperse, the smaller droplets will have larger solubility when compared with the larger ones (due to curvature effects). With time, the smaller droplets disappear and their molecules diffuse to the bulk and become deposited on the larger droplets. With time the droplet size distribution shifts to larger values.

Two general methods may be applied to reduce Ostwald ripening [1]:

(i) Addition of a second disperse phase component which is insoluble in the continuous medium (e.g. squalane). In this case partitioning between different droplet
sizes occurs, with the component having low solubility expected to be concentrated in the smaller droplets. During Ostwald ripening in a two component system, equilibrium is established when the difference in chemical potential between
different sized droplets (which results from curvature effects) is balanced by the
difference in chemical potential resulting from partitioning of the two components. This effect reduces further growth of droplets.

(ii) Modification of the interfacial film at the O/W interface: reduction in γ results
in a reduction of Ostwald ripening rate. By using surfactants that are strongly
adsorbed at the O/W interface (i.e. polymeric surfactants) and which do not desorb during ripening (by choosing a molecule that is insoluble in the continuous
phase) the rate could be significantly reduced. An increase in the surface dilational modulus ε (= $d\gamma/d\ln A$) and a decrease in γ would be observed for the
shrinking drop and this tends to reduce further growth [1].

A–B–A block copolymers such as PHS–PEO–PHS (which is soluble in the oil droplets
but insoluble in water) can be used to achieve the above effect. Similar effects can
also be obtained using a graft copolymer of hydrophobically modified inulin, namely
INUTEC® SP1 (ORAFTI, Belgium). This polymeric surfactant adsorbs with several alkyl
chains (which may dissolve in the oil phase) leaving loops and tails of strongly hydrated inulin (polyfructose) chains. The molecule has limited solubility in water and
hence it resides at the O/W interface. These polymeric emulsifiers enhance the Gibbs
elasticity, thus significantly reducing the Ostwald ripening rate.

Coalescence refers to the process of thinning and disruption of the liquid film between the droplets which may be present in a creamed or sedimented layer, in a floc or
simply during droplet collision, with the result of fusion of two or more droplets into
larger ones. This process of coalescence results in a considerable change in the droplet
size distribution, which shifts to larger sizes. The limiting case for coalescence is the
complete separation of the emulsion into two distinct liquid phases. The thinning and
disruption of the liquid film between the droplets is determined by the relative magnitudes of the attractive versus repulsive forces. To prevent coalescence, the repulsive
forces must exceed the van der Waals attraction, thus preventing film rupture.

Several methods may be applied to achieve the above effects:

(i) Use of mixed surfactant films. In many cases using mixed surfactants, say anionic
and nonionic or long chain alcohols, can reduce coalescence as a result of several
effects: high Gibbs elasticity; high surface viscosity; hindered diffusion of surfactant molecules from the film.

(ii) Formation of lamellar liquid crystalline phases at the O/W interface. Surfactant
or mixed surfactant film can produce several bilayers that "wrap" the droplets.
As a result of these multilayer structures, the potential drop is shifted to longer

distances, thus reducing the van der Waals attraction. For coalescence to occur, these multilayers have to be removed "two-by-two" and this forms an energy barrier preventing coalescence.

(iii) Use of polymeric surfactants that produce high steric repulsion which results in a high disjoining pressure that reduces or eliminates coalescence.

An example of a very effective stabilizer for O/W emulsions is INUTEC SP1 based on a graft copolymer AB_n with A being polyfructose and B are several alkyl groups grafted on the polyfructose chain [3]. This polymeric surfactant produces enhanced steric stabilization both in water and high electrolyte concentrations as described in Chapter 7. The high stability observed using INUTEC® SP1 is related to its strong hydration both in water and in electrolyte solutions. Emulsions of Isopar M/water and cyclomethicone/water were prepared using INUTEC SP1. 50/50 (v/v) O/W emulsions were prepared and the emulsifier concentration was varied from 0.25 to 2 (w/v) % based on the oil phase. 0.5 (w/v) % emulsifier was sufficient for the stabilization of these 50/50 (v/v) emulsions [3]. The emulsions were stored at room temperature and 50 °C and optical micrographs were taken at intervals of time (for a year) in order to check the stability. Emulsions prepared in water were very stable, showing no change in droplet size distribution over more than a one year period and this indicated absence of coalescence.

The second example is a W/O emulsion stabilized using an A–B–A block copolymer of poly(12-hydroxystearic acid) (PHS) (the A chains) and poly(ethylene oxide) (PEO) (the B chain): PHS–PEO–PHS. The PEO chain (that is soluble in the water droplets) forms the anchor chain, whereas the PHS chains form the stabilizing chains. PHS is highly soluble in most hydrocarbon solvents and is strongly solvated by its molecules. The structure of the PHS–PEO–PHS block copolymer is shown in Fig. 11.1 and its conformation of the polymeric at the W/O interface is schematically shown in Fig. 11.2.

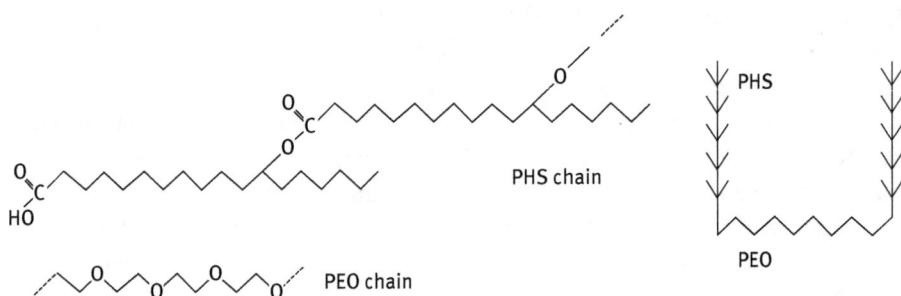

Fig. 11.1: Schematic representation of the structure of PHS–PEO–PHS block copolymer.

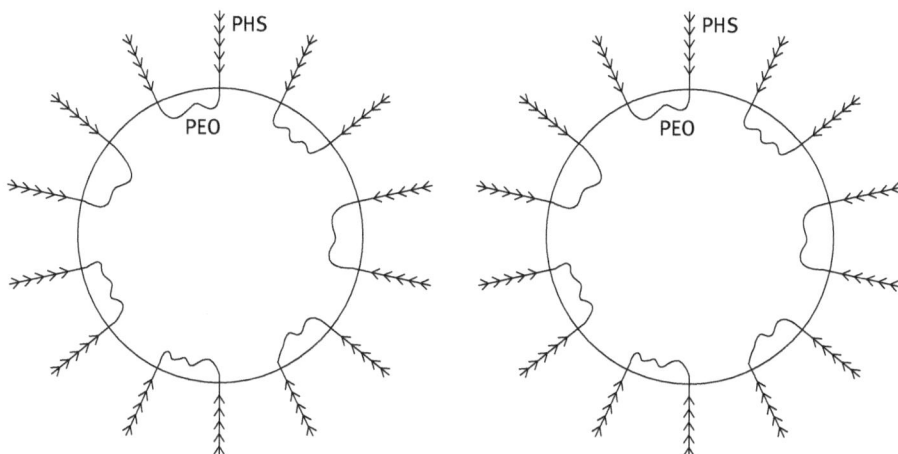

Fig. 11.2: Conformation of PHS–PEO–PHS polymeric surfactant at the W/O interface.

11.3 Use of polymeric surfactants for stabilizing nanoemulsions in cosmetics

Nanoemulsions are transparent or translucent systems in the size range 20–200 nm [2]. Whether the system is transparent or translucent depends on the droplet size, the volume fraction of the oil and the refractive index difference between the droplets and the medium. Nanoemulsions having diameters < 50 nm appear transparent when the oil volume fraction is < 0.2 and the refractive index difference between the droplets and the medium is not large. With increasing droplet diameter and oil volume fraction, the system may appear translucent and at higher oil volume fractions the system may become turbid.

Nanoemulsions are only kinetically stable. They have to be distinguished from microemulsions (that cover the size range 5–50 nm) that are mostly transparent and thermodynamically stable. The long-term physical stability of nanoemulsions (with no apparent flocculation or coalescence) makes them unique and they are sometimes referred to as "approaching thermodynamic stability". The inherently high colloid stability of nanoemulsions can be well understood from a consideration of their steric stabilization (when using nonionic surfactants and/or polymers) and how this is affected by the ratio of the adsorbed layer thickness to droplet radius as discussed in Chapter 5.

Unless adequately prepared (to control the droplet size distribution) and stabilized against Ostwald ripening (that occurs when the oil has some finite solubility in the continuous medium), nanoemulsions may show an increase in droplet size and an initially transparent system may become turbid on storage.

The attraction of nanoemulsions for application in personal care and cosmetics is due to the following advantages [2]:

(i)   The very small droplet size causes a large reduction in the gravity force and Brownian motion may be sufficient to overcome gravity. This means that no creaming or sedimentation occurs on storage.

(ii)   The small droplet size also prevents any flocculation of the droplets. Weak flocculation is prevented and this enables the system to remain dispersed with no separation.

(iii)   The small droplets also prevent their coalescence, since these droplets are nondeformable and hence surface fluctuations are prevented. In addition, the significant surfactant film thickness (relative to droplet radius) prevents any thinning or disruption of the liquid film between the droplets.

(iv)   Nanoemulsions are suitable for efficient delivery of active ingredients through the skin. The large surface area of the emulsion system allows rapid penetration of actives.

(v)   Due to their small size, nanoemulsions can penetrate through the "rough" skin surface and this enhances penetration of actives.

(vi)   The transparent nature of the system, their fluidity (at reasonable oil concentrations) as well as the absence of any thickeners may give them a pleasant aesthetic character and skin feel.

(vii)   Unlike microemulsions (which require a high surfactant concentration, usually in the region of 20 % and higher), nanoemulsions can be prepared using reasonable surfactant concentration. For a 20 % O/W nanoemulsion, a surfactant concentration in the region of 5–10 % may be sufficient.

(viii) The small size of the droplets allows them to deposit uniformly on substrates. Wetting, spreading and penetration may be also enhanced as a result of the low surface tension of the whole system and the low interfacial tension of the O/W droplets.

(ix)   Nanoemulsions can be applied for delivery of fragrants which may be incorporated in many personal care products. This could also be applied in perfumes when an alcohol-free formulation is desired.

(x)   Nanoemulsions may be applied as a substitute for liposomes and vesicles (which are much less stable) and it is possible in some cases to build lamellar liquid crystalline phases around the nanoemulsion droplets.

The inherently high colloid stability of nanoemulsions when using polymeric surfactants is due to their steric stabilization. The mechanism of steric stabilization was discussed in Chapter 5. As was shown in Fig. 5.4, the energy-distance curve shows a shallow attractive minimum at separation distance comparable to twice the adsorbed layer thickness 2δ. This minimum decreases in magnitude as the ratio between adsorbed layer thickness to droplet size increases, as was shown in Fig. 5.5 of Chapter 5.

With nanoemulsions the ratio of adsorbed layer thickness to droplet radius (δ/R) is relatively large (0.1–0.2) when compared with macroemulsions.

These systems approach thermodynamic stability against flocculation and/or coalescence. The very small size of the droplets and the dense adsorbed layers ensure lack of deformation of the interface, lack of thinning and disruption of the liquid film between the droplets and hence coalescence is also prevented.

One of the main problems with nanoemulsions is Ostwald ripening, which results from the difference in solubility between small and large droplets [2]. The difference in chemical potential of dispersed phase droplets between different sized droplets was given by Lord Kelvin:

$$c(r) = c(\infty) \exp\left(\frac{2\gamma V_m}{rRT}\right), \tag{11.1}$$

where $c(r)$ is the solubility surrounding a particle of radius r, $c(\infty)$ is the bulk phase solubility and V_m is the molar volume of the dispersed phase. The quantity $(2\gamma V_m/RT)$ is termed the characteristic length. It has an order of $\approx 1\,nm$ or less, indicating that the difference in solubility of a $1\,\mu m$ droplet is of the order of 0.1 % or less.

Theoretically, Ostwald ripening should lead to condensation of all droplets into a single drop (i.e. phase separation). This does not occur in practice since the rate of growth decreases with increasing droplet size.

For two droplets of radii r_1 and r_2 (where $r_1 < r_2$),

$$\frac{RT}{V_m} \ln\left[\frac{c(r_1)}{c(r_2)}\right] = 2\gamma\left(\frac{1}{r_1} - \frac{1}{r_2}\right). \tag{11.2}$$

Equation (11.2) shows that the greater the difference between r_1 and r_2, the higher the rate of Ostwald ripening.

Ostwald ripening can be quantitatively assessed from plots of the cube of the radius versus time t [2]:

$$r^3 = \frac{8}{9}\left[\frac{c(\infty)\gamma V_m}{\rho RT}\right]t, \tag{11.3}$$

where D is the diffusion coefficient of the disperse phase in the continuous phase.

Ostwald ripening can be reduced by incorporating a second component that is insoluble in the continuous phase (e.g. squalane) [2]. In this case significant partitioning between different droplets occurs; the component with low solubility in the continuous phase is expected to be concentrated in the smaller droplets. During Ostwald ripening in a two component disperse phase system, equilibrium is established when the difference in chemical potential between different sized droplets (which results from curvature effects) is balanced by the difference in chemical potential resulting from partitioning of the two components. If the secondary component has zero solubility in the continuous phase, the size distribution will not deviate from the initial one (the growth rate is equal to zero). In the case of limited solubility of the secondary component, the distribution is the same as governed by equation (11.3), i.e. a mixture growth rate is obtained which is still lower than that of the more soluble component.

The above method is of limited application since one requires a highly insoluble oil as the second phase which is miscible with the primary phase.

Another method for reducing Ostwald ripening depends on modifying the interfacial film at the O/W interface [2]. According to equation (11.3), a reduction in γ results in a reduction of Ostwald ripening. However, this alone is not sufficient since one has to reduce γ by several orders of magnitude. It has been suggested that by using surfactants which are strongly adsorbed at the O/W interface (i.e. polymeric surfactants) and which do not desorb during ripening, the rate could be significantly reduced. An increase in the surface dilational modulus and a decrease in γ would be observed for the shrinking drops. The difference in γ between the droplets would balance the difference in capillary pressure (i.e. curvature effects).

To achieve the above effect it is useful to use A–B–A block copolymers that are soluble in the oil phase and insoluble in the continuous phase. A strongly adsorbed polymeric surfactant that has limited solubility in the aqueous phase can also be used. Polymeric surfactants are expected to significantly reduce Ostwald ripening due to the high interfacial elasticity produced by the adsorbed polymeric surfactant molecules [2]. To test this hypothesis, several nanoemulsions were formulated using a graft copolymer of hydrophobically modified inulin. The inulin backbone consists of polyfructose with a degree of polymerization > 23. This hydrophilic backbone is hydrophobically modified by attachment of several C_{12} alkyl chains [2]. The polymeric surfactant (with a trade name of INUTEC® SP1) adsorbs with several alkyl chains that can be soluble in the oil phase or strongly attached to the oil surface, leaving the strongly hydrated hydrophilic polyfructose loops and tails "dangling" in the aqueous phase. These hydrated loops and tails (with a hydrodynamic thickness > 5 nm) provide effective steric stabilization.

Oil/water (O/W) nanoemulsions were prepared by two step emulsification processes. In the first step, an O/W emulsion was prepared using a high speed stirrer, namely an Ultra-Turrax [2]. The resulting coarse emulsion was subjected to high pressure homogenization using a Microfluidizer (Microfluidics, USA). In all cases, the pressure used was 700 bar and homogenization was carried out for 1 min. The z-average droplet diameter was determined using PCS measurements as discussed before.

Figure 11.3 shows plots of r^3 versus t for nanoemulsions of the hydrocarbon oils that were stored at 50 °C. It can be seen that both paraffinum liquidum with low and high viscosity give almost a zero slope, indicating absence of Ostwald ripening in this case. This is not surprising since both oils have very low solubility and the hydrophobically modified inulin, INUTEC® SP1, strongly adsorbs at the interface giving high elasticity that reduces both Ostwald ripening and coalescence. However, with the more soluble hydrocarbon oils, namely isohexadecane, there is an increase in r^3 with time, giving a rate of Ostwald ripening of 4.1×10^{-27} m^3 s^{-1}. The rate for this oil is almost three orders of a magnitude lower than that obtained with a nonionic surfactant, namely laureth-4 (C_{12}-alkylchain with 4 mol ethylene-oxide) when stored at 50 °C. This clearly shows the effectiveness of INUTEC® SP1 in reducing Ostwald ripening. This

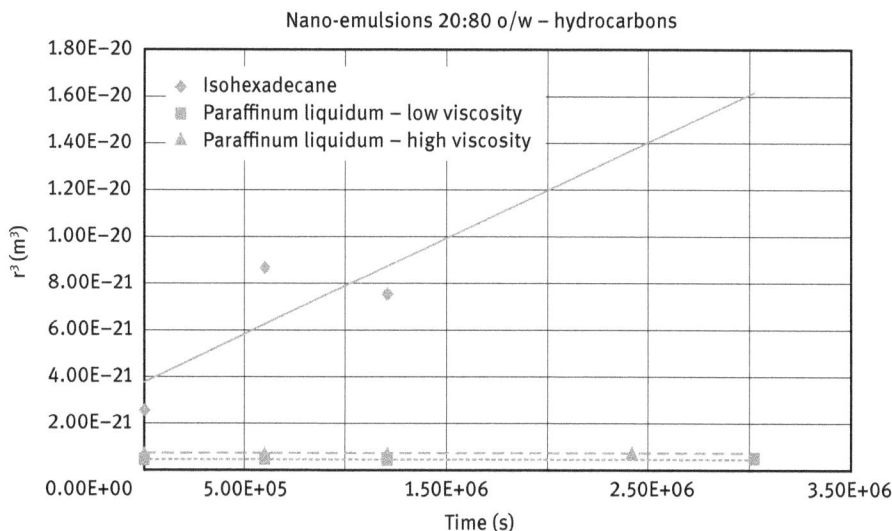

Fig. 11.3: r^3 versus t for nanoemulsions based on hydrocarbon oils.

reduction can be attributed to the enhancement of the Gibbs dilational elasticity [2] which results from the multipoint attachment of the polymeric surfactant with several alkyl groups to the oil droplets. This results in a reduction of the molecular diffusion of the oil from the smaller to the larger droplets.

Figure 11.4 shows the results for the isopropylalkylate O/W nanoemulsions. As with the hydrocarbon oils, there is a significant reduction in the Ostwald ripening rate with increasing alkyl chain length of the oil. The rate constants are 1.8×10^{-27}, 1.7×10^{-27} and 4.8×10^{-28} m^3 s^{-1} respectively.

Figure 11.5 shows the r^3 versus t plots for nanoemulsions based on natural oils. In all cases, the Ostwald ripening rate is very low. However, a comparison between squalene and squalane shows that the rate is relatively higher for squalene (unsaturated oil) when compared with squalane (with lower solubility). The Ostwald ripening rate for these natural oils is given in Tab. 11.1.

Tab. 11.1: Ostwald ripening rates for nanoemulsions based on natural oils.

Oil	Ostwald ripening rate (m^3 s^{-1})
Squalene	2.9×10^{-28}
Squalane	5.2×10^{-30}
Ricinus communis	3.0×10^{-29}
Macadamia ternifolia	4.4×10^{-30}
Buxis chinensis	≈ 0

Nano-emulsions 20:80 o/w – isopropyl alkylate

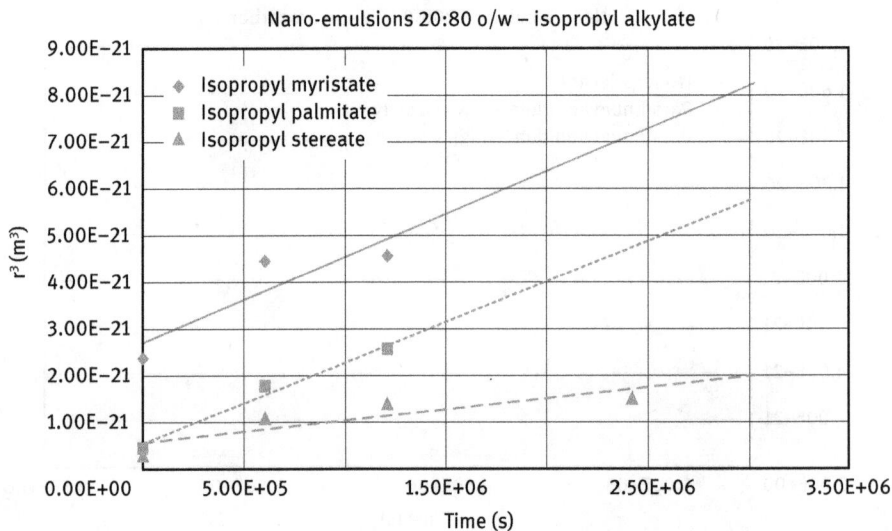

Fig. 11.4: r^3 versus t for nanoemulsions based on isopropylalkylate.

Nano-emulsions 20:80 o/w – natural oils

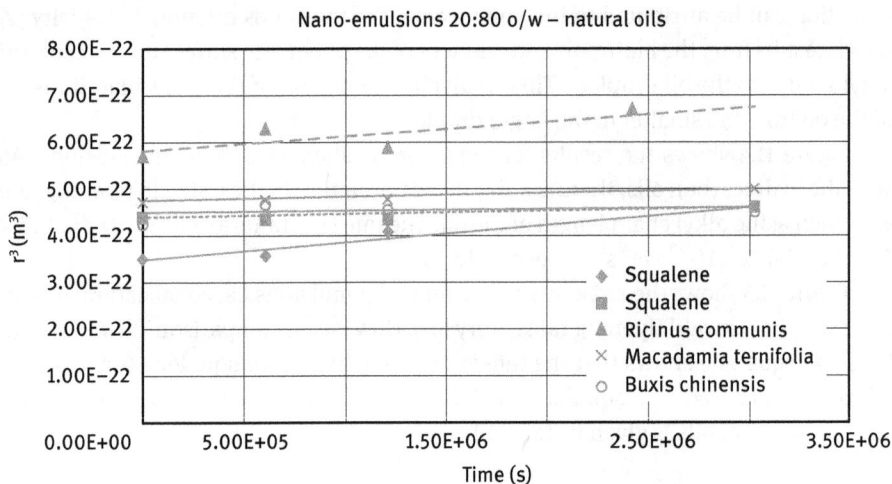

Fig. 11.5: r^3 versus t for nanoemulsions based on natural oils.

Figure 11.6 shows the results based on silicone oils. Both dimethicone and phenyl trimethicone give an Ostwald ripening rate close to zero, whereas cyclopentasiloxane gives a rate of 5.6×10^{-28} m^3 s^{-1}.

Figure 11.7 shows the results for nanoemulsions based on esters and the Ostwald ripening rates are given in Tab. 11.2. C$_{12-15}$ alkylbenzoate seems to give the highest rate.

Tab. 11.2: Ostwald ripening rates for nanoemulsions based on esters.

Oil	Ostwald ripening rate (m^3 s^{-1})
Butyl stearate	1.8×10^{-28}
Caprylic capric triglyceride	4.9×10^{-29}
Cetearyl ethylhexanoate	1.9×10^{-29}
Ethylhexyl palmitate	5.1×10^{-29}
Cetearyl isononanoate	1.8×10^{-29}
C$_{12-15}$ alkyl benzoate	6.6×10^{-28}

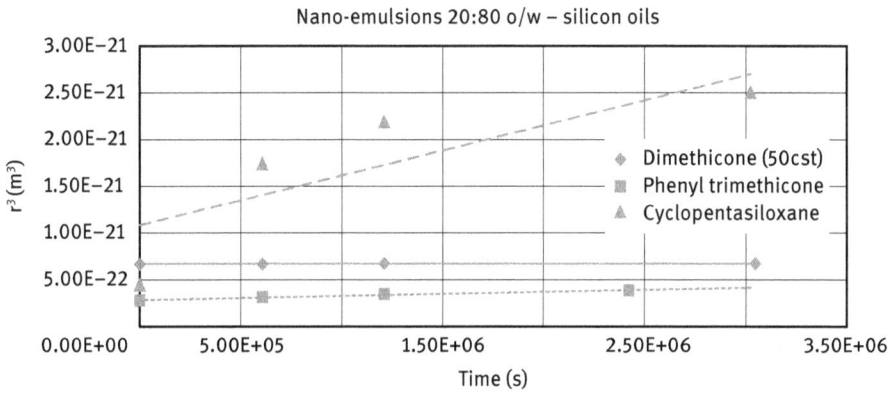

Fig. 11.6: r^3 versus t for nanoemulsions based on silicon oils.

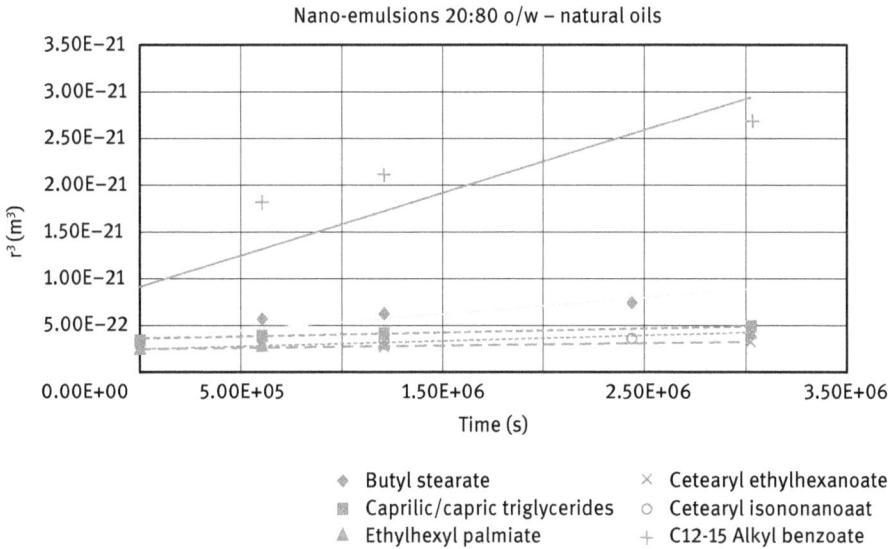

Fig. 11.7: r^3 versus t for nanoemulsions based on esters.

Figure 11.8 gives a comparison for two nanoemulsions based on polydecene, a highly insoluble nonpolar oil and PPG-15 stearyl ether which is relatively more polar. Polydecene gives a low Ostwald ripening rate of 6.4×10^{-30} m^3 s^{-1} which is one order of magnitude lower than that of PPG-15 stearyl ether (5.5×10^{-29} m^3 s^{-1}).

The influence of adding glycerol (which is sometimes added to personal care formulations as a humectant), which can be used to prepare transparent nanoemulsions

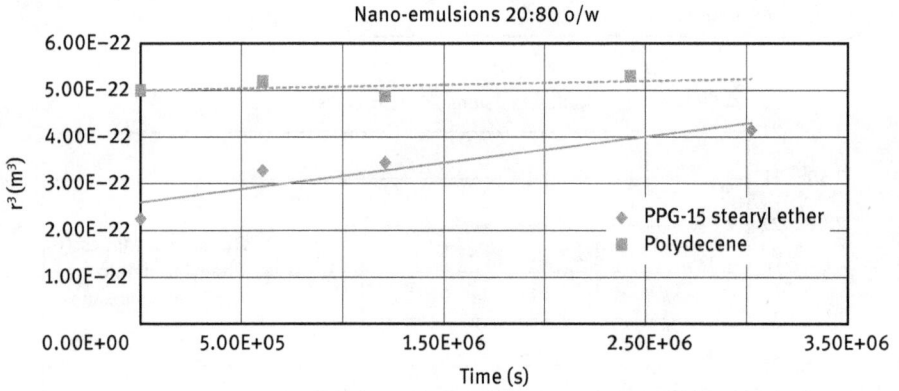

Fig. 11.8: r^3 versus t for nanoemulsions based on PPG-15 stearyl ether and polydecene.

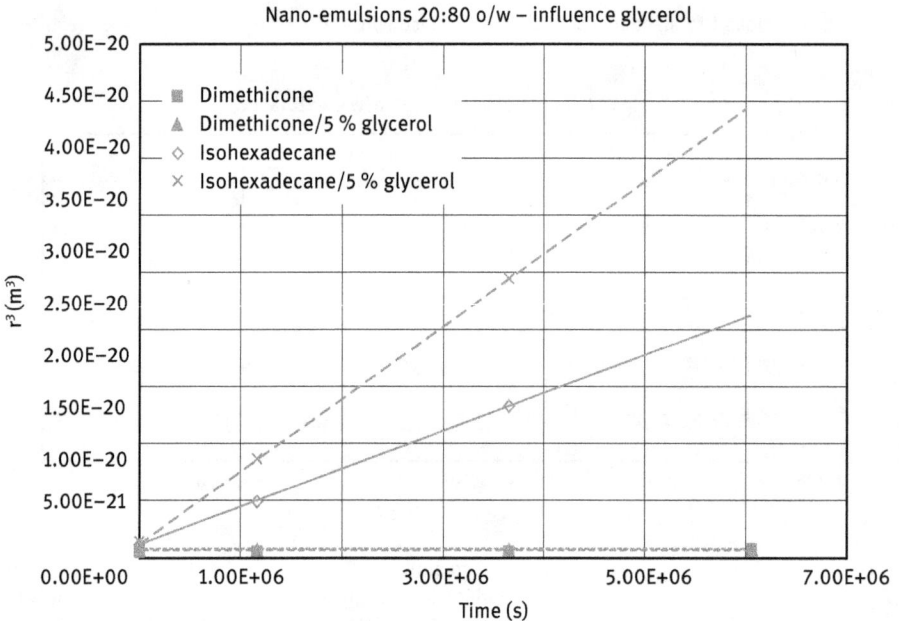

Fig. 11.9: Influence of glycerol on the Ostwald ripening rate of nanoemulsions.

(by matching the refractive index of the oil and the aqueous phase), on the Ostwald ripening rate is shown in Fig. 11.9. With the more insoluble silicone oil, addition of 5 % glycerol does not show an increase in the Ostwald ripening rate, whereas for the more soluble isohexadecane oil, glycerol increases the rate.

It can be seen that hydrophobically modified inulin, HMI (INUTEC® SP1), reduces the Ostwald ripening rate of nanoemulsions when compared with nonionic surfactants such as laureth-4. This is due to the strong adsorption of INUTEC® SP1 at the oil-water interface (by multipoint attachment) and enhancement of the Gibbs dilational elasticity, both reducing the diffusion of oil molecules from the smaller to the larger droplets [2]. The present study also showed a large influence of the nature of the oil-phase, with the more soluble and more polar oils giving the highest Ostwald ripening rate. However in all cases, when using INUTEC® SP1, the rates are reasonably low allowing one to use this polymeric surfactant in formulating nanoemulsions for personal care applications.

11.4 Stabilizing nonaqueous dispersions for sunscreens

Sunscreens are formulations that are used for UV protection [2]. The actives employed in these preparations are of two basic types: organics, which can absorb UV radiation of specific wavelengths due to their chemical structure and inorganics, which both absorb and scatter UV radiation. Inorganics have several benefits over organics in that they are capable of absorbing over a broad spectrum of wavelengths and they are mild and nonirritant. Both of these advantages are becoming increasingly important as the demand for *daily* UV protection against both UVB (wavelength 290–320 nm) and UVA (wavelength 320–400 nm) radiation increases. Since UVB is much more effective than UVA at causing biological damage, solar UVB contributes about 80 % towards a sunburn reaction, with solar UVA contributing the remaining 20 %.

The ability of fine particle inorganics to absorb radiation depends upon their refractive index. For inorganic semiconductors such as titanium dioxide and zinc oxide, this is a complex number indicating their ability to absorb light. The band gap in these materials is such that UV light up to around 405 nm can be absorbed. They can also scatter light due to their particulate nature and their high refractive indices make them particularly effective scatterers. Both scattering and absorption depend critically on particle size [2]. Particles of around 250 nm for example are very effective at scattering visible light and TiO_2 of this particle size is the most widely used white pigment. At smaller particle sizes, absorption and scattering maxima shift to the UV region, and at 30–50 nm UV attenuation is maximized.

The benefits of a predispersion of inorganic sunscreens are widely acknowledged. However, it requires an understanding of the nature of colloidal stabilization in order to optimize this predispersion (for both UV attenuation and stability) and exceed the performance of powder-based formulations. Dispersion rheology and its dependence on interparticle interactions is a key factor in this optimization. Optimization of sunscreen actives however does not end there; an appreciation of the end application is crucial to maintaining performance. Formulators need to incorporate the particulate actives into an emulsion, mousse or gel with due regard to aesthetics (skin feel and transparency), stability and rheology.

Both dispersion stability and dispersion rheology depend upon adsorbed amount Γ and steric layer thickness δ (which in turn depends on oligomer molecular weight M_n and solvency χ) as discussed in detail in Chapter 5. In order to optimize formulation, the adsorption strength χ^s must also be considered. The nature of interaction between particles, dispersant, emulsifiers and thickeners must be considered with regard to competitive adsorption and/or interfacial stability if a formulation is to deliver its required protection when spread on the skin.

TiO_2 and ZnO absorb and scatter UV light. They provide a broad spectrum and they are inert and safe to use. Larger particles scatter visible light and they cause whitening. The scattering and absorption depend on the refractive index (which depends on the chemical nature), the wavelength of light and the particle size and shape distribution. The total attenuation is maximized in UVB for 30–50 nm particles as illustrated in Fig. 11.10 which shows the effect of particle size on UV attenuation.

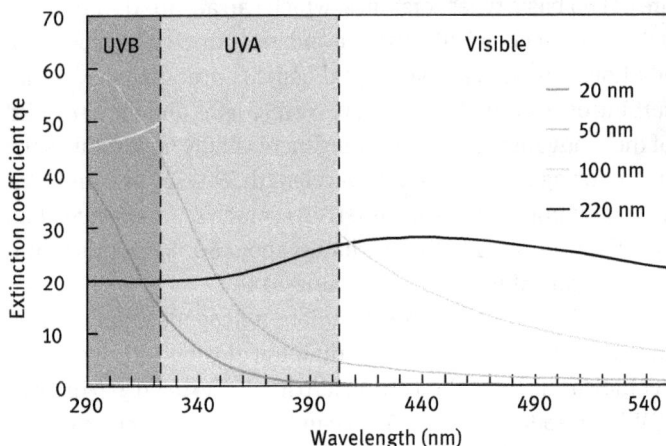

Fig. 11.10: Effect of particle size on UVA and UVB absorption.

The performance of any sunscreen formulation is defined by a number referred to as the sun protection factor (SPF) as described before [2].

To keep the particles well dispersed (as single particles) high steric repulsion is required to overcome strong van der Waals attraction and this shows the importance of using polymeric surfactants as described in Chapter 5.

The most effective molecules are the A–B, A–B–A block or BA_n graft polymeric surfactants [2] where B refers to the anchor chain. For a hydrophilic particle this may be a carboxylic acid, an amine or phosphate group or other larger hydrogen bonding type block such as polyethylene oxide. The A chains are referred to as the stabilizing chains which should be highly soluble in the medium and strongly solvated by its molecules. For nonaqueous dispersions, the A chains could be polypropylene oxide, a long chain alkane, oil soluble polyester or polyhydroxystearic acid (PHS).

One of the most useful concepts for assessing solvation of any polymer by the medium is to use Hildebrand's solubility parameter, δ^2, which is related to the heat of vaporization, ΔH, by the following equation [4]:

$$\delta^2 = \frac{\Delta H - RT}{V_M},\tag{11.4}$$

where V_M is the molar volume of the solvent.

Hansen [5] first divided Hildebrand's solubility parameter into three terms as follows:

$$\delta^2 = \delta_d^2 + \delta_p^2 + \delta_h^2,\tag{11.5}$$

where δ_d, δ_p and δ_h correspond to London dispersion effects, polar effects and hydrogen bonding effects, respectively.

Hansen and Beerbower [6] developed this approach further and proposed a stepwise approach such that theoretical solubility parameters can be calculated for any solvent or polymer based upon its component groups. In this way we can arrive at theoretical solubility parameters for dispersants and oils. In principle, solvents with a similar solubility parameter to the polymer should also be a good solvent for it (low χ).

For sterically stabilized dispersions, the resulting energy-distance curve often shows a shallow minimum, G_{min}, at particle-particle separation distance h comparable to twice the adsorbed layer thickness δ. The depth of this minimum depends on the particle size R, Hamaker constant A and adsorbed layer thickness δ. At constant R and A, G_{min} decreases with increasing δ/R. This is illustrated in Fig. 11.11.

When δ becomes smaller than 5 nm, G_{min} may become deep enough to cause weak flocculation. This is particularly the case with concentrated dispersions since the entropy loss on flocculation becomes very small and a small G_{min} would be sufficient to cause weak flocculation ($\Delta G_{flocc} < 0$). This can be explained by considering

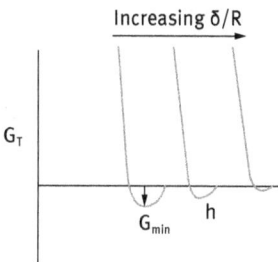

Fig. 11.11: Schematic representation of energy-distance curves at increasing δ/R ratios.

the free energy of flocculation [2]:

$$\Delta G_{flocc} = \Delta H_{flocc} - T\Delta S_{flocc}. \tag{11.6}$$

Since for concentrated dispersions ΔS_{flocc} is very small, then ΔG_{flocc} depends only on the value of ΔH_{flocc}. This in turn depends on G_{min}, which is negative. In other words, ΔG_{flocc} becomes negative causing weak flocculation. This will result in a three-dimensional coherent structure with a measurable yields stress. This weak gel can be easily redispersed by gentle shaking or mixing. However, the gel will prevent any separation of the dispersion on storage. So, we can see that the interaction energies also determine the dispersion rheology.

At high solids content and for dispersions with larger δ/R, viscosity is also increased by steric repulsion. With a dispersion consisting of very small particles, as is the case with UV attenuating TiO_2, significant rheological effects can be observed even at moderate volume fraction of the dispersion. This is due to the much higher effective volume fraction of the dispersion compared with the core volume fraction due to the adsorbed layer.

Let us for example consider a 50 % w/w TiO_2 dispersion with a particle radius of 20 nm and with a 3000 molecular weight stabilizer giving an adsorbed layer thickness of ≈ 10 nm. The effective volume fraction is given by [2]:

$$\phi_{eff} = \phi \left[1 + \frac{\delta}{R} \right]^3$$

$$= \phi[1 + 10/20]^3$$

$$\approx 3\phi \tag{11.7}$$

The effective volume fraction can be three times that of the core particle volume fraction. For a 50 % solids (w/w) TiO_2 dispersion, the core volume fraction ϕ is ~ 0.25 (taking an average density of 3 g cm^{-3} for the TiO_2 particles) which means that ϕ_{eff} is about 0.75, which is sufficient to fill the whole dispersion space producing a highly viscous material. It is important therefore to choose the minimum δ for stabilization.

In the case of steric stabilization as employed in these oil dispersions, the important success criteria for well stabilized but handleable dispersions are:
(i)   complete coverage of the surface – high Γ adsorbed amount);
(ii) strong adsorption (or "anchoring") of the chains to the surface;
(iii) effective stabilizing chain, chain well solvated, $\chi < 0.5$ and adequate (but not too large) steric barrier δ.

However, a colloidally stable dispersion does not guarantee a stable and optimized final formulation. TiO_2 particles are always surface modified in a variety of ways in order to improve dispersability and compatibility with other ingredients. It is important that we understand the impact these surface treatments may have upon the dispersion and more importantly upon the final formulation. As will be disused below, TiO_2

is actually formulated into a suspoemulsion, i.e. a suspension in an emulsion. Many additional ingredients are added to ensure cosmetic elegance and function. The emulsifiers used are structurally and functionally not very different to the dispersants used to optimize the fine particle inorganics. Competitive adsorption may occur with some partial desorption of a stabilizer from one or other of the available interfaces. Thus one requires strong adsorption (which should be irreversible) of the polymer to the particle surface.

Dispersions of surface modified TiO_2 (Tab. 11.1) in alkyl benzoate and hexamethyltetracosane (squalane) were prepared at various solid loadings using a polymeric/oligomeric polyhydroxystearic acid (PHS) surfactant of molecular weight 2500 (PHS2500) and 1000 (PHS1000) [2]. For comparison, results were also obtained using a low molecular weight (monomeric) dispersant, namely isostearic acid, ISA. The titania particles had been coated with alumina and/or silica. The electron micrograph in Fig. 11.12 shows the typical size and shape of these rutile particles. The surface area and particle size of the three powders used are summarized in Tab. 11.3.

The dispersions of the surface-modified TiO_2 powder, dried at 110 °C, were prepared by milling (using a horizontal bead mill) in polymer solutions of different concentrations for 15 minutes and were then allowed to equilibrate for more than 16 hours at room temperature before making the measurements.

100 nm

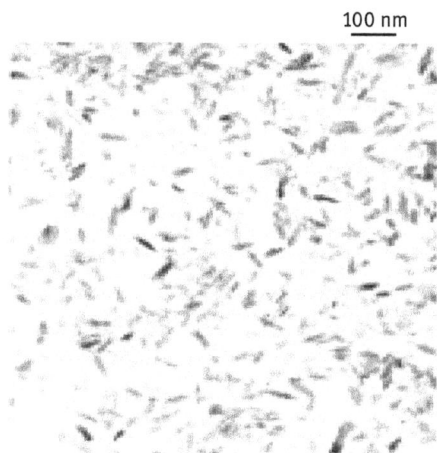

Fig. 11.12: Transmission electron micrograph of titanium dioxide particles.

Tab. 11.3: Surface modified TiO_2 powders.

Powder	Coating	Surface Area* / (m²/g)	Particle size** / nm
A	Alumina/silica	95	40–60
B	Alumina/stearic acid	70	30–40
C	Silica/stearic acid	65	30–40

* BET N2, ** equivalent sphere diameter, X-ray disc centrifuge

The adsorption isotherms were obtained by preparing dispersions of 30 w/w % TiO$_2$ at different polymer concentration (C$_0$, mg/l). The particles and adsorbed dispersant were removed by centrifugation at 20 000 rpm ($\approx$ 48 000 g) for 4 hours, leaving a clear supernatant. The concentration of the polymer in the supernatant was determined by acid value titration. The adsorption isotherms were calculated by mass balance to determine the amount of polymer adsorbed at the particle surface (Γ, mg m^2) of a known mass of particulate material (m g) relative to that equilibrated in solution (C$_e$, mg/l):

$$\Gamma = \frac{(C_0 - C_e)}{mA_s}. \tag{11.8}$$

The surface area of the particles (A$_s$, m^2/g) was determined by BET nitrogen adsorption method.

Dispersions of various solid loadings were obtained by milling at progressively increasing TiO$_2$ concentration at an optimum dispersant/solids ratio [2]. The dispersion stability was evaluated by viscosity measurement and by attenuation of UV-vis radiation. The viscosity of the dispersions was measured by subjecting the dispersions to an increasing shear stress, from 0.03 Pa to 200 Pa over 3 min at 25 °C using a Bohlin CVO rheometer. It was found that the dispersions exhibited shear thinning behaviour and high zero shear viscosity, identified from the plateau region at low shear stress (where viscosity was apparently independent of the applied shear stress). UV-vis attenuation was determined by measuring transmittance of radiation between 250 nm and 550 nm. Samples were prepared by dilution with a 1 w/v % solution of dispersant in cyclohexane to approximately 20 mg/l and placed in a 1 cm pathlength cuvette in a UV-vis spectrophotometer. The sample solution extinction ε (l g^{-1} cm^{-1}) was calculated from Beer's law (equation (11.9)):

$$\varepsilon = \frac{A}{cl}, \tag{11.9}$$

where A is absorbance, c is concentration of attenuating species (g/l), l is pathlength (cm). The dispersions of powders B and C were finally incorporated into typical water-in-oil sunscreen formulations at 5 % solids with an additional 2 % of organic active (butyl methoxy dibenzoyl methane) and assessed for efficacy, SPF (sun protection factor) as well as stability (visual observation, viscosity). SPF measurements were made on an Optometrics SPF-290 analyser fitted with an integrating sphere.

Figure 11.13 shows the adsorption isotherms of ISA, PHS1000 and PHS2500 on TiO$_2$ in alkylbenzoate (Fig. 11.13 (a)) and in squalane (Fig. 11.13 (b)). The adsorption of the low molecular weight ISA from alkylbenzoate is of low affinity (Langmuir type) indicating reversible adsorption (possibly physisorption). In contrast, the adsorption isotherms for PHS100 and PHS2500 are of the high affinity type, indicating irreversible adsorption and possible chemisorption due to acid-base interaction. From squalane, all adsorption isotherms show high affinity type and they show higher adsorption values when compared with the results using alkylbenzoate. This reflects the difference in solvency of the dispersant by the medium as will be discussed below.

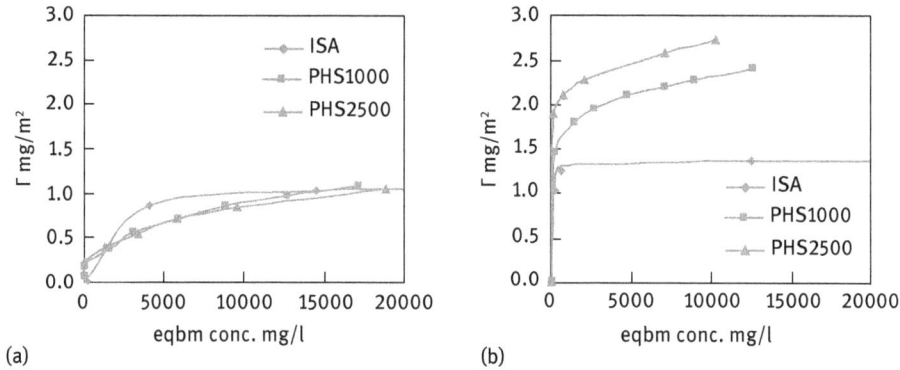

Fig. 11.13: Adsorption isotherms in alkylbenzoate (a) and in squalane (b).

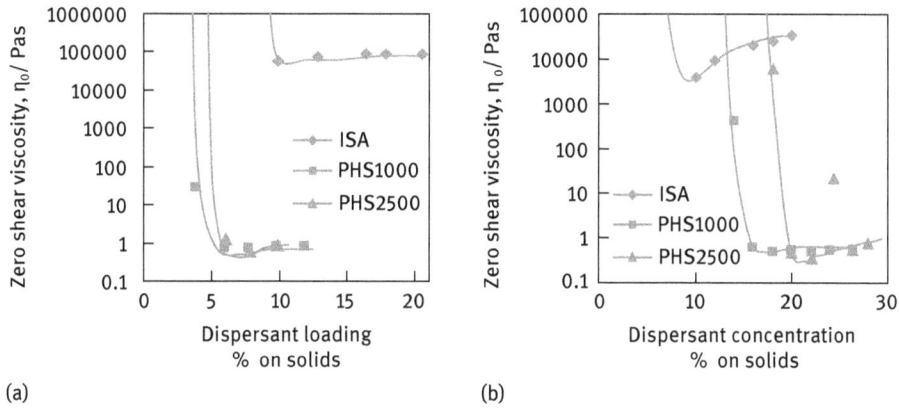

Fig. 11.14: Dispersant demand curve in alkylbenzoate (left) and squalane (right).

Figure 11.14 shows the variation in zero shear viscosity with dispersant loading % for a 40 % dispersion. It can be seen that the zero shear viscosity decreases very rapidly with increasing dispersant loading and eventually the viscosity reaches a minimum at an optimum loading that depends on the solvent used as well as the nature of the dispersant.

With the molecular dispersant ISA, the minimum viscosity that could be reached at high dispersant loading was very high (several orders of magnitude more than the optimized dispersions) indicating poor dispersion of the powder in both solvents. Even reducing the solids content of TiO_2 to 30 % did not result in a low viscosity dispersion. With PHS1000 and PHS2500, a low minimum viscosity could be reached at 8–10 % dispersant loading in alkylbenzoate and 18–20 % dispersant loading in squalane. In the latter case the dispersant loading required for reaching a viscosity minimum is higher for the higher molecular weight PHS.

The quality of the dispersion was assessed using UV-vis attenuation measurements. At very low dispersant concentration a high solids dispersion can be achieved by simple mixing, but the particles are aggregated as demonstrated by the UV-vis curves (Fig. 11.15). These large aggregates are not effective as UV attenuators. As the PHS dispersant level is increased, UV attenuation is improved and above 8 wt % dis-

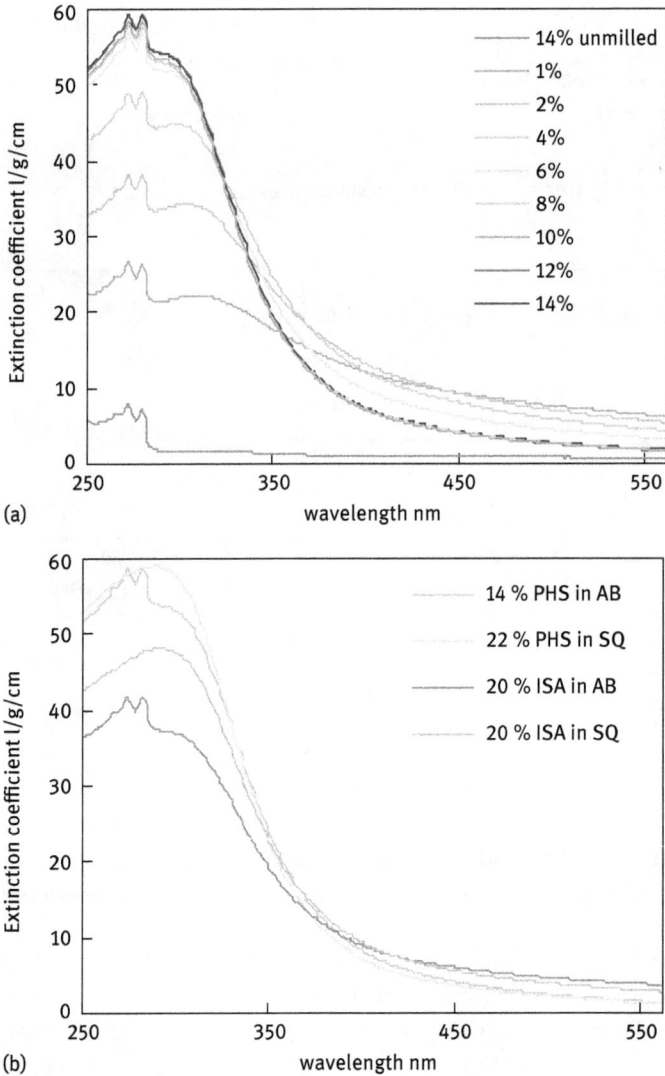

Fig. 11.15: UV-vis attenuation (a) for milled dispersions with 1–14 % PHS1000 dispersant and unmilled at 14 % dispersant on solids; and (b) for dispersions in squalane (SQ) and in alkyl-benzoate (AB) using 20 % isostearic acid (ISA) as dispersant compared to optimized PHS1000 dispersions in the same oils.

persant on particulate mass, optimized attenuation properties (high UV, low visible attenuation) are achieved (for the PHS1000 in alkyl benzoate). However, milling is also required to break down the aggregates into their constituent nanoparticles and a simple mixture which is unmilled has poor UV attenuation even at 14 % dispersant loading.

The UV-vis curves obtained when monomeric isostearic acid was incorporated as a dispersant (Fig. 11.15) indicate that these molecules do not provide a sufficient barrier to aggregation, resulting in relatively poor attenuation properties (low UV, high visible attenuation).

The steric layer thickness δ could be varied by altering the dispersion medium and hence the solvency of the polymer chain. This had a significant effect upon dispersion rheology. Solid loading curves (Fig. 11.16 (a) and (b)) demonstrate the differences in effective volume fraction due to the adsorbed layer (equation (11.7)).

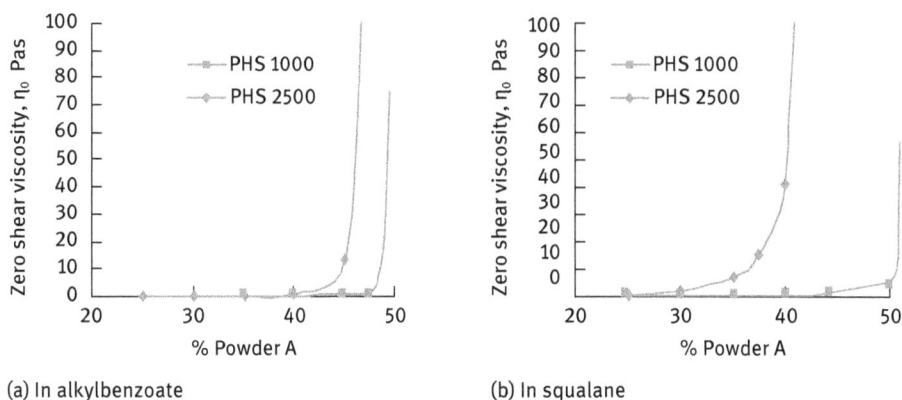

(a) In alkylbenzoate   (b) In squalane

Fig. 11.16: Zero shear viscosity dependency on solid loadings (a) in alkylbenzoate and (b) in squalane.

In the poorer solvent case (squalane), the effective volume fraction and adsorbed layer thickness showed a strong dependency on molecular weight with solid loadings becoming severely limited above 35 % for the higher molecular weight whereas $\approx 50\%$ could be reached for the lower molecular weight polymer. In alkyl benzoate no strong dependency was seen with both systems achieving more than 45 % solids. Solids weight fraction above 50 % resulted in very high viscosity dispersions in both solvents.

The same procedure described above enabled optimized dispersion of equivalent particles with alumina and silica inorganic coatings (powders B and C). Both particles additionally had the same level of organic (stearate) modification. These optimized dispersions were incorporated into water-in-oil formulations and their stability/efficacy monitored by visual observation and SPF measurements (Tab. 11.4).

Tab. 11.4: Sunscreen emulsion formulations from dispersions of powders B and C.

Emulsion	Visual observation	SPF	Emulsifier level
Powder B emulsion 1	Good homogenous emulsion	29	2.0 %
Powder C emulsion 1	Separation, inhomogeneous	11	2.0 %
Powder C emulsion 2	Good homogeneous emulsion	24	3.5 %

The formulation was destabilized by the addition of the powder C dispersion and poor efficacy was achieved despite an optimized dispersion before formulation. When emulsifier concentration was increased from 2 to 3.5 % (emulsion 2) the formulation became stable and efficacy was restored.

The anchor of the chain to the surface (described qualitatively through χ^s) is very specific and this could be illustrated by silica-coated particles which showed lower adsorption of the PHS (Fig. 11.17).

In addition, when a quantity of emulsifier was added to an optimized dispersion of powder C (silica surface) the acid value of the equilibrium solution was seen to rise, indicating some displacement of the PHS2500 by the emulsifier.

The dispersant demand curves (Fig. 11.15 (a) and (b)) and solid loading curves (Fig. 11.16 (a) and (b)) show that one can reach a stable dispersion using PHS1000 or PHS2500 both in alkylbenzoate and in squalane. This can be understood in terms

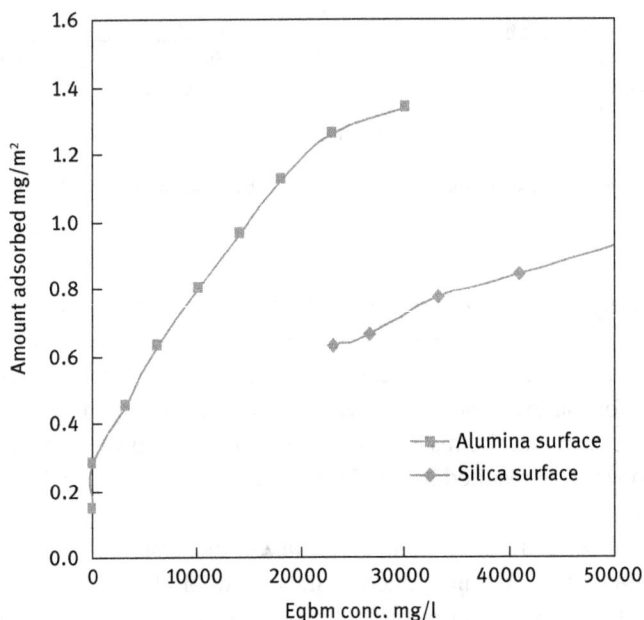

Fig. 11.17: Adsorption isotherms for PHS2500 on powder B (alumina surface) and powder C (silica surface).

of the stabilization produced when using these polymeric dispersants. Addition of sufficient dispersant enables coverage of the surface and results in a steric barrier preventing aggregation due to van der Waals attraction. Both molecular weight oligomers were able to achieve stable dispersions. The much smaller molecular weight "monomer", isostearic acid, however, is insufficient to provide this steric barrier and dispersions were aggregated, leading to high viscosities, even at 30 % solids. UV-vis curves confirm that these dispersions are not fully dispersed since their full UV potential is not realized (Fig. 11.15). Even at 20 % isostearic acid the dispersions are seen to give a lower E_{max} and increased scattering at visible wavelengths. indicating a partially aggregated system.

The differences between alkylbenzoate and squalane observed in the optimum dispersant concentration required for maximum stability can be understood by examining the adsorption isotherms in Fig. 11.13 (a) and (b). The nature of the steric barrier depends on the solvency of the medium for the chain, and is characterized by the Flory–Huggins interaction parameter χ. Information on the value of χ for the two solvents can be obtained from solubility parameter calculations (equation (11.2)). The results of these calculations are given in Tab. 11.5 for PHS, alkylbenzoate and squalane.

Tab. 11.5: Hansen and Beerbower solubility parameters for the polymer and both solvents.

	δ_T	δ_d	δ_p	δ_h	$\Delta\delta_T$
PHS	19.00	18.13	0.86	5.60	
Alkyl benzoate	17.01	19.13	1.73	4.12	1.99
Squalane	12.9	15.88	0	0	6.1

It can be seen that both PHS and alkylbenzoate have polar and hydrogen bonding contributions to the solubility parameter δ_T. In contrast, squalane which is nonpolar has only a dispersion component to δ_T. The difference in the total solubility parameter $\Delta\delta_T$ value is much smaller for alkylbenzoate when compared with squalane. Thus one can expect that alkylbenzoate is a better solvent for PHS when compared with squalane. This explains the higher adsorption amounts of the dispersants in squalane when compared with alkylbenzoate (Fig. 11.13). The PHS finds adsorption at the particle surface energetically more favourable than remaining in solution. The adsorption values at the plateau for PHS in squalane ($> 2\,mg\,m^{-2}$ for PHS1000 and $> 2.5\,mg\,m^{-2}$ for PHS2500) is more than twice the value obtained in alkylbenzoate ($1\,mg\,m^{-2}$ for both PHS1000 and PHS2500).

It should be mentioned, however, that both alkylbenzoate and squalane will have χ values less than 0.5, i.e., good solvent conditions and a positive steric potential. This is consistent with the high dispersion stability produced in both solvents. However, the relative difference in solvency for PHS between alkylbenzoate and squalane is expected to have a significant effect on the conformation of the adsorbed layer. In squalane, a poorer solvent for PHS, the polymer chain is denser when compared to the polymer layer in alkylbenzoate. In the latter case a diffuse layer that is typical for polymers in good solvents is produced. This is illustrated in Fig. 11.18 (a) which shows a higher hydrodynamic layer thickness for the higher molecular weight PHS2500. A schematic representation of the adsorbed layers in squalane is shown in Fig. 11.18 (b), which also shows a higher thickness for the higher molecular weight PHS2500.

PHS 1000   PHS 2500   PHS 1000   PHS 2500

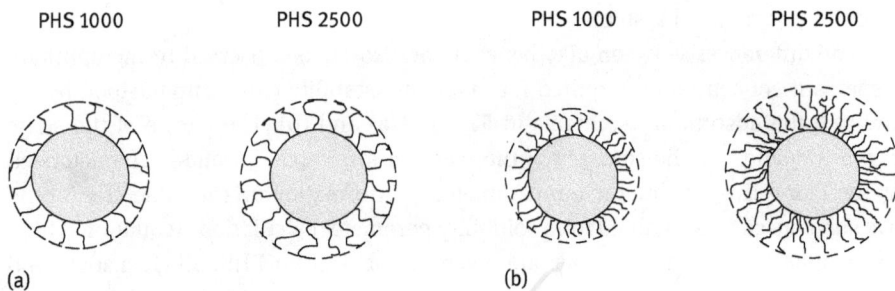

(a)   (b)

Fig. 11.18: (a) (left): Well solvated polymer results in diffuse adsorbed layers (alkylbenzoate). (b) (right): Polymers are not well solvated and form dense adsorbed layers (squalane).

In squalane, the dispersant adopts a close packed conformation with little solvation and high amounts are required to reach full surface coverage ($\Gamma > 2$ mg m^2). It seems also that in squalane the amount of adsorption depends much more on the molecular weight of PHS than in the case of alkylbenzoate. It is likely that with the high molecular weight PHS2500 in squalane the adsorbed layer thickness can reach higher values when compared with the results in alkylbenzoate. This larger layer thickness increases the effective volume fraction and this restricts the total solids that can be dispersed. This is clearly shown from the results of Fig. 11.16, which shows a rapid increase in zero shear viscosity at a solid loadings > 35 %. With the lower molecular weight PHS1000, with smaller adsorbed layer thickness, the effective volume fraction is lower and high solid loadings (≈ 50 %) can be reached. The solid loadings that can be reached in alkylbenzoate when using PHS2500 is higher (≈ 40 %) than that obtained in squalane. This implies that the adsorbed layer thickness of PHS2500 is smaller in alkylbenzene when compared with the value in squalane as schematically shown in Fig. 11.18. The solid loading with PHS1000 in alkylbenzene is similar to that in squalane, indicating a similar adsorbed layer thickness in both cases.

The solid loading curves demonstrate that with an extended layer such as that obtained with the higher molecular weight (PHS2500) the maximum solid loading becomes severely limited as the effective volume fraction (equation (11.7)) is increased.

In squalane, the monomeric dispersant, isostearic acid, shows a high affinity adsorption isotherm with a plateau adsorption of $1\,\mathrm{mg\,m^2}$ but this provides an insufficient steric barrier (δ/R too small, Fig. 11.4) to ensure colloidal stability.

Most sunscreen formulations consist of an oil-in-water (O/W) emulsion into which the particles are incorporated. These active particles can be in either the oil phase, or the water phase, or both as illustrated in Fig. 11.19. For a sunscreen formulation based on a W/O emulsion, the added nonaqueous sunscreen dispersion mostly stays in the oil continuous phase.

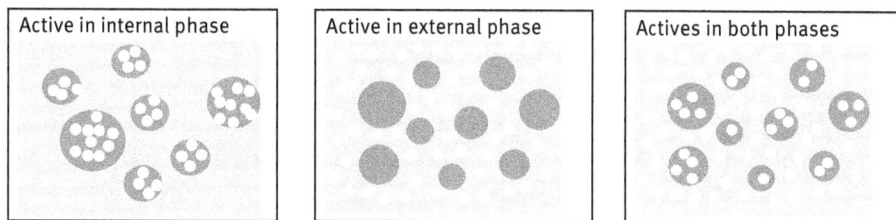

Fig. 11.19: Schematic representation of the location of active particles in sunscreen formulations.

On addition of the sunscreen dispersion to an emulsion to produce the final formulation, one has to consider the competitive adsorption of the dispersant/emulsifier system. In this case the strength of adsorption of the dispersant to the surface modified TiO_2 particles must be considered. As shown in Fig. 11.17, the silica coated particles (C) show lower PHS2500 adsorption compared to the alumina coated particles (B). However, the dispersant demand for the two powders to obtain a colloidally stable dispersion was similar in both cases (12–14 % PHS2500). This appears at first sight to indicate similar stabilities. However, when added to a water-in-oil emulsion prepared using an A–B–A block copolymer of PHS–PEO–PHS as emulsifier, the system based on the silica coated particles (C) became unstable, showing separation and coalescence of the water droplets. The SPF performance also dropped drastically from 29 to 11. In contrast, the system based on alumina coated particles (B) remained stable showing no separation as illustrated in Tab. 11.4. These results are consistent with the stronger adsorption (higher χ^s) of PHS2500 on the alumina coated particles. With the silica coated particles, it is likely that the PHS–PEO–PHS block copolymer becomes adsorbed on the particles, thus depleting the emulsion interface from the polymeric emulsifier and this is the cause of coalescence. It is well known that molecules based on PEO can adsorb on silica surfaces [2]. By addition of more emulsifier (increasing its concentration from 2 to 3.5 %) the formulation remained stable as is illustrated in Tab. 11.2. This final set of results demonstrates how a change in surface coating can alter the adsorption strength which can have consequences for the final formulation. The same optimization process used for powder A enabled stable dispersions to be formed from powders B and C. Dispersant demand curves showed optimized dispersion rheology at similar added dispersant levels of 12–14 % PHS2500. To the dispersion scientist these appeared to be stable TiO$_2$ dispersions. However, when the optimized dispersions were formulated into the external phase of a water-in-oil emulsion differences were observed and alterations in formulation were required to ensure emulsion stability and performance.

11.5 Use of polymeric surfactants for stabilizing liposome and vesicles

Liposomes are spherical phospholipid liquid crystalline phases (smectic mesophases) that are simply produced by dispersion of phospholipid (such as lecithin) in water by simple shaking [2]. This results in the formation of multilayer structures consisting of several bilayers of lipids (several μm). When sonicated, these multilayer structures produce unilamellar structures (with size range of 25–50 nm) that are referred to as vesicles. A schematic picture of liposomes and vesicles is given in Fig. 11.20.

Glycerol containing phospholipids are used for the preparation of liposomes and vesicles. The structure of some lipids is shown in Fig. 11.21. The most widely used lipid for cosmetic formulations is phosphatidylcholine that can be obtained from eggs or soybean. In most preparations, a mixture of lipids is used to obtain the most optimum structure. These liposome bilayers can be considered as mimicking models of biological membranes. They can solubilize both lipophilic active ingredients in the lipid bilayer phase, as well as hydrophilic molecules in the aqueous layers between the lipid bilayers and in the inner aqueous phase. For example, addition of liposomes to cosmetic formulations can be applied for enhancing the penetration of anti-wrinkle agents [2]. They will also form lamellar liquid crystalline phases and they do not disrupt the stratum corneum. No facilitated transdermal transport is possible, thus eliminating skin irritation. Phospholipid liposomes can be used as in vitro indicators for studying skin irritation by surfactants.

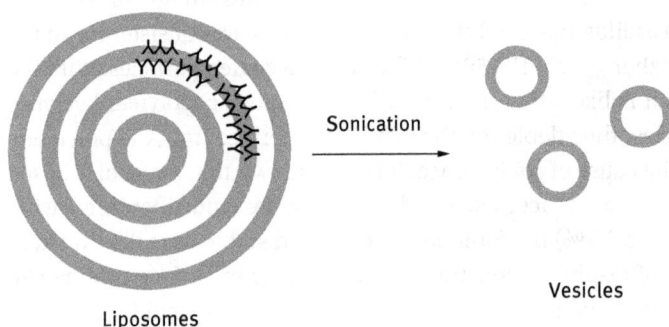

Fig. 11.20: Schematic representation of liposomes and vesicles.

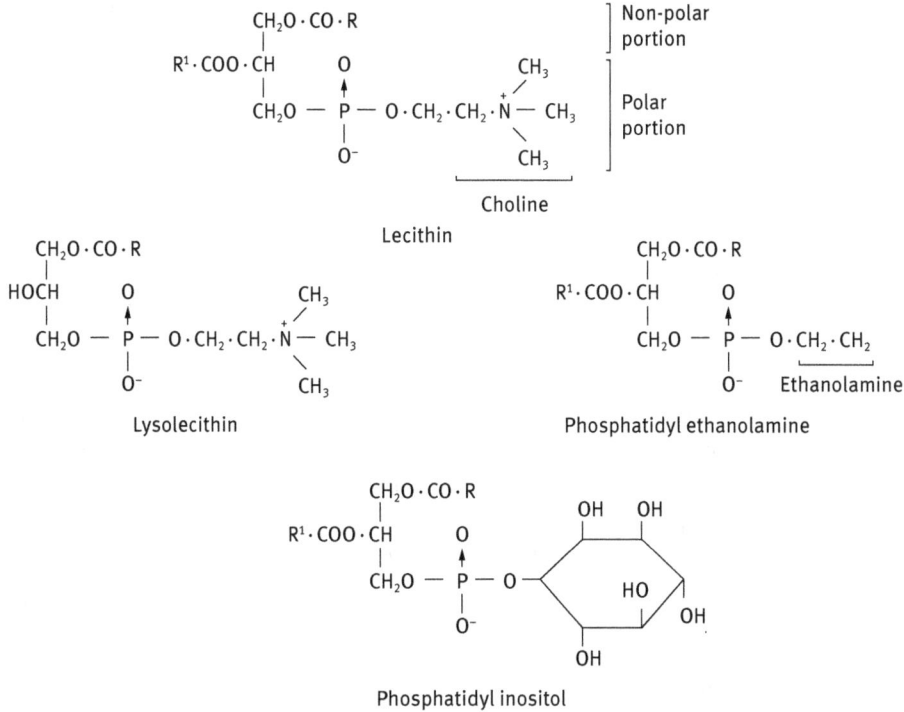

Fig. 11.21: Structure of lipids.

The nomenclature for liposomes is far from being clear; it is now generally accepted that "All types of lipid bilayers surrounding an aqueous phase are in the general category of liposomes" [2]. The term "liposome" is usually reserved for vesicles composed, even partly, by phospholipids. The more generic term "vesicle" is to be used to describe any structure consisting of one or more bilayers of various other surfactants. In general, the terms "liposome" and "phospholipid vesicle" are used interchangeably. Liposomes are classified in terms of the number of bilayers, as Multilamellar Vesicles (MLVs > 400 nm), Large Unilamellar Vesicles (LUVs > 100 nm) and Small Unilamellar Vesicles (SUVs < 100 nm). Other types reported are the Giant Vesicles (GV), which are unilamellar vesicles of diameter between 1–5 μm and Large Oligolamellar Vesicles (LOV), where a few vesicles are entrapped in the LUV or GV.

For phospholipid molecules containing two hydrocarbon chains of 16–18 carbon atoms per chain, the volume of the hydrocarbon part of the molecule is double the volume of a single-chain molecule, while the optimum surface area for its head group is of the same order as that of a single-chain surfactant ($a_0 \approx 0.5-0.7$ nm^2). Thus the only way for this double-chain surfactant is to form aggregates of the bilayer sheet or the close bilayer vesicle type. This can be explained using the critical packing parameter concept (CPP) [2]. The CPP is a geometric expression given by the ratio of the cross-sectional area of the hydrocarbon tail(s), a, to that of the head group, a_0. a is equal to the volume of the hydrocarbon chain(s), v, divided by the critical chain length, l_c, of the hydrocarbon tail. Thus the CPP is given by [2]

$$CPP = \frac{v}{a_0 l_c}. \tag{11.10}$$

Regardless of shape, any aggregated structure should satisfy the following criterion: no point within the structure can be farther from the hydrocarbon-water surface than l_c which is roughly equal to, but less than the fully extended length l of the alkyl chain.

For a spherical micelle, the radius $r = l_c$ and from simple geometry $CPP = v/a_0 l_c \leq 1/3$. Once $v/a_0 l_c > 1/3$, spherical micelles cannot be formed and when $1/2 \geq CPP > 1/3$ cylindrical micelles are produced. When the $CPP > 1/2$ but < 1, vesicles are produced. These vesicles will grow until $CPP \approx 1$, when planer bilayers will start forming.

The bilayer sheet lipid structure is energetically unfavourable to the spherical vesicle, because of the lower aggregation number of the spherical structure. Without the introduction of packing constrains, the vesicles should shrink to such a small size that they would actually form micelles. For double-chain amphiphiles three considerations are important:

(i)   an optimum, a_0 (almost the same as that for single-chain surfactants), must be achieved by considering the various opposing forces;
(ii) structures with minimum aggregation number N must be formed;
(iii) aggregates into bilayers must be the most favoured structure.

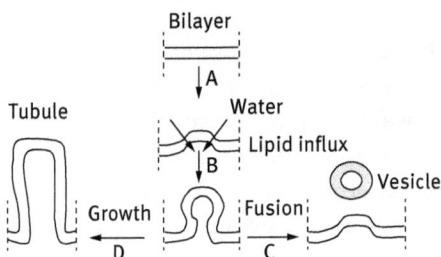

Fig. 11.22: Bilayer vesicle and tubule formation.

A schematic picture of the formation of bilayer vesicle and tubule structures is shown in Fig. 11.22. Steps A and B are energetically favourable. Step C is governed by packing constraints and thermodynamics in terms of the least aggregation number. The spherical vesicle is an equilibrium state of the aggregate in water and it is certainly more favoured over extended bilayers.

The main drawback of applications of liposomes in cosmetic formulations is their metastability. On storage, the liposomes tend to aggregate and fuse to form larger polydisperse systems and finally the system reverses into a phospholipid lamellar phase in water. This process takes place relatively slowly because of the slow exchange between the lipids in the vesicle and the monomers in the surrounding medium. Therefore, it is essential to investigate both the chemical and physical stability of the liposomes. The process of aggregation can be examined by obtaining measurements of particle size as a function of time. Maintenance of the vesicle structure can be assessed using freeze fracture and electron microscopy.

Several methods have been applied to increase the rigidity and physicochemical stability of the liposome bilayer of which the following methods are the most commonly used: hydrogenation of the double bonds within the liposomes; polymerization of the bilayer using synthesized polymerizable amphiphiles and inclusion of cholesterol to rigidify the bilayer [2].

Other methods to increase the stability of the liposomes include modification of the liposome surface, for example by physical adsorption of polymeric surfactants onto the liposome surface (e.g. proteins and block copolymers). Another approach is to covalently bond the macromolecules to the lipids and subsequently form vesicles. A third method is to incorporate the hydrophobic segments of the polymeric surfactant within the lipid bilayer. This latter approach has been successfully applied using an A–B–A block copolymer of polyethylene oxide (A) and polypropylene oxide (PPO), namely Poloxamers (Pluronics) [2]. Two different techniques of adding the copolymer were attempted [2]. In the first method (A), the block copolymer was added after formation of the vesicles. In the second method, the phospholipid and copolymer are first mixed together and this is followed by hydration and formation of SUV vesicles. These two methods are briefly described below.

The formation of small unilamellar vesicles (SUVs) was carried out by sonication of 2 % w/w of the hydrated lipid (for about 4 hours). This produced SUV vesicles with a mean vesicle diameter of 45 nm (polydispersity index of 1.7–2.4). This was followed by the addition of the block copolymer solution and dilution of 100 times to obtain a lipid concentration of 0.02 % (method A). In the second method, (method I) SUV vesicles were prepared in the presence of the copolymer at the required molar ratio.

In method (A), the hydrodynamic diameter increases with increasing block copolymer concentration, particularly those with high PEO content, reaching a plateau at a certain concentration of the block copolymer. The largest increase in hydrodynamic diameter (from ≈ 43 nm to ≈ 48 nm) was obtained using Pluronic F127 (that contains a molar mass of 8330 PPO and molar mass of 3570 PEO). In method (I), the mean vesicle diameter showed a sharp increase with an increase in % w/w copolymer, reaching a maximum at a certain block copolymer concentration, after which any further increase in polymer concentration showed a sharp reduction in average diameter. For example with Pluronic F127, the average diameter increased from ≈ 43 nm to ≈ 78 nm at 0.02 w/w % block copolymer and then it decreased sharply with a further increase in polymer concentration reaching ≈ 45 nm at 0.06 w/w % block copolymer. This reduction in average diameter at high polymer concentration is due to the presence of excess micelles of the block copolymer.

Method (A)   Method (I)

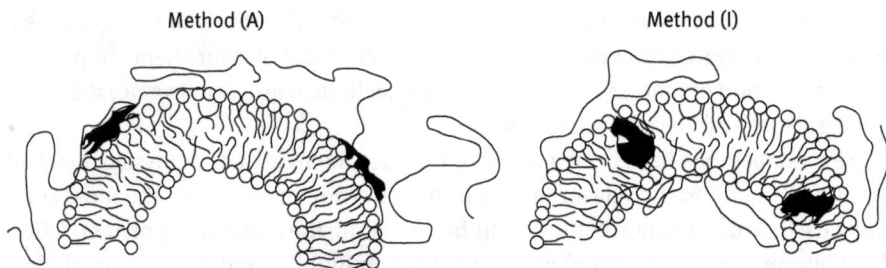

Fig. 11.23: Schematic representation of vesicle structure in the presence of triblock copolymer prepared using method (A) and (I) [4].

A schematic representation of the structure of the vesicles obtained on addition of the block copolymer using methods (A) and (I) is shown in Fig. 11.23. With method (A), the triblock copolymer is adsorbed on the vesicle surface by both PPO and PEO blocks. These "flat" polymer layers are prone to desorption due to the weak binding onto the phospholipid surface. In contrast, with the vesicles prepared using method (I) the polymer molecules are more strongly attached to the lipid bilayer with PPO segments "buried" in the bilayer environment surrounded by the lipid fatty acids. The PEO chains remain at the vesicle surface, free to dangle in solution and attain the preferred conformation. The resulting sterically stabilized vesicles [(I) system] have several advantages over the (A) system with the copolymer simply coating their outer surface. The anchoring of the triblock copolymer using method (I) results in irreversible adsorption and lack of desorption. This is confirmed by dilution of both systems. With (A), dilution of the vesicles results in reduction of the diameter to its original bare liposome system, indicating polymer desorption. In contrast, dilution of the vesicles prepared by method (I) showed no significant reduction in diameter size, indicating strong anchoring of the polymer to the vesicle. A further advantage of constructing the vesicles with bilayer-associated copolymer molecules is the possibility of increased rigidity of the lipid-polymer bilayer [2].

11.6 Polymeric surfactants in multiple emulsions

Multiple emulsions are complex systems of "emulsions of emulsions" [2]. Two main types can be distinguished:
(i)   Water-in-Oil-in-Water (W/O/W) multiple emulsions, where the dispersed oil droplets contain emulsified water droplets.
(ii) Oil-in-Water-in-Oil (O/W/O) multiple emulsions, where the dispersed water droplets contain emulsified oil droplets.

The most commonly used multiple emulsions in cosmetics are the W/O/W. The W/O/W multiple emulsion may be considered as water/water emulsions in which the internal water droplets are separated by an "oily layer" (membrane). The internal droplets could also consist of a polar solvent such as glycol or glycerol which may contain a dissolved or dispersed active ingredient. The O/W/O multiple emulsion can be considered as an oil/oil emulsion separated by an "aqueous layer" (membrane).

Due to the oily liquid or aqueous membrane formed, multiple emulsions ensure complete protection of the entrapped active ingredient used in many cosmetic systems (e.g. anti-wrinkle agents) and controlled release of this active ingredient from the internal to the external phase. In addition, multiple emulsions offer several advantages such as protection of fragile ingredients, separation of incompatible ingredients, prolonged hydration of the skin and in some cases formation of firm gelled structure. Furthermore, a pleasant skin feels like that of O/W emulsion combined with the well-known moisturizing properties of W/O emulsions are obtained with W/O/W multiple emulsions. Multiple emulsions can be usefully applied for controlled release by controlling the rate of the breakdown process of the multiple emulsions on application. Initially, one prepares a stable multiple emulsions (with a shelf life of two years) which on application breaks down in a controlled manner, thus releasing the active ingredient in a similarly controlled manner (slow or sustained release).

For applications in personal care and cosmetics, a wider range of surfactants can be used provided these molecules satisfy some essential criteria such as lack of skin irritation, lack of toxicity on application and safety to the environment (biodegradability of the molecule is essential in this case).

Two main criteria are essential for the preparation of stable multiple emulsions:

(i) Two emulsifiers with low and high HLB numbers. Emulsifier 1 should prevent coalescence of the internal water droplets, preferably a polymeric surfactant that produces a viscoelastic film which also reduces water transport. The secondary emulsifier should also produce an effective steric barrier at the O/W interface to prevent any coalescence of the multiple emulsion droplet.

(ii) Optimum osmotic balance. This is essential to reduce water transport. This is achieved by addition of electrolytes or nonelectrolytes. The osmotic pressure in the external phase should be slightly lower than that of the internal phase to compensate for curvature effects.

These multiple emulsions are usually prepared in a two-stage process. For example, a W/O/W multiple emulsion is formulated by first preparing a W/O emulsion using a polymeric surfactant with a low HLB number (5–6) using a high speed mixer (e.g. an Ultra-Turrax or Silverson). The resulting W/O emulsion is further emulsified in aqueous solution containing a polymeric surfactant with a high HLB number (9–12) using a low speed stirrer (e.g. a paddle stirrer). A schematic representation of the preparation of multiple emulsions is given in Fig. 11.24.

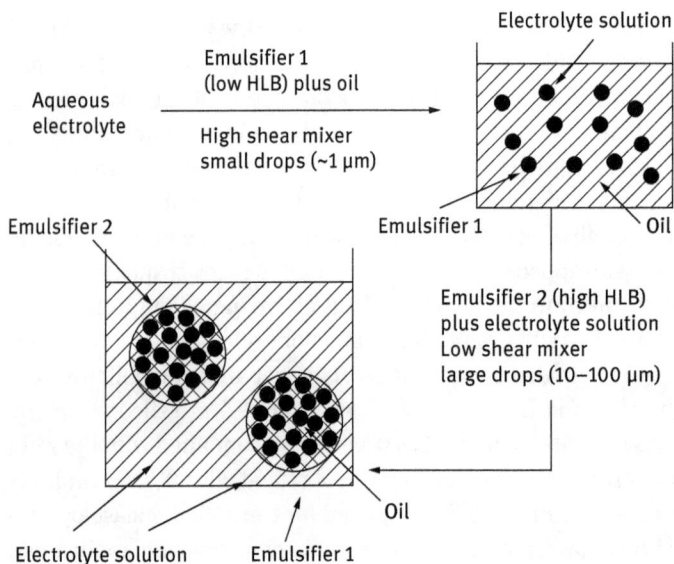

Fig. 11.24: Scheme for preparing a W/O/W multiple emulsion.

The yield of the multiple emulsion can be determined using dialysis for W/O/W multiple emulsions. A water soluble marker is used and its concentration in the outside phase is determined:

$$\% \text{ multiple emulsion} = \frac{C_i}{C_i + C_e} \times 100, \qquad (11.11)$$

where C_i is the amount of marker in the internal phase and C_e is the amount of marker in the external phase. It has been suggested that if a yield of more than 90 % is required, the lipophilic (low HLB) surfactant used to prepare the primary emulsion must be ≈ 10 times higher in concentration than the hydrophilic (high HLB) surfactant.

The oils that can be used for the preparation of multiple emulsions must be cosmetically acceptable (no toxicity). Most convenient oils are vegetable oils such as soybean or safflower oil. Paraffinic oils with no toxic effect may be used. Also some polar oils such as isopropyl myristate can be applied; silicone oils can also be used. The low HLB emulsifiers (for the primary W/O emulsion) are mostly the sorbitan esters (Spans), but these may be mixed with other polymeric emulsifiers such as silicone emulsifiers. The high HLB surfactant can be chosen from the Tween series, although the block copolymers PEO–PPO–PEO (Poloxamers or Pluronics) may give much better stability. The polymeric surfactant INUTEC® SP1 can also give much higher stability. For controlling the osmotic pressure of the internal and external phases, electrolytes such as NaCl or nonelectrolytes such as sorbitol may be used.

In most cases, a "gelling agent" is required, both for the oil and the outside external phase. For the oil phase, fatty alcohols may be used. For the aqueous continuous phase, one can use the same "thickeners" that are used in emulsions, e.g. hydroxyethyl cellulose, xanthan gum, alginates, carrageenans, etc. Sometimes liquid crystalline phases are applied to stabilize the multiple emulsion droplets. These can be generated using a nonionic surfactant and long chain alcohol. "Gel" coating around the multiple emulsion droplets may also be formed to enhance stability.

As an illustration, a typical formulation of a W/O/W multiple emulsion is described below, using two different thickeners, namely Keltrol (xanthan gum from Kelco) and Carbopol 980 (a crosslinked polyacrylate gel produced by BF Goodrich). These thickeners were added to reduce creaming of the multiple emulsion. A two-step process was used in both cases.

The primary W/O emulsion was prepared using an A–B–A block copolymer (where A is poly(hydroxystearic acid), PHS, and B is poly(ethylene oxide), PEO), i.e. PHS–PEO–PHS. 4 g of PHS–PEO–PHS were dissolved in 30 g of a hydrocarbon oil. For quick dissolution, the mixture was heated to 75 °C. The aqueous phase consisted of 65.3 g water, 0.7 g $MgSO_4 \cdot 7H_2O$ and a preservative. This aqueous solution was also heated to 75 °C. The aqueous phase was added to the oil phase slowly while stirring intensively using a high speed mixer. The W/O emulsion was homogenized for 1 min and allowed to cool to 40–45 °C followed by further homogenization for another minute and stirring was continued until the temperature reached ambient.

The primary W/O emulsion was emulsified in an aqueous solution containing the polymeric surfactant PEO–PPO–PEO, namely Pluronic PEF127. 2 g of the polymeric surfactant were dissolved in 16.2 g water containing a preservative by stirring at 5 °C. 0.4 g $MgSO_4 \cdot 7H_2O$ were then added to the aqueous polymeric surfactant solution. 60 g of the primary W/O emulsion were slowly added to the aqueous PFE127 solution while stirring slowly at 700 rpm (using a paddle stirrer). An aqueous Keltrol solution was prepared by slowly adding 0.7 g Keltrol powder to 20.7 g water, while stirring. The resulting thickener solution was further stirred for 30–40 min until a homogeneous gel was produced. The thickener solution was slowly added to the multiple emulsion while stirring at low speed (400 rpm) and the whole system was homogenized for 1 minute followed by gentle stirring at 300 rpm until the thickener completely dispersed in the multiple emulsion (about 30 min stirring was sufficient). The final system was investigated using optical microscopy to ensure that a multiple emulsion was produced. The formulation was left standing for several months and the droplets of the multiple emulsion were investigated using optical microscopy (see below). The rheology of the multiple emulsion was also measured at various intervals to ensure that the consistency of the product remained the same on long storage.

The second multiple emulsion was made under the same conditions as above except using Carbopol 980 as a thickener (gel). In this case, no $MgSO_4$ was added, since the Carbopol gel was affected by electrolytes. The aqueous PEF127 polymeric surfactant solution was made by dissolving 2 g of the polymer in 23 g water. 15 g of 2% master gel of Carbopol were added to the PEF127 solution while stirring until the Carbopol was completely dispersed. 60 g of the primary W/O emulsion were slowly added to the aqueous solution of PEF127/Carbopol solution, while stirring thoroughly at 700 rpm. Triethanolamine was added slowly, while gently stirring until the pH of the system reached 6.0–6.5.

Another example of a W/O/W multiple emulsion was prepared using two polymeric surfactants. A W/O emulsion was prepared using an A–B–A block copolymer of PHS–PEO–PHS. This emulsion was prepared using a high speed mixer giving droplet sizes in the region of 1 µm. The W/O emulsion was then emulsified in an aqueous solution of hydrophobically modified inulin (INUTEC® SP1) using low speed stirring to produce multiple emulsion droplets in the range 10–100 µm. The osmotic balance was achieved using 0.1 mol dm^{-3} MgCl$_2$ in the internal water droplets and outside continuous phase. The multiple emulsion was stored at room temperature and 50 °C and photomicrographs were taken at various time intervals. The multiple emulsion was very stable for several months. A photomicrograph of the W/O/W multiple emulsion is shown in Fig. 11.25.

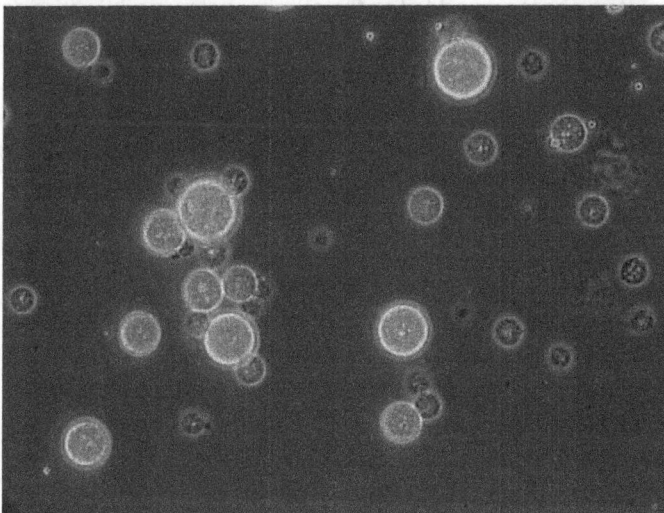

Fig. 11.25: Photomicrograph of the W/O/W multiple emulsion.

An O/W/O multiple emulsion was made by first preparing a nanoemulsion using INUTEC® SP1. The nanoemulsion was then emulsified into an oil solution of PHS–PEO–PHS using a low speed stirrer. The O/W/O multiple emulsion was stored at room temperature and 50 °C and photomicrographs taken at various time intervals. The O/W/O multiple emulsion was stable for several months both at room temperature and 50 °C. A photomicrograph of the O/W/O multiple emulsion is shown in Fig. 11.26.

A schematic representation of a W/O/W multiple emulsion drop is shown in Fig. 11.27.

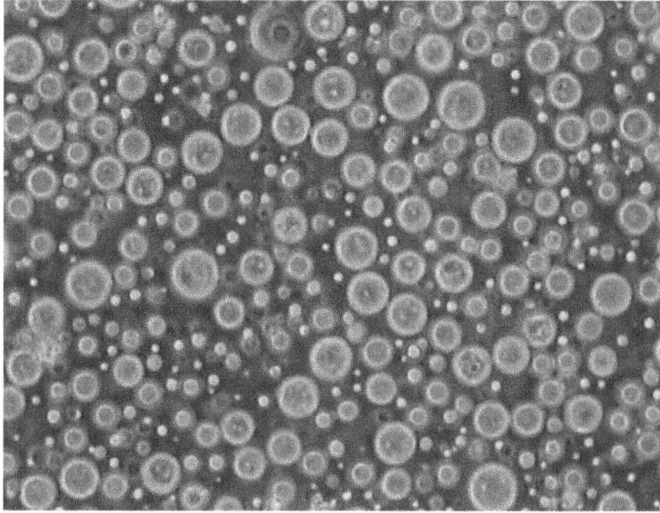

Fig. 11.26: Photomicrograph of the O/W/O multiple emulsion.

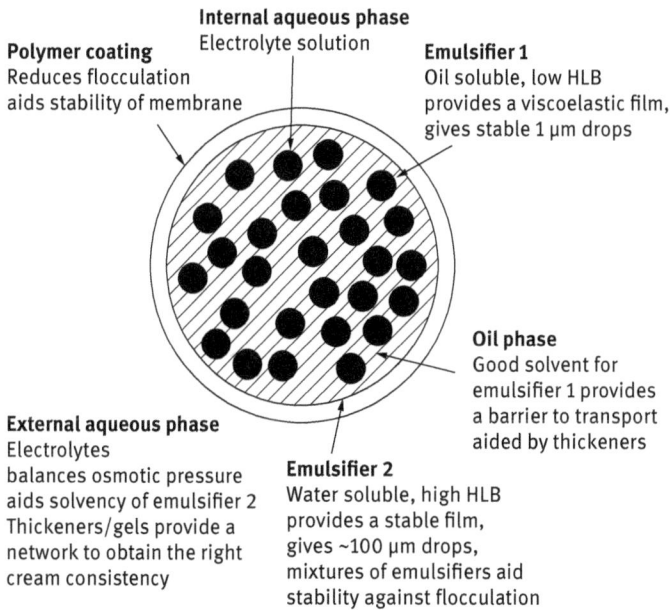

Internal aqueous phase
Electrolyte solution

Polymer coating
Reduces flocculation
aids stability of membrane

Emulsifier 1
Oil soluble, low HLB
provides a viscoelastic film,
gives stable 1 µm drops

Oil phase
Good solvent for
emulsifier 1 provides
a barrier to transport
aided by thickeners

External aqueous phase
Electrolytes
balances osmotic pressure
aids solvency of emulsifier 2
Thickeners/gels provide a
network to obtain the right
cream consistency

Emulsifier 2
Water soluble, high HLB
provides a stable film,
gives ~100 µm drops,
mixtures of emulsifiers aid
stability against flocculation

Fig. 11.27: Schematic picture of a multiple emulsion drop.

References

[1] Tadros, Th. F., "Emulsions", De Gruyter, Germany (2016).

[2] Tadros, Th. F., "Formulation of Cosmetics and Personal Care", De Gruyter, Germany (2016).

[3] Tadros, Th. F., Vandamme, A, Levecke, B., Booten, K. and Stevens, C. V., Advances Colloid Interface Sci., 108–109, 207 (2004).

[4] Hildebrand, J. H., "Solubility of Non-Electrolytes", 2nd edition, Reinhold, New York (1936).

[5] Hansen, C. M., J. Paint Technol., 39, 104–117, 505–514 (1967).

[6] Hansen, C. M. and Beerbower, A., in "Handbook of Solubility Parameters and other Cohesion Parameters", A. F. M. Barton (ed.), CRC Press, Boca Raton, FL (1983).

12 Polymeric surfactants in paints and coatings

12.1 Introduction

Polymeric surfactants play major roles in the formulation of paints and coatings of which the following are worth mentioning: use of block and graft copolymers in emulsion and suspension polymerization for the preparation of polymer colloids that are used in paints and coatings; use in nonaqueous dispersion polymerization (NAD), for dispersions used in nonaqueous paints; stabilizing preformed polymer colloids and pigments that are used in coatings, etc. [1]. The main advantages of using A–B, A–B–A block and BA_n graft (where B is the "anchor" chain and A is the stabilizing chain) stems from the strong adsorption of the polymeric surfactant at the interface, where B is chosen to be highly insoluble in the medium and strongly attached to the surface of the particle or droplet by multipoint attachment with several B chains. The A chain is chosen to be highly soluble in the medium and strongly solvated by its molecules, thus providing effective steric stabilization [1, 2]. This strong repulsion arises from the unfavourable mixing of the A chain (when these are in good solvent conditions) and loss of configurational entropy on considerable overlap of the A chains [3].

In the following I will give some examples illustrating the use of polymeric surfactants for stabilizing dispersions used in paints and coatings.

12.2 Use of polymeric surfactants in emulsion polymerization

This subject has been discussed in detail in Chapter 9 and only a summary is given here. Block and graft copolymers (polymeric surfactants) are expected to be better stabilizers when compared to simple surfactants. The use of these polymeric surfactants in emulsion polymerization and the stabilization of the resulting polymer particles is crucial. Most aqueous emulsion and dispersion polymerization reported in the literature is based on a few commercial products with a broad molecular weight distribution and varying block composition. The results obtained from these studies could not establish what effect the structural features of the block copolymer has on their stabilizing ability and effectiveness in polymerization. Fortunately, model block copolymers with well-defined structures could be synthesized and their role in emulsion polymerization has been investigated using model polymers and model latexes.

A series of well-defined A–B block copolymers of polystyrene-block-polyethylene oxide (PS–PEO) were synthesized [1, 4] and used for emulsion polymerization of styrene. These molecules are "ideal" since the polystyrene block is compatible with the polystyrene formed and thus it forms the best anchor chain. The PEO chain (the stabilizing chain) is strongly hydrated with water molecules and it extends into the aqueous phase forming the steric layer necessary for stabilization (see Chapter 5).

DOI 10.1515/9783110487282-013

However, the PEO chain can become dehydrated at high temperature (due to the breaking of hydrogen bonds), thus reducing the effective steric stabilization. Thus, the emulsion polymerization should be carried out at temperatures well below the theta (θ)-temperature of PEO (see Chapter 5).

Five block copolymers were synthesized [1, 4] with various molecular weights of the PS and PEO blocks. The molecular weight of the polystyrene block and the resulting PS–PEO polymer was determined using gel permeation chromatography. The mole percent of ethylene oxide and the percent of PEO in the block were determined using H^1 NMR spectroscopy. The molecular weight of the blocks varied from M_n = 1000–7000 for PS and M_w = 3000–9000 for PEO. These five block copolymers were used for emulsion polymerization of styrene at 50 °C (well below the θ-temperature of PEO). The results indicated that for efficient anchoring the PS block need not be more than 10 monomer units. The PEO block should have a $M_w \geq 3000$. However, the ratio of the two blocks is very important; for example if the wt % of PEO is ≤ 3000, the molecule becomes insoluble in water (not sufficiently hydrophilic) and no polymerization could occur when using this block copolymer. In addition, the 50 % PEO block could produce a latex but this was unstable and coagulated at 35 % conversion. It became clear from these studies that the percent PEO in the block copolymer plays an important role and this should exceed 75 %. However, the overall molecular weight of the block copolymer is also very important. For example, if one uses a PS block with M_n = 7000, the PEO molecular weight has to be 21 000, which is too high and may result in bridging flocculation unless one prepares a very dilute latex.

Another systematic study of the effect of block copolymer on emulsion polymerization was carried out using blocks of poly(methylmethacrylate)-block-polyethylene oxide (PMMA–PEO) for the preparation of PMMA latexes. The ratio and molecular weight of PMMA to PEO in the block copolymer was varied. Ten different PMMA–PEO blocks were synthesized [4] with M_n for PMMA varying between 400 and 2500. The M_w of PEO was varied between 750 and 5000. The recipe for MMA polymerization consisted of 100 monomer, 800 g water, 20 g PMMA–PEO block copolymer and 0.5 g potassium persulphate. Polymerization was carried out at 45 °C, which is well below the θ-temperature of PEO. The rate of polymerization, R_p, was calculated by using latex samples drawn from the reaction mixture at various time intervals (the amount of latex was determined gravimetrically). The particle size of each latex sample was determined by dynamic light scattering (Photon Correlation Spectroscopy, PCS). The number of particles, N, in each case was calculated from the weight of the latex and the z-average diameter. The results obtained were used to study the effect of the anchoring group PMMA, molecular weight, the effect of PEO molecular weight and the effect of the total molecular weight of the block copolymer.

The results of the systematic study of varying the PMMA and PEO block molecular weight, the percent PEO in the chain as well as the overall molecular weight clearly show the effect of these factors on the resulting latex. For example, when using a block copolymer with 400 molecular weight of PMMA and 750 molecular weight of PEO (i.e.

containing 65 wt % PEO) the resulting latex had fewer particles when compared with the other surfactants. The most dramatic effect was obtained when the PMMA molecular weight was increased to 900 while keeping the PEO molecular weight (750) the same. This block copolymer contains only 46 wt % PEO and it became insoluble in water due to the lack of hydrophilicity. The latex produced was unstable and it collapsed at the early stage of polymerization. The PEO molecular weight of 750 is insufficient to provide effective steric stabilization (see Chapter 5). By increasing the molecular weight of PEO to 2000 or 5000 while keeping the PMMA molecular weight at 400 or 800, a stable latex was produced with a small particle diameter and large number of particles. The best results were obtained by keeping the molecular weight of PMMA at 800 and that of PEO at 2000. This block copolymer gave the highest conversion rate, the smallest particle diameter and the largest number of particles. It is interesting to note that by increasing the PEO molecular weight to 5000 while keeping the PMMA molecular weight at 800, the rate of conversion decreased, the average diameter increased and the number of particles decreased when compared with the results obtained using 2000 molecular weight for PEO. It seems that when the PEO molecular weight is increased, the hydrophilicity of the molecule increased (86 wt % PEO) and this reduced the efficiency of the copolymer. It seems that by increasing hydrophilicity of the block copolymer and its overall molecular weight the rate of adsorption of the polymer to the latex particles and its overall adsorption strength may have decreased. The effect of the overall molecular weight of the block copolymer and its overall hydrophilicity have a great influence on latex production. Increasing the overall molecular weight of the block copolymer above 6200 resulted in a reduction in the rate of conversion, an increase in the particle diameter and a reduction in the number of latex particles. The worst results were obtained with an overall molecular weight of 7500 while reducing the PEO wt % in which case particles with 322 nm diameter were obtained and the number of latex particles was significantly reduced.

The importance of the affinity of the anchor chain (PMMA) to the latex particles was investigated by using different monomers [1, 4]. For example, when using styrene as the monomer the resulting latex was unstable and it showed the presence of coagulum. This can be attributed to the lack of chemical compatibility of the anchor chain (PMMA) and the polymer to be stabilized, namely polystyrene. This clearly indicates that block copolymers of PMMA–PEO are not suitable for emulsion polymerization of styrene. However, when using vinyl acetate monomer, where the resulting poly(vinyl acetate) latex should have strong affinity to the PMMA anchor, no latex was produced when the reaction was carried out at 45 °C. It was speculated that the water solubility of the vinyl acetate monomer resulted in the formation of oligomeric chain radicals which could exist in solution without nucleation. Polymerization at 60 °C, which did nucleate particles, was found to be controlled by chain transfer of the vinyl acetate radical with the surfactant, resulting in broad molecular weight distributions

Emulsion polymerization of MMA using triblock copolymers was carried using PMMA-block-PEO-PMMA using blocks with the same PMMA molecular weight (800 or 900) while varying the PEO molecular weight from 3400 to 14 000 in order to vary the loop size. Although the rate of polymerization was not affected by the loop size, the particles with the smallest diameter were obtained with the 10 000 molecular weight PEO. Comparing the results obtained using the triblock copolymer with those obtained using diblock copolymer (while keeping the PMMA block molecular weight the same) showed the same rate of polymerization. However, the average particle diameter was smaller and the total number of particles larger when using the diblock copolymer. This clearly shows the higher efficacy of the diblock copolymer when compared with the triblock copolymer [4].

The first systematic study of the effect of graft copolymers on emulsion polymerization was carried out by using well characterized graft copolymers with different backbone and side chain lengths [4]. Several grafts of poly(p-methyl styrene)-graft-polyethylene oxide, (PMSt)–(PEO)$_n$, were synthesized and used in styrene emulsion polymerization. Three different PMSt chain lengths (with molecular weight of 750, 2000 and 5000) and three different PEO chain lengths were prepared. In this way the structure of the amphipathic graft copolymer could be changed in three different ways:

(i)   three different PEO graft chain lengths;
(ii) three different backbone chain lengths with the same weight % PEO;
(iii) four different wt % PEO grafts.

Emulsion polymerization of styrene or methylmethacrylate was carried out using hydrophobically modified inulin (INUTEC® SP1) [1] and this showed an optimum weight ratio of (INUTEC®)/monomer of 0.0033 for PS and 0.001 for PMMA particles. The (initiator)/(monomer) ratio was kept constant at 0.00125. The monomer conversion was higher than 85 % in all cases. Latex dispersions of PS reaching 50 % and of PMMA reaching 40 % could be obtained using such a low concentration of INUTEC® SP1. The stability of the latexes was determined by determining the critical coagulation concentration (ccc) using CaCl$_2$. The ccc was low (0.0175–0.05 mol dm^{-3}), but this was higher than that for the latex prepared without surfactant. Post addition of INUTEC® SP1 resulted in a large increase in the ccc.

As with the emulsions, the high stability of the latex when using INUTEC® SP1 is due to the strong adsorption of the polymeric surfactant on the latex particles and formation of strongly hydrated loops and tails of polyfructose that provide effective steric stabilization. Evidence for the strong repulsion produced when using INUTEC® SP1 was obtained from atomic force microscopy investigations [1] in which the force between hydrophobic glass spheres and hydrophobic glass plate, both containing an adsorbed layer of INUTEC® SP1, was measured as a function of distance of separation both in water and in the presence of various Na$_2$SO$_4$ concentrations.

12.3 Use of polymeric surfactants for stabilizing preformed latex particles

For this purpose, polystyrene (PS) latexes were prepared using surfactant-free emulsion polymerization [1]. Two latexes with z-average diameter of 427 and 867 (as measured using Photon Correlation Spectroscopy, PCS) that are reasonably monodisperse were prepared. Two polymeric surfactants, namely Hypermer CG-6 and Atlox 4913 (UNIQEMA, UK) were used. Both are of the graft ("comb") type consisting of polymethylmethacrylate/polymethacrylic acid (PMMA/PMA) backbone with methoxy-capped polyethylene oxide (PEO) side chains (M = 750 Da). Hypermer CG-6 is the same graft copolymer as Atlox 4913 but it contains a higher proportion of methacrylic acid in the backbone. The average molecular weight of the polymer is ≈ 5000 Da. Figure 12.1 shows a typical adsorption isotherm of Atlox 4913 on the two latexes. Similar results were obtained for Hypermer CG-6 but the plateau adsorption was lower ($1.2\,\mathrm{mg\,m^{-2}}$ compared with $1.5\,\mathrm{mg\,m^{-2}}$ for Atlox 4913). It is likely that the backbone of Hypermer CG-6 that contains more PMA is more polar and hence less strongly adsorbed. The amount of adsorption was independent of particle size.

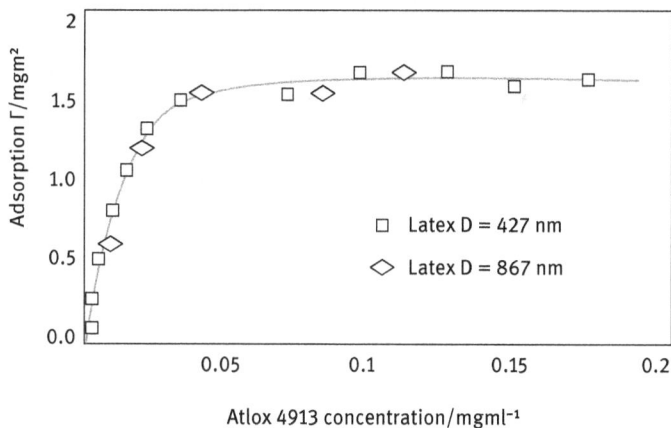

Fig. 12.1: Adsorption isotherms of Atlox 4913 on the two latexes at 25 °C.

The influence of temperature on adsorption is shown in Fig. 12.2. The amount of adsorption increases with increasing temperature. This is due to the poorer solvency of the medium for the PEO chains. The PEO chains become less hydrated at higher temperature and the reduction in solubility of the polymer enhances adsorption.

The adsorbed layer thickness of the graft copolymer on the latexes was determined using rheological measurements. Steady state (shear stress $\delta-\gamma$ $\sigma-\gamma$ shear rate) measurements were carried out and the results were fitted to the Bingham equa-

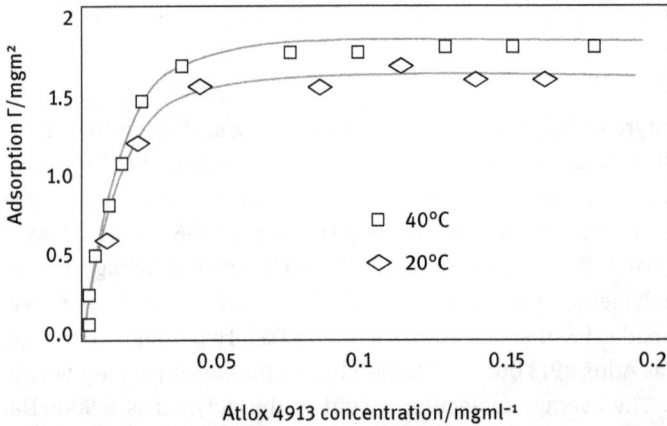

Fig. 12.2: Effect of temperature on adsorption of Atlox 4913 on PS.

tion to obtain the yield value σ_β and the high shear viscosity η of the suspension [5]:

$$\sigma = \sigma_\beta + \eta\dot{\gamma}. \tag{12.1}$$

As an illustration, Fig. 12.3 shows a plot of σ_β versus volume fraction ϕ of the latex for Atlox 4913. Similar results were obtained for latexes stabilized using Hypermer CG-6.

At any given volume fraction, the smaller latex has higher σ_β when compared to the larger latex. This is due to the higher ratio of adsorbed layer thickness to particle radius, Δ/R, for the smaller latex. The effective volume fraction of the latex ϕ_{eff} is

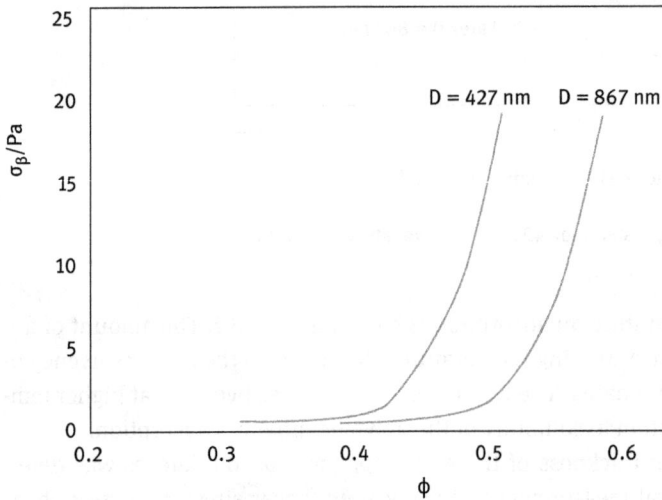

Fig. 12.3: Variation of yield stress with latex volume fraction for Atlox 4913.

related to the core volume fraction ϕ by the equation,

$$\phi_{eff} = \phi \left[1 + \frac{\Delta}{R} \right]^3. \tag{12.2}$$

As discussed before, ϕ_{eff} can be calculated from the relative viscosity η_r using the Dougherty–Krieger equation [6, 7],

$$\eta_r = \left[1 - \left(\frac{\phi_{eff}}{\phi_p} \right) \right]^{-[\eta]\phi_p}, \tag{12.3}$$

where ϕ_p is the maximum packing fraction.

The maximum packing fraction, ϕ_p, can be calculated using the following empirical equation:

$$\frac{(\eta_r^{1/2} - 1)}{\phi} = \left(\frac{1}{\phi_p} \right) (\eta^{1/2} - 1) + 1.25. \tag{12.4}$$

The results showed a gradual decrease of adsorbed layer thickness Δ with increasing volume fraction ϕ. For the latex with diameter D of 867 nm and Atlox 4913, Δ decreased from 17.5 nm at $\phi = 0.36$ to 6.5 nm at $\phi = 0.57$. For Hypermer CG-6 with the same latex, Δ decreased from 11.8 nm at $\phi = 0.49$ to 6.5 nm at $\phi = 0.57$. The reduction of Δ with increasing ϕ may be due to overlap and/or compression of the adsorbed layers as the particles come close to each other at higher volume fraction of the latex.

The stability of the latexes was determined using viscoelastic measurements [5]. For this purpose, dynamic (oscillatory) measurements were used to obtain the storage modulus G^*, the elastic modulus G' and the viscous modulus G'' as a function of strain amplitude γ_0 and frequency ω (rad s^{-1}). The method relies on application of a sinusoidal strain or stress and the resulting stress or strain is measured simultaneously. For a viscoelastic system, the strain and stress sine waves oscillate with the same frequency but out of phase. From the time shift, Δt, and ω one can obtain the phase angle shift, δ.

The ratio of the maximum stress, σ_0, to the maximum strain, γ_0, gives the complex modulus $|G^*|$:

$$|G^*| = \frac{\sigma_0}{\gamma_0}. \tag{12.5}$$

$|G^*|$ can be resolved into two components: storage (elastic) modulus G', the real component of the complex modulus; and loss (viscous) modulus G'', the imaginary component of the complex modulus. The complex modulus can be resolved into G' and G'' using vector analysis and the phase angle shift δ,

$$G' = |G^*| \cos \delta \tag{12.6}$$

$$G'' = |G^*| \sin \delta \tag{12.7}$$

G' is measured as a function of electrolyte concentration and/or temperature to assess the latex stability. As an illustration, Fig. 12.4 shows the variation of G' with temperature for latex stabilized with Atlox 4913 in the absence of any added electrolyte and

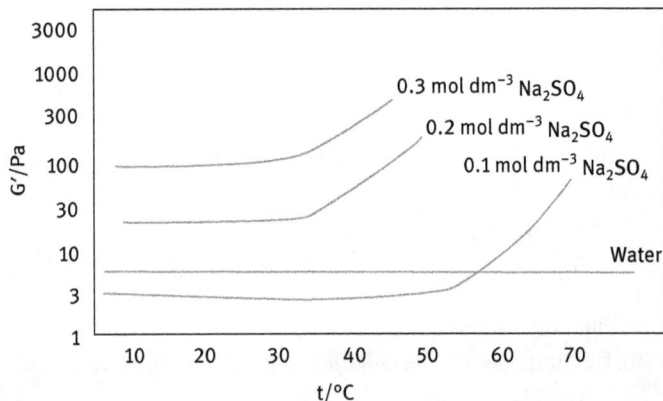

Fig. 12.4: Variation of G' with temperature in water and at various Na$_2$SO$_4$ concentrations.

in the presence of 0.1, 0.2 and 0.3 mol dm^{-3} Na$_2$SO$_4$. In the absence of electrolyte, G' showed no change with temperature up to 65 °C. In the presence of 0.1 mol dm^{-3} Na$_2$SO$_4$, G' remained constant up to 40 °C above which G' increased with a further increase in temperature. This temperature is denoted as the critical flocculation temperature (CFT). The CFT decreases with increasing electrolyte concentration reaching $\approx$ 30 °C in 0.2 and 0.3 mol dm^{-3} Na$_2$SO$_4$. This reduction in CFT with increasing electrolyte concentration is due to the reduction in solvency of the PEO chains with increasing electrolyte concentrations. The latex stabilized with Hypermer CG-6 gave relatively higher CFT values when compared with that stabilized using Atlox 4913.

12.4 Dispersion polymerization

This method is usually applied in the preparation of nonaqueous latex dispersions and hence it is referred to as NAD [8]. The method has also been adapted to prepare aqueous latex dispersions by using an alcohol-water mixture [9, 10].

In the NAD process the monomer, normally an acrylic, is dissolved in a nonaqueous solvent, normally an aliphatic hydrocarbon, and an oil soluble initiator and a stabilizer (to protect the resulting particles from flocculation, sometimes referred to as "protective colloid") are added to the reaction mixture. The most successful stabilizers used in NAD are block and graft copolymers. These block and graft copolymers are assembled in a variety of ways to provide the molecule with an "anchor chain" and a stabilizing chain. The anchor chain should be sufficiently insoluble in the medium and have a strong affinity to the polymer particles produced. In contrast, the stabilizing chain should be soluble in the medium and strongly solvated by its molecules to provide effective steric stabilization (see Chapter 5). The length of the anchor and stabilizing chains has to be carefully adjusted to ensure strong adsorption (by multipoint attachment of the anchor chain to the particle surface) and a sufficiently "thick" layer of the stabilizing chain prevents close approach of the particles to a distance where the van der Waals attraction becomes strong. The criteria for effective steric stabilization by block and graft copolymers are discussed in detail in Chapter 5. Several configurations of block and graft copolymers are possible, as illustrated in Fig. 12.5.

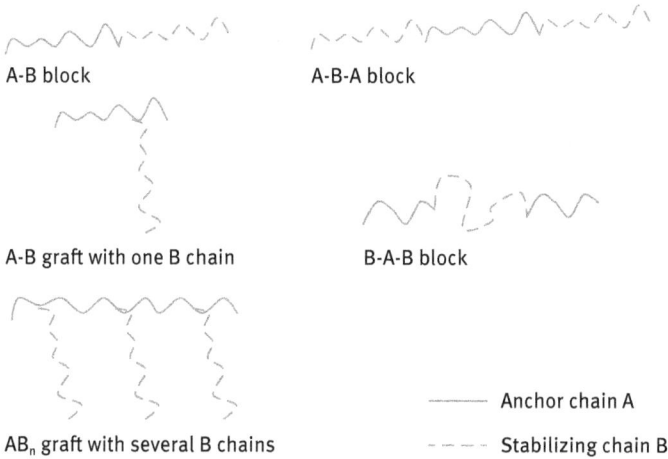

A-B block

A-B-A block

A-B graft with one B chain

B-A-B block

AB$_n$ graft with several B chains

———— Anchor chain A

– – – – Stabilizing chain B

Fig. 12.5: Configurations of block and graft copolymers.

Typical preformed graft stabilizers based on poly(12-hydroxystearic acid) (PHS) are simple to prepare and effective in NAD polymerization. Commercial 12-hydroxystearic acid contains 8–15 % palmitic and stearic acids which limits the molecular weight during polymerization to an average of 1500–2000. This oligomer may be converted to a "macromonomer" by reacting the carboxylic group with glycidyl methacrylate. The macromonomer is then copolymerized with an equal weight of methyl methacrylate (MMA) or similar monomer to give a "comb" graft copolymer with an average molecular weight of 10 000–20 000. The graft copolymer contains on average 5–10 PHS chains pendent from a polymeric anchor backbone of PMMA. This graft copolymer can stabilize latex particles of various monomers. The major limitation of the monomer composition is that the polymer produced should be insoluble in the medium used.

Two main criteria must be considered in the process of dispersion polymerization:

(i)   the insolubility of the formed polymer in the continuous phase;

(ii) the solubility of the monomer and initiator in the continuous phase.

Initially, dispersion polymerization starts as a homogeneous system but after sufficient polymerization, the insolubility of the resulting polymer in the medium forces the chain to precipitate. Initially, polymer nuclei are produced which then grow to polymer particles. These are stabilized against aggregation by the block or graft copolymer that is added to the continuous phase before the process of polymerization starts. It is essential to choose the right block or graft copolymer which should have a strong anchor chain A and good stabilizing chain B as schematically represented in Fig. 12.5.

Dispersion polymerization may be considered a heterogeneous process which may include emulsion, suspension, precipitation and dispersion polymerization. In dispersion and precipitation polymerization, the initiator must be soluble in the continuous phase, whereas in emulsion and suspension polymerization the initiator is chosen to be soluble in the disperse phase of the monomer.

NAD polymerization is carried in two steps:

(i) Seed stage: the diluent, a portion of the monomer, a portion of dispersant and initiator (azo or peroxy type) are heated to form an initial low-concentration fine dispersion.

(ii) Growth stage: the remaining monomer together with more dispersant and initiator are then fed over the course of several hours to complete the growth of the particles.

A small amount of transfer agent is usually added to control the molecular weight. Excellent control of particle size is achieved by proper choice of the designed dispersant and correct distribution of dispersant between the seed and growth stages. NAD acrylic polymers are applied in automotive thermosetting polymers and hydroxy monomers may be included in the monomer blend used.

Two main factors must be considered when considering the long-term stability of a nonaqueous polymer dispersion. The first and very important factor is the nature of the "anchor chain" A. As mentioned above, this should have a strong affinity to the produced latex and in most cases it can be designed to be "chemically" attached to the polymer surface. Once this criterion is satisfied, the second important factor in determining the stability is the solvency of the medium for the stabilizing chain B. The solvency of the medium is characterized by the Flory–Huggins interaction parameter χ. Three main conditions can be identified: $\chi < 0.5$ (good solvent for the stabilizing chain); $\chi > 0.5$ (poor solvent for the stabilizing chain); and $\chi = 0.5$ (referred to as the θ-solvent). Clearly, to maintain stability of the latex dispersion, the solvent must be better than a θ-solvent. The solvency of the medium for the B chain is affected by addition of a nonsolvent and/or temperature changes. It is, therefore, essential to determine the critical volume fraction (CFV) of a nonsolvent above which flocculation (sometimes referred to as incipient flocculation) occurs. One should also determine the critical flocculation temperature at any given solvent composition, below which flocculation occurs. The correlation between CFV or CFT and the flocculation of the nonaqueous polymer dispersion has been demonstrated by Napper [3] who investigated the flocculation of poly(methyl methacrylate) dispersions stabilized by poly(12-hydroxy stearic acid) or poly(n-lauryl methacrylate-co-glycidyl methacrylate) in hexane by adding a nonsolvent such as ethanol or propanol and cooling the dispersion. The dispersions remained stable until the addition of ethanol transformed the medium to a θ-solvent for the stabilizing chains in solution. However, flocculation did occur under conditions of slightly better than θ-solvent for the chains. The same was found for the CFT which was 5–15 K above the θ-temperature. This difference was accounted for by the polydispersity of the polymer chains. The θ-condition is usually determined by cloud point measurements and the least soluble component will precipitate first giving values that are lower than the CFV or higher than the CFT.

The process of dispersion polymerization has been applied in many cases using completely polar solvents such as alcohol or alcohol-water mixtures [9]. The results obtained showed completely different behaviour when compared with dispersion polymerization in nonpolar media. For example, results obtained by Lok and Ober [9] using styrene as monomer and hydroxypropyl cellulose as stabilizer showed a linear increase of particle diameter with increasing weight percent of the monomer. There was no region in monomer concentration where instability occurred (as has been observed for the dispersion polymerization of methyl methacrylate in aliphatic hydrocarbons). Replacing water in the continuous phase with 2-methoxyethanol, Lok and Ober [9] were able to grow large, monodisperse particles up to 15 μm in diameter. They concluded from these results that the polarity of the medium is the controlling factor in the formation of particles and their final size. The authors suggested a mechanism in which the polymeric surfactant molecule grafts to the polystyrene chain, forming a physically anchored stabilizer (nuclei). These nuclei grow to form the polymer particles. Paine [10] carried out dispersion polymerization of styrene by systematically increasing the alcohol chain length from methanol to octadecanol and using hydroxypropyl cellulose as stabilizer. The results showed an increase in particle diameter with increasing number of carbon atoms in the alcohol, reaching a maximum when hexanol was used as the medium, after which there was a sharp decrease in the particle diameter with any further increase in the number of carbon atoms in the alcohol. Paine explained his results in terms of the solubility parameter of the dispersion medium. The largest particles are produced when the solubility parameter of the medium is closest to those of styrene and hydroxypropyl cellulose.

References

[1] Tadros, Th. F., "Colloids in Paints", Wiley-VCH, Germany (2010).
[2] Tadros, Th. F., "Interfacial Phenomena and Colloid Stability", De Gruyter, Germany (2015).
[3] Napper, D. H., "Polymeric Stabilisation of Colloidal Dispersions", Academic Press, London (1983).
[4] Piirma, I., "Polymeric Surfactants", Surfactant Science Series, No. 42, Marcel Dekker, New York, (1992).
[5] Tadros, Th. F., "Rheology of Dispersions", Wiley-VCH, Germany (2010).
[6] Krieger, I. M., and Dougherty, T. J., Trans. Soc. Rheol, 3, 137 (1959).
[7] Krieger, I. M., Advances Colloid and Interface Sci., 3, 111 (1972).

[8]   Barrett, K. E. J. (ed.), "Dispersion Polymerization in Organic Media", John Wiley & Sons Ltd, Chichester (1975).

[9]   Lok, K. P. and Ober, C. K., Can. J. Chem., 63, 209 (1985).

[10]   Paine, A. J., J. Polymer Sci., Part A, 28, 2485 (1990).

13 Polymeric surfactants in agrochemicals

13.1 Introduction

Polymeric surfactants play an important role in the formulation of many agrochemicals of which suspension concentrates (SCs), emulsion concentrates (EWs), suspoemulsions (mixtures of suspensions and emulsions) and oil-based formulations are worth mentioning [1–3]. With SCs, one needs to prepare a highly concentrated suspension (with volume fraction $\phi > 0.4$) and this requires the use of block and graft copolymers [1–3]. With EWs, polymeric surfactants are essential to prevent flocculation, Ostwald ripening and coalescence [1–3]. With suspoemulsions polymeric surfactants are essential to prevent homo- and heteroflocculation as well as coalescence of the emulsion [1–3]. Oil-based formulations require the use of polymeric surfactants that provide effective steric repulsion, since in this case electrostatic repulsion is ineffective due to the low dielectric constant of the oil [1–3]. A summary of the applications of polymeric surfactants in the above mentioned formulations is given below.

13.2 Polymeric surfactants in suspension concentrates (SCs)

The formulation of agrochemicals as dispersions of solids in aqueous solution (to be referred to as suspension concentrates or SCs) has attracted considerable attention [1–3]. Such formulations are a natural replacement for wettable powders (WPs). Although wettable powders are simple to formulate they are not the most convenient for the farmer. Apart from being dusty (and occupying a large volume due to their low bulk density), they tend to settle fast in the spray tank and they do not provide optimum biological efficiency as a result of the large particle size of the system. In addition, one cannot incorporate the necessary adjuvants (mostly surfactants) in the formulation. These problems are overcome by formulating the agrochemical as an aqueous SC.

Several advantages may be quoted for SCs. Firstly, one may control the particle size by controlling the milling conditions and proper choice of the dispersing agent. Secondly, it is possible to incorporate high concentrations of surfactants in the formulation which is sometimes essential for enhancing wetting, spreading and penetration. Stickers may also be added to enhance adhesion and in some cases to provide slow release.

In recent years there has been considerable research into the factors that govern the stability of suspension concentrates [4]. The theories of colloid stability could be applied to predict the physical states of these systems on storage.

Suspension concentrates are usually formulated using a wet milling process which requires the addition of a surfactant/dispersing agent [4]. This agent should satisfy the following criteria:

DOI 10.1515/9783110487282-014

(i) It is a good wetting agent for the agrochemical powder (both external and internal surfaces of the powder aggregates or agglomerates must be spontaneously wetted).
(ii) It is a good dispersing agent to break such aggregates or agglomerates into smaller units and subsequently help in the milling process (one usually aims at a dispersion with a volume mean diameter of 1–2 μm).
(iii) It should provide good stability in the colloid sense (this is essential for maintaining the particles as individual units once formed).

Powerful dispersing agents, usually polymeric surfactants, are particularly important for the preparation of highly concentrated suspensions (sometimes require for seed dressing). Any flocculation will cause a rapid increase in the viscosity of the suspension and this makes the wet milling of the agrochemical a difficult process.

Dry powders of organic compounds usually consist of particles of various degrees of complexity, depending on the isolation stages and the drying process. Generally, the particles in a dry powder form aggregates (in which the particles are joined together with their crystal faces) or agglomerates (in which the particles touch at edges or corners) forming a looser, more open structure. It is essential in the dispersion process to wet the external as well as the internal surfaces and displace the air entrapped between the particles. This is usually achieved by the use of surface active agents of the ionic or nonionic type. In some cases, polymeric surfactants or polyelectrolytes may be efficient in this wetting process. This may be the case since these polymeric surfactants contain a very wide distribution of molecular weights and the low molecular weight fractions may act as efficient wetting agents. For efficient wetting, the molecules should lower the surface tension of water and they should diffuse fast in solution and become quickly adsorbed at the solid/solution interface.

Let us consider an agrochemical powder with surface area A. Before the powder is dispersed in the liquid it has a surface tension γ_{SV} and after immersion in the liquid it has a surface tension γ_{SL}. The work of dispersion, W_d, is simply given by the difference in adhesion or wetting tension of SL and SV [4],

$$W_d = A(\gamma_{SL} - \gamma_{SV}) = -A\gamma_{LV} \cos \theta, \tag{13.1}$$

where γ_{LV} is the liquid surface tension and θ is the equilibrium contact angle at the wetting line. It is clear from equation (13.1) that if $\theta < 90°$, $\cos \theta$ is positive and W_d is negative, i.e. wetting of the powder is spontaneous. Since surfactants are added in sufficient amounts ($\gamma_{dynamic}$ is lowered sufficiently), spontaneous dispersion is the rule rather than the exception.

Wetting of the internal surface requires penetration of the liquid into the channels between and inside the agglomerates. The process is similar to forcing a liquid through fine capillaries as discussed in detail in Chapter 3. To enhance the rate of penetration, γ_{LV} has to be made as high as possible, θ as low as possible and η as low as possible. For dispersion of powders into liquids, one should use surfactants that lower θ

while not reducing γ_{LV} too much. The viscosity of the liquid should also be kept at a minimum. Thickening agents (such as polymers) should not be added during the dispersion process. It is also necessary to avoid foam formation during the dispersion process.

For the dispersion of aggregates and agglomerates into smaller units one requires high speed mixing, e.g. a Silverson mixer. In some cases the dispersion process is easy and the capillary pressure may be sufficient to break up the aggregates and agglomerates into primary units. The process is aided by the surfactant which becomes adsorbed on the particle surface. However, one should be careful during the mixing process not to entrap air (foam) which causes an increase in the viscosity of the suspension and prevents easy dispersion and subsequent grinding. If foam formation becomes a problem, one should add anti-foaming agents such as polysiloxane anti-foaming agents [4].

After completion of the dispersion process, the suspension is transferred to a ball or bead mill for size reduction. Milling or comminution (the generic term for size reduction) is a complex process and there is little fundamental information on its mechanism. For the breakdown of single crystals into smaller units, mechanical energy is required [4]. In a bead mill, for example, this energy is supplied by impaction of the glass beads with the particles. As a result, permanent deformation of the crystals and crack initiation occur. This will eventually lead to the fracture of the crystals into smaller units. However, since the milling conditions are random, it is inevitable that some particles receive impacts that are far in excess of those required for fracture, whereas others receive impacts that are insufficient to fracture them. This makes the milling operation grossly inefficient and only a small fraction of the applied energy is actually used in comminution. The rest of the energy is dissipated as heat, vibration, sound, interparticulate friction, friction between the particles and beads, and elastic deformation of unfractured particles. For these reasons, milling conditions are usually established by a trial and error procedure. Of particular importance is the effect of various surface active agents and macromolecules on the grinding efficiency. As a result of adsorption of surfactants at the solid/liquid interface, the surface energy at the boundary is reduced and this facilitates the process of deformation or destruction. The adsorption of the surfactant at the solid/solution interface in cracks facilitates their propagation. The surface energy manifests itself in destructive processes on solids, since the generation and growth of cracks and separation of one part of a body from another is directly connected with the development of new free surface. Thus, as a result of adsorption of surface active agents at structural defects on the surface of the crystals, fine grinding is facilitated. In the extreme case where there is a very great reduction in surface energy at the sold/liquid boundary, spontaneous dispersion may take place with the result of the formation of colloidal particles ($< 1\,\mu m$).

When considering the stability of agrochemical suspension concentrates one must distinguish between the colloid stability and the overall physical stability. Colloid stability implies absence of an aggregation between the particles which requires the presence of an energy barrier that is produced by electrostatic [5, 6], steric repulsion [7, 8] or a combination of the two (electrosteric). This is achieved by the use of powerful dispersing agents, e.g. surfactants of the ionic or nonionic type, nonionic polymers or polyelectrolytes. These dispersing agents must be strongly adsorbed onto the particle surfaces and fully cover them. With ionic surfactants, irreversible flocculation is prevented by the repulsive force generated from the presence of an electrical double layer at the particle solution interface. Depending on the conditions, this repulsive force can be made sufficiently large to overcome the ubiquitous van der Waals attraction between the particles, at intermediate distances of separation. With nonionic surfactants and polymeric surfactants, repulsion between the particles is

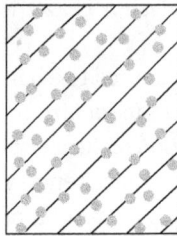

(a) Stable colloidal suspension

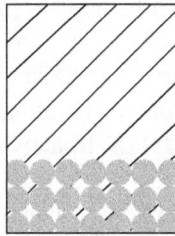

(b) Stable coarse sisoension (uniform size)

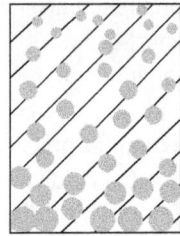

(c) Stable coarse sisoension (size distribution)

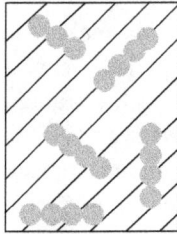

(d) Coagulated suspension (chain aggregates)

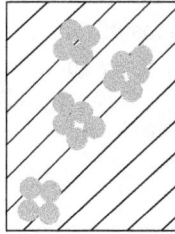

(e) Coagulated suspension (compact clusters)

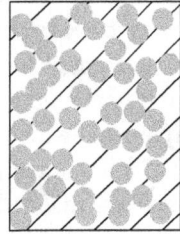

(f) Coagulated suspension (open structure)

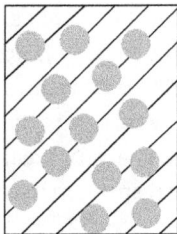

(g) Weakly flocculated structure

(h) Bridging flocculation

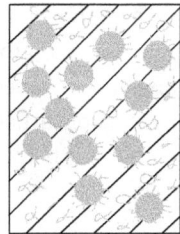

(i) Depletion flocculation

Fig. 13.1: States of the suspension.

ensured by the steric interaction of the adsorbed layers on the particle surfaces (see Chapter 5). With polyelectrolytes, both electrostatic and steric repulsion exist. Physical stability implies absence of sedimentation and/or separation, ease of dispersion on shaking and/or dilution in the spray tanks. To achieve overall physical stability, one may apply controlled and reversible flocculation methods and/or use a rheology modifier.

To distinguish between colloid stability/instability and physical stability one must consider the state of the suspension on standing as schematically illustrated in Fig. 13.1. These states are determined by:

(i)   magnitude and balance of the various interaction forces, electrostatic repulsion, steric repulsion and van der Waals attraction;
(ii) particle size and shape distribution;
(iii) density difference between disperse phase and medium which determines the sedimentation characteristics;
(iv) conditions and prehistory of the suspension, e.g. agitation which determines the structure of the flocs formed (chain aggregates, compact clusters, etc.);
(v)   presence of additives, e.g. high molecular weight polymers that may cause bridging or depletion flocculation.

These states may be described in terms of three different energy-distance curves:

(a) Electrostatic, produced for example by the presence of ionogenic groups on the surface of the particles, or adsorption of ionic surfactants.
(b) Steric, produced for example by adsorption of nonionic surfactants or polymers.
(c) Electrostatic and steric (electrosteric) as for example produced by polyelectrolytes. These are illustrated below in Fig. 13.2.

States (a)–(c) in Fig. 13.1 correspond to a suspension that is stable in the colloid sense. The stability is obtained as a result of net repulsion due to the presence of extended double layers (i.e. at low electrolyte concentration), the result of steric repulsion pro-

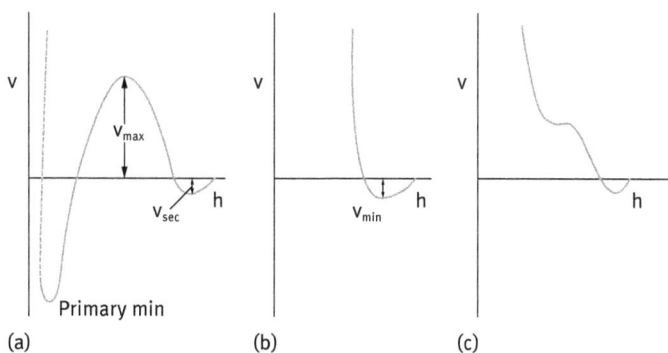

Fig. 13.2: Energy-distance curves for three stabilization mechanisms.

ducing adsorption of nonionic surfactants or polymers, or the result of a combination of double layer and steric repulsion (electrosteric). State (a) represents the case of a suspension with small particle size (submicron) where Brownian diffusion overcomes the gravity force producing uniform distribution of the particles in the suspension, i.e.

$$kT > (4/3)\pi R^3 \Delta\rho gh, \tag{13.2}$$

where k is the Boltzmann constant, T is the absolute temperature, R is the particle radius, $\Delta\rho$ is the buoyancy (difference in density between the particles and the medium), g is the acceleration due to gravity and h is the height of the container.

A good example of the above case is a nanosuspension with particle size well below 1 μm that is stabilized by an ionic surfactant or nonionic surfactant or polymer. This suspension will show no separation on storage for long periods of time [9].

States (b) and (c) in Fig. 13.1 represent the case of suspensions where the particle size range is outside the colloid range (> 1 μm). In this case, gravity force exceeds Brownian diffusion. With state (b), the particles are uniform and they will settle under gravity forming a hard sediment (technically referred to as "clay" or "cake". The repulsive forces between the particles allow them to move past each other until they reach small distances of separation (that are determined by the location of the repulsive barrier). Due to the small distances between the particles in the sediment, it is very difficult to redisperse the suspension by simple shaking.

With case (c) consisting of a wide distribution of particle sizes, the sediment may contain larger proportions of the larger size particles, but still a hard "clay" is produced. These "clays" are dilatant (i.e. shear thickening) and they can be easily detected by inserting a glass rod in the suspension. Penetration of the glass rod into these hard sediments is very difficult.

States (d)–(f) in Fig. 13.1 represent the case for coagulated suspensions which either have a small repulsive energy barrier or where it is completely absent. State (d) represents the case of coagulation under no stirring conditions. In this case chain aggregates are produced that will settle under gravity forming a relatively open structure. State (e) represents the case of coagulation under stirring conditions whereby compact aggregates are produced that will settle faster than the chain aggregates and the sediment produced is more compact. State (f) represents the case of coagulation at high volume fraction of the particles, ϕ. In this case whole particles will form a "one-floc" structure that is formed from chains and cross chains that extend from one wall in the container to the other. Such a coagulated structure may undergo some compression (consolidation) under gravity leaving a clear supernatant liquid layer at the top of the container. This phenomenon is referred to as syneresis.

State (g) in Fig. 13.1 represents the case of weak and reversible flocculation. This occurs when the secondary minimum in the energy-distance curve (Fig. 13.2(a)) is deep enough to cause flocculation. This can occur at moderate electrolyte concentrations, in particular with larger particles. The same occurs with sterically and electrosterically stabilized suspensions (Fig. 13.2(b) and (c)). This occurs when the adsorbed layer thickness is not very large, particularly with large particles. The minimum depth required for causing weak flocculation depends on the volume fraction of the suspension. The higher the volume fraction, the lower the minimum depth required for weak flocculation.

The above flocculation is weak and reversible, i.e. on shaking the container re-dispersion of the suspension occurs. On standing, the dispersed particles aggregate to form a weak "gel". This process (referred to as sol-gel transformation) leads to reversible time dependency of viscosity (thixotropy). On shearing the suspension, viscosity decreases and when the shear is removed, viscosity is recovered. This phenomenon is applied in paints. On application of the paint (by a brush or roller), the gel is fluidized, allowing uniform coating of the paint. When shearing is stopped, the paint film recovers its viscosity and this avoids any dripping.

State (h) in Fig. 13.1 represents the case where the particles are not completely covered by the polymer chains. In this case, simultaneous adsorption of one polymer chain on more than one particle occurs, leading to bridging flocculation. If the polymer adsorption is weak (low adsorption energy per polymer segment), the flocculation could be weak and reversible. In contrast, if the adsorption of the polymer is strong, tough flocs are produced and the flocculation is irreversible. The latter phenomenon is used for solid/liquid separation, e.g. in water and effluent treatment.

Case (i) represents a phenomenon referred to as depletion flocculation, produced by addition of "free" nonadsorbing polymer [10]. In this case, the polymer coils cannot approach the particles to a distance Δ (that is determined by the radius of gyration of free polymer, R_G), since the reduction of entropy on close approach of the polymer coils is not compensated by an adsorption energy. The suspension particles will be surrounded by a depletion zone with thickness Δ. Above a critical volume fraction of the free polymer, ϕ_p^+, the polymer coils are "squeezed out" from between the particles and the depletion zones begin to interact. The interstices between the particles are now free from polymer coils and hence an osmotic pressure is exerted outside the particle surface (the osmotic pressure outside is higher than in between the particles) resulting in weak flocculation [10].

The magnitude of the depletion attraction free energy, G_{dep}, is proportional to the osmotic pressure of the polymer solution, which in turn is determined by ϕ_p and molecular weight M. The range of depletion attraction is proportional to the thickness of the depletion zone, Δ, which is roughly equal to the radius of gyration, R_G, of the free polymer [12].

Ostwald ripening (crystal growth) occurs as a result of the difference in solubility between the small and large crystals [3, 4]. The smaller crystals have a higher solubility than the larger ones. On storage of the suspension, the smaller crystals dissolve and become deposited on the larger ones. This leads to a shift in the particle size distribution to larger values. This results in an enhancement of the sedimentation of the particles, increased flocculation and reduced bio-efficacy of the agrochemical.

Another mechanism for crystal growth is related to polymorphic changes in solutions, and again the driving force is the difference in solubility between the two polymorphs. In other words, the less soluble form grows at the expense of the more soluble phase. This is sometimes also accompanied by changes in the crystal habit. Thus, prevention of crystal growth or at least reducing it to an acceptable level is essential in most suspension concentrates. Many surfactants and polymers may act as crystal growth inhibitors if they adsorb strongly on the crystal faces, thus preventing solute deposition. However, the choice of an inhibitor is still an art and there are not many rules that can be used for selecting crystal growth inhibitors.

One of the most effective polymeric surfactant that is used in SCs is the amphipathic graft copolymer consisting of a polymeric backbone B (polystyrene or polymethyl methacrylate) and several A chains ("teeth") such as polyethylene oxide. This graft copolymer is sometimes referred to as a "comb" stabilizer. This copolymer is usually prepared by grafting a macromonomer such methoxy polyethylene oxide methacrylate with polymethyl methacrylate.

BA_n graft copolymer based on polymethylmethacrylate (PMMA) backbone (with some polymethacrylic acid) on which several PEO chains (with average molecular weight of 750) are grafted (e.g. TERSPERSE® 2500 supplied by Huntsman, Belgium) is shown below.

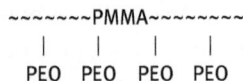

```
~~~~~~~PMMA~~~~~~~~
   |    |    |    |
  PEO  PEO  PEO  PEO
```

Another example is styrene/maleic anhydride copolymer with grafted pendants chains (PEO/PPO) (e.g. TERSPERSE® 2612, Huntsman), schematically shown in Fig. 13.3.

Graft copolymers based on polysaccharides have been developed for the stabilization of disperse systems. One of the most useful graft copolymers are those based on inulin obtained from chicory roots. It is a linear polyfructose chain (with a degree of polymerization > 23) with a glucose end. The latter molecule is used to prepare a series of graft copolymers by random grafting of alkyl chains (using alky isocyanate) on the inulin backbone. The first molecule of this series is INUTEC® SP1 (Beneo-Remy, Belgium) that is obtained by random grafting of C_{12} alkyl chains. It has an average molecular weight of ≈ 5000 Da and its structure is given in Fig. 13.4. The molecule is schematically illustrated in Fig. 13.5, which shows the hydrophilic polyfructose chain (backbone) and the randomly attached alkyl chains.

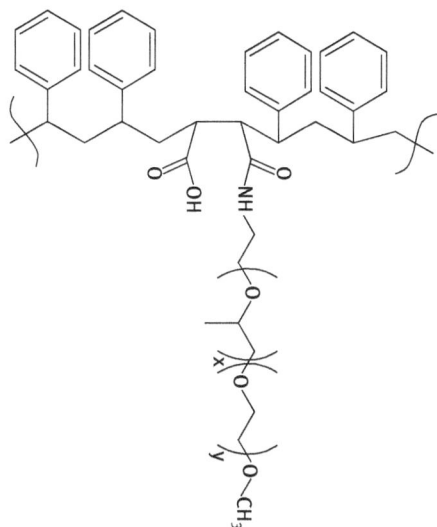

Fig. 13.3: Schematic structure of TERESPERSE® 2612.

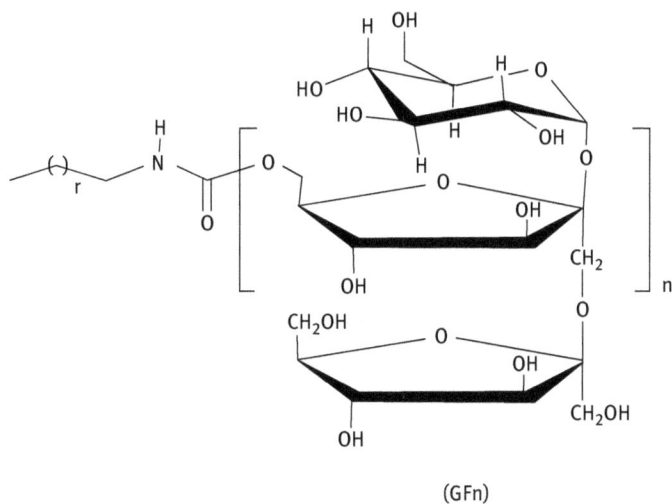

(GFn)

Fig. 13.4: Structure of INUTEC® SP1.

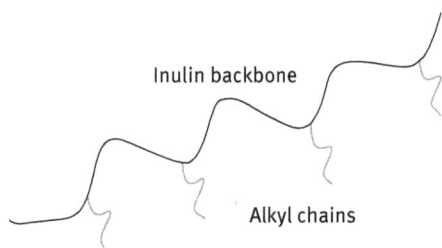

Inulin backbone

Alkyl chains

Fig. 13.5: Schematic representation of INUTEC® SP1 polymeric surfactant.

The main advantages of INUTEC® SP1 as a stabilizer for suspensions are:

(i) Strong adsorption to the particle by multipoint attachment with several alkyl chains. This ensures lack of desorption and displacement of the molecule from the interface.

(ii) Strong hydration of the linear polyfructose chains both in water and in the presence of high electrolyte concentrations and high temperatures. This ensures effective steric stabilization.

13.3 Emulsion concentrates (EWs)

Many agrochemicals are formulated as oil-in-water (O/W) emulsion concentrates (EWs) [3]. These systems offer many advantages over the more traditionally used emulsifiable concentrates (ECs). By using an O/W system one can reduce the amount of oil in the formulation since in most cases a small proportion of oil is added to the agrochemical oil (if this has a high viscosity) before emulsification. In some cases, if the agrochemical oil has a low to medium viscosity, one can emulsify the active ingredient directly into water. With many agrochemicals with low melting point, which are not suitable for the preparation of a suspension concentrate, one can dissolve the active ingredient in a suitable oil and the oil solution is then emulsified into water. Aqueous-based EWs produce fewer hazards to the operator reducing any skin irritation. In addition, in most cases EWs are less phytotoxic to plants when compared with ECs. The O/W emulsion is convenient for incorporating water soluble adjuvants (mostly surfactants). EWs can also be less expensive when compared to ECs, since a lower surfactant concentration is used to produce the emulsion and also one replaces a great proportion of oil by water. The only drawback to EWs when compared to ECs is the need to use high speed stirrers and/or homogenizers to obtain the required droplet size distribution. In addition, EWs require control and maintenance of their physical stability. EWs are only kinetically stable and one has to control the breakdown processes that occur on storage such as creaming or sedimentation, flocculation, Ostwald ripening, coalescence and phase inversion.

Polymeric surfactants are used in EWs to eliminate flocculation, Ostwald ripening and coalescence. This is due to the strong steric repulsion produced by these molecules which eliminates flocculation and coalescence, as well as enhancing the Gibbs dilational elasticity that is essential for reducing Ostwald ripening.

Four main criteria are necessary when using polymeric surfactants for EWs:

(i) Complete coverage of the droplets by the stabilizing chains.

(ii) Firm attachment (strong anchoring) of the chains to the droplets.
  This requires the chains to be insoluble in the medium and soluble in the oil. However, this is incompatible with stabilization which requires a chain that is soluble in the medium and strongly solvated by its molecules. These conflicting requirements are solved by the use of A–B, A–B–A block or BA_n graft copolymers (B is the "anchor" chain and A is the stabilizing chain(s)):

B~~~~~~~~~~~~A   A~~~~~B~~~~~A   ~~~~~~~B~~~~~~~

                                         | | | | |

                                        A   A   A   A   A

Examples for the B chains for O/W emulsions are polystyrene, polymethylmeth-acrylate, polypropylene oxide and alkyl polypropylene oxide. For the A chain(s), polyethylene oxide (PEO) or polyvinyl alcohol are good examples.

(iii) Thick adsorbed layers. The adsorbed layer thickness should be in the region of 5–10 nm. This means that the molecular weight of the stabilizing chains could be in the region of 1000–5000.

(iv) The stabilizing chain should be maintained in good solvent conditions ($\chi < 0.5$) under all conditions of temperature changes on storage.

The driving force for Ostwald ripening is the difference in solubility between the small and large droplets (the smaller droplets have higher Laplace pressure and higher solubility than the larger ones). As a result molecular diffusion of oil molecules occurs from the smaller to the larger droplets.

Several methods can be applied to reduce Ostwald ripening:

(i) Addition of a second disperse phase component which is insoluble in the continuous medium (e.g. squalane). In this case partitioning between different droplet sizes occurs, with the component having low solubility expected to be concentrated in the smaller droplets. During Ostwald ripening in a two component system, equilibrium is established when the difference in chemical potential between different sized droplets (which results from curvature effects) is balanced by the difference in chemical potential resulting from the partitioning of the two components. This effect reduces further growth of droplets.

(ii) Modification of the interfacial film at the O/W Interface. A reduction in γ results in a reduction of the Ostwald ripening rate. By using surfactants that are strongly adsorbed at the O/W interface (i.e. polymeric surfactants) and which do not desorb during ripening (by choosing a molecule that is insoluble in the continuous phase) the rate could be significantly reduced [11]. An increase in the surface dilational modulus ε ($= d\gamma/d \ln A$) and a decrease in γ would be observed for the shrinking drop and this tends to reduce further growth. A–B–A block copolymers such as PHS–PEO–PHS (which is soluble in the oil droplets but insoluble in water) can be used to achieve the above effect. This polymeric emulsifier enhances the Gibbs elasticity and causes reduction of γ to very low values.

Emulsion coalescence occurs as a result of the thinning and disruption of the liquid film between the droplets. When two emulsion droplets come into close contact in a floc or creamed layer or during Brownian diffusion, thinning and disruption of the liquid film may occur resulting in eventual rupture. On close approach of the droplets, film thickness fluctuations may occur. Alternatively, the liquid surfaces undergo some fluctuations forming surface waves. The surface waves may grow in amplitude and the apices may join as a result of the strong van der Waals attraction (at the apex, the film thickness is the smallest). The same applies if the film thins to a small value (critical thickness for coalescence).

A very useful concept was introduced by Deryaguin [12] who suggested that a "disjoining pressure", $\pi(h)$, is produced in the film which balances the excess normal pressure,

$$\pi(h) = P(h) - P_0 \tag{13.3}$$

where $P(h)$ is the pressure of a film with thickness h and P_0 is the pressure of a sufficiently thick film such that the net interaction free energy is zero.

$\pi(h)$ may be equated to the net force (or energy) per unit area acting across the film

$$\pi(h) = -\frac{dG_T}{dh}, \tag{13.4}$$

where G_T is the total interaction energy in the film.

$\pi(h)$ is made up of three contributions due to electrostatic repulsion (π_E), steric repulsion (π_S) and van der Waals attraction (π_A),

$$\pi(h) = \pi_E + \pi_S + \pi_A. \tag{13.5}$$

To produce a stable film, $\pi_E + \pi_S > \pi_A$ and this is the driving force for preventing coalescence which can be achieved by two mechanisms and their combination:
(i) Increased repulsion, both electrostatic and steric.
(ii) Dampening of the fluctuation by enhancing the Gibbs elasticity.

In general, smaller droplets are less susceptible to surface fluctuations and hence coalescence is reduced. This explains the high stability of nanoemulsions.

Several methods may be applied to achieve the above effects:
(i) Use of mixed surfactant films. In many cases using mixed surfactants, say anionic and nonionic or long chain alcohols can reduce coalescence as a result of several effects: high Gibbs elasticity; high surface viscosity; hindered diffusion of surfactant molecules from the film.
(ii) Formation of lamellar liquid crystalline phases at the O/W interface. This mechanism was suggested by Friberg and co-workers [13], who suggested that surfactant or mixed surfactant film can produce several bilayers that "wrap" the droplets. As a result of these multilayer structures, the potential drop is shifted to longer distances thus reducing the van der Waals attraction. For coalescence to occur, these multilayers have to be removed "two-by-two" and this forms an energy barrier preventing coalescence.

13.4 Suspoemulsions

Suspoemulsions are mixtures of suspensions and emulsions that are applied in many agrochemical formulations. In suspoemulsions two active ingredients are formulated with one as an aqueous suspension and the other as an oil/water emulsion [3]. Two main types can be distinguished:
(i) a system in which the solid particles and the emulsion droplets remain as separate entities;
(ii) a system in which the solid particles are dispersed in the oil droplets.

The first system is the one that is commonly applied in agrochemical formulations.

As mentioned above, with suspoemulsions two active ingredients are formulated together which offers convenience to the farmer and also may result in synergism in biological efficacy. A wider spectrum of disease control may be achieved, particularly with many fungicides and herbicides. With many suspoemulsions an adjuvant that enhances the biological efficacy is added.

The formulation of suspoemulsions is not an easy task; one may produce a stable suspension and emulsion separately but when these are mixed they become unstable due to the following interactions:
(i) Homoflocculation of the suspension particles. This can happen if the dispersing agent used for preparing the suspension is not strongly adsorbed and hence it becomes displaced by the emulsifier which is more strongly adsorbed but not a good stabilizer for the suspension particles.
(ii) Emulsion coalescence. This can happen if the emulsifier is not strongly adsorbed at the O/W or W/O interface, resulting in its partial or complete displacement by the suspension dispersant which is not a good emulsion stabilizer. This results in coalescence of the emulsion droplets with ultimate separation of oil (for O/W) or water (for W/O).
(iii) Heteroflocculation between the oil droplets and suspension particles. The latter may be partially wetted by the oil and they may reside at the O/W interface (this is particularly the case if the oil droplets are much larger than the suspension particles). Heteroflocculation can also occur with suspension particles dispersed in a W/O emulsion.
(iv) Phase transfer and crystallization. This happens when the suspension particles have some solubility in the oil phase. The small suspension particles, which have higher solubility than the larger ones (due to curvature effects), may become dissolved in the oil phase and they become recrystallized onto the larger suspension particles (a form of Ostwald ripening process). Large and sometimes needle-shaped crystals may be produced as a result of crystal habit modification (that sometimes occurs with Ostwald ripening).

The use of strongly "anchored" dispersants and emulsifiers, i.e. polymeric surfactants, is crucial for reducing the interaction between the particles and the droplets. The interaction can also be significantly reduced by the addition of rheology modifiers such as hydroxyethyl cellulose (HEC) or xanthan gum. These thickeners produce a "three-dimensional" gel network by the overlap of the polymer coils of HEC or the double helices of xanthan gum. Apart from their effect in reducing creaming and sedimentation by producing a high residual viscosity (at low shear rates), these polymers will also prevent trapping of the oil droplets in the suspension and the suspension particles in the emulsion.

Heteroflocculation results from the competitive adsorption between the dispersant and emulsifier, particularly when these are not strongly anchored to the surfaces. Displacement of some or all of the dispersant by the emulsifier and vice versa may result in attraction between the particles and droplets. The repulsive barrier is weakened in both cases. If the particles are partially wetted by the oil, they may reside at the O/W interface if the oil droplets are sufficiently large.

The above processes of attraction may continue for long periods of time and ultimately the suspoemulsion becomes physically unstable. Any flocculation will result in entrapment of the liquid between the particles in floc structure and this causes a significant increase in the viscosity of the system.

Competitive adsorption may be reduced by using the same surfactant for dispersing the solid and emulsifying the oil. A better method for reducing competitive adsorption is to use a polymeric surfactant that is strongly and irreversibly adsorbed on the suspension particles and emulsion droplets, such as the graft copolymer of polymethylmethacrylate backbone with several polyethylene oxide chains grafted on it. This graft copolymer (Atlox 4913) has a weight average molecular weight of $\approx 20\,000$ and it adsorbs strongly and irreversibly on hydrophobic particles, e.g. polystyrene latex and most agrochemical suspensions. By using the above graft copolymer as dispersant and an A–B–A block copolymer of PEO (A) and polypropylene oxide (PPO) (B) as emulsifier one can obtain very stable suspoemulsions. A good polymeric stabilizer is INUTEC® SP1 (ORAFTI, Belgium), described in Fig. 13.3, that consists of an inulin (linear polyfructose with degree of polymerization >23) chain on which several alkyl chains are grafted. This polymeric surfactant adsorbs on hydrophobic particles and emulsion droplets by multipoint attachment with several alkyl groups leaving strongly hydrated loops and tails of polyfructose that provide an effective steric barrier.

Coalescence of the emulsion droplets on storage accelerates the instability of the suspoemulsion. Large oil droplets can induce heteroflocculation with the suspension particles residing at the O/W interface. Emulsion coalescence can be reduced by one or more of the following methods:
(i)   Reduction of droplet size by using a high pressure homogenizer.
(ii) Use of an effective emulsion stabilizer such as INUTEC® SP1.

(iii) Incorporation of an oil soluble polymeric surfactant such as Atlox 4912 (or Arlacel P135 which can be used in cosmetics). This is an A–B–A block copolymer of
   polyhydroxy stearic acid (PHS, A) and PEO (B).

The criteria for preparing a stable suspoemulsion are:
(i)   Use a strongly adsorbed ("anchored") dispersant by multipoint attachment of a
       block or graft copolymer.
(ii)   Use a polymeric stabilizer for the emulsion (also with multipoint attachment),
       e.g. INUTEC® SP1.
(iii) Prepare the suspension and emulsion separately and allow enough time for com-
       plete adsorption (equilibrium).
(iv)   Use low shear when mixing the suspension and emulsion.
(v)   When dissolving an active in an oil (e.g. with many agrochemicals), choose an
       oil in which the suspension particles are insoluble and also the oil should not
       wet the particles.
(vi)   Increase dispersant and emulsifier concentrations to ensure that the lifetime of
       any bare patches produced during collision is very short.
(vii) Reduce emulsion droplet size by using a high pressure homogenizer. Smaller
       droplets are less deformable and coalescence is prevented. In addition, accu-
       mulation of the suspension particles at the O/W interface is prevented.
(viii) Use a rheology modifier such as HEC or xanthan gum that produces a viscoelastic
       solution that prevents creaming or sedimentation and prevents entrapment of
       the oil droplets in between the suspension particles or the suspension particles
       in between the emulsion droplets.
(ix)   If possible, it is preferable to use a higher volume fraction of the oil when com-
       pared to the suspension. In many cases, flocculation is more rapid at higher solid
       volume fractions. The emulsion oil phase volume can be increased by incorpo-
       rating an inert oil.

When one wishes to prepare a suspoemulsion with a high volume fraction, ϕ, of
suspension and emulsion (e.g. $\phi > 0.4$) it is preferable to emulsify the oil directly
into a prepared suspension. To do so, one prepares the suspension first (e.g. by bead
milling) using the polymeric dispersant. The suspension is left to equilibrate for sufficient time (preferably overnight) to ensure complete adsorption of the polymer. The
polymeric emulsifier is then added and the oil is emulsified into the SC, for example
using a Silverson or Ultra-Turrax. Overmixing, which may result in orthokinetic (or
shear) flocculation and dilatancy, must be avoided and the whole system should be
cooled as much as possible during emulsification.

Preventing crystallization is by far the most serious instability problem with suspoemulsions (particularly with many agrochemicals). It arises from the partial solubility of the suspension particles in the oil droplets. The process is accelerated at higher
temperatures and also on temperature cycling. As mentioned above, smaller particles

will have higher solubility than larger ones. This is due to the fact that the higher the curvature, the higher the solubility. On storage, the smaller particles will dissolve in the oil and they recrystallize on the larger particles which may be at the vicinity of the O/W interface. Some crystal habit modification may be produced and large plates or needles are formed which can reach several µm.

Several procedures may be applied to inhibit recrystallization:

(i)   Diluting the oil phase with another inert oil in which the particles are insoluble.
(ii) Use of a strongly adsorbed polymeric surfactant such as Atlox 4913 or INU-TEC® SP1 that prevents entry of the suspension particles into the oil droplets.
(iii) Addition of electrolytes in the continuous phase. This has the effect of enhancing the polymeric surfactant adsorption thus preventing particle entry into the oil droplets.
(iv) Use of crystal growth inhibitors: e.g. flat dye molecules which are insoluble in the oil and are strongly adsorbed on the particle surface. This prevents particle entry into the oil droplets.
(v)   Use of analogues of the solid active ingredient (having the same basic structure) that are insoluble in the oil and become incorporated on the surface of the solid particles.
(vi) Use of thickeners such as HEC and xanthan gum. This will increase the low shear rate viscosity of the medium and hence slow down the diffusion of the small particles thus preventing their entry into the oil droplets. These thickeners can produce gels in the continuous phase that is viscoelastic and this can prevent particle diffusion.

13.5 Oil-based suspension concentrates

Oil-based suspensions are currently used for the formulation of many agrochemicals, in particular those which are chemically unstable in aqueous media [3]. These suspensions allow one to use oils (such as methyl oleate) which may enhance the biological efficacy of the active ingredient. In addition, one may incorporate water insoluble adjuvants in the formulation. The most important criterion for the oil used is to have minimum solubility of the active ingredient, otherwise Ostwald ripening or crystal growth will occur on storage.

The oil-based suspension has to be diluted into water to produce an oil-in-water emulsion. A self-emulsifiable system has to be produced and this requires the presence of the appropriate surfactants for self-emulsification. The surfactants used for self-emulsification should not interfere with the dispersing agent that is used to stabilize the suspension particles in the nonaqueous media. Displacement of the dispersing agent with the emulsifiers can lead to flocculation of the suspension.

To prevent sedimentation of the particles (since the density of the active ingredient is higher than that of the oil in which it is dispersed), an appropriate rheology modifier (antisettling agent) that is effective in the nonaqueous medium must be incorporated in the suspension. This rheology modifier should not interfere with the self-emulsification process of the oil-based suspension.

Two main types of nonaqueous suspensions may be distinguished:

(i) Suspensions in polar media such as alcohol, glycols, glycerol, esters. These media have a relative permittivity $\varepsilon_r > 10$. In this case double layer repulsion plays an important role, in particular when using ionic dispersing agents.

(ii) Suspensions in nonpolar media, $\varepsilon_r < 10$, such as hydrocarbons (paraffinic or aromatic oils) which can have a relative permittivity as low as 2. In this case charge separation and double layer repulsion are not effective and hence one has to depend on the use of dispersants (polymeric surfactants) that produce steric stabilization.

The most effective dispersants are polymeric surfactants which may be classified into two main categories:

(i) homopolymers and

(ii) block and graft copolymers.

The homopolymers adsorb as random coils with tail-train-loop configurations. In most cases there is no specific interaction between the homopolymer and the particle surface and it seldom can provide effective stabilization. The block and graft copolymers can provide effective stabilization providing they satisfy the following criteria:

(i) Strong adsorption of the dispersant to the particle surface. This can be provided by a block B that is chosen to be insoluble in the nonaqueous medium and has some affinity to the particle surface. In the case when the affinity to the surface is not strong, one can rely on "rejection anchoring" whereby the insoluble B chain is rejected towards the surface as a result of its insolubility in the nonaqueous medium.

(ii) Strongly solvated A blocks that provide effective steric stabilization as a result of their unfavourable mixing and loss of entropy when the particles approach each other in the suspension.

(iii) A reasonably thick adsorbed layer to prevent any strong flocculation.

Based on the above principles various polymeric surfactants have been designed for suspensions in nonaqueous medium. One of the most effective stabilizing chains in nonaqueous media is poly(12-hydroxystearic acid) (PHS) that has a molar mass in the region of 1000–2000 Da. This chain is strongly solvated in most hydrocarbon oils (paraffinic or aromatic oils). It is also strongly solvated in many esters such as methyl oleate that are commonly used for oil-based suspensions. For the B chain one can choose a polar chain such as polyethylene imine or polyvinylpyrrolidone which is insoluble in most oils. The B chain could also be polystyrene or polymethylmethacrylate which is insoluble in aliphatic hydrocarbons and may have some affinity to the hydrophobic agrochemical particle.

References

[1]   Tadros, Th. F., Advances in Colloid and Interface Science, 12, 141 (1980).
[2]   Tadros, Th. F., "Surfactants in Agrochemicals", Marcel Dekker, NY (1994).
[3]   Tadros, Th. F., "Colloids in Agrochemicals", Wiley-VCH, Germany (2009).
[4]   Tadros, Th. F., "Dispersion of Powders in Liquids and Stabilisation of Suspensions", Wiley-VCH, Germany (2012).
[5]   Kruyt, H. R. (ed.), "Colloid Science", Vol. I, Elsevier, Amsterdam (1952).
[6]   Verwey, E. J. W. and Overbeek, J. Th. G., "Theory of Stability of Lyophobic Colloids", Elsevier, Amsterdam (1948).
[7]   Tadros, Th. F., "Polymer Adsorption and Dispersion Stability", in "The Effect of Polymers on Dispersion Properties", Th. F. Tadros (ed.), Academic Press, London (1981).
[8]   Napper, D. H., "Polymeric Stabilisation of Colloidal Dispersions", Academic Press, London (1981).
[9]   Tadros, Th. F., "Nanodispersions", De Gruyter, Germany (2016).
[10] Asakura, S. and Oosawa, F., J. Phys. Chem., 22, 1255 (1954); J. Polym. Sci., 33, 183 (1958).
[11] Walstra, P. in "Encyclopedia of Emulsion Technology", Vol. 4, P. Becher (ed.), Marcel Dekker, NY (1996).
[12] Deryaguin, B. V. and Scherbaker, R. L., Kolloid Zh., 23, 33 (1961).
[13] Friberg, S., Jansson, P. O. and Cederberg, E., J. Colloid Interface Sci., 55, 614 (1976).

14 Polymeric surfactants in the food industry

14.1 Introduction

The use of surfactants in the food industry has been known for many centuries. Naturally occurring surfactants such as lecithin from egg yolk and various proteins from milk are used for the preparation of many food products such as mayonnaise, salad creams, dressings, desserts, etc. Later, polar lipids such as monoglycerides have been introduced as emulsifiers for food products. More recently, synthetic surfactants such as sorbitan esters and their ethoxylates and sucrose esters have been used in food emulsions. For example, esters of monostearate or mono-oleate with organic carboxylic acids, e.g. citric acid are used as anti-spattering agents in margarine for frying [1].

Many foods are colloidal systems, containing particles of various kinds. The particles may remain as individual units suspended in the medium, but in most cases aggregation of these particles takes place forming three-dimensional structures, generally referred to as "gels". These aggregation structures are determined by the interaction forces between the particles that are determined by the relative magnitudes of attractive (van der Waals forces) and repulsive forces. The latter can be electrostatic or steric in nature depending on the composition of the food formulation. It is clear that the repulsive interactions will be determined by the nature of the surfactant present in the formulation. Such surfactants can be ionic or polar in nature, or they may be polymeric in nature. The latter are sometimes added not only to control the interaction between particles or droplets in the food formulation, but also to control the consistency (rheology) of the system [1].

Many food formulations contain mixtures of surfactants (emulsifiers) and hydrocolloids. The interaction between the surfactant and polymer molecule plays a major role in the overall interaction between the particles or droplets, as well as the bulk rheology of the whole system. Such interactions are complex and require fundamental studies of their colloidal properties. As will be discussed later, many food products contain proteins that are used as emulsifiers. The interaction between proteins and hydrocolloids is also very important in determining the interfacial properties and bulk rheology of the system. In addition, the proteins can also interact with the emulsifiers present in the system and this interaction requires particular attention [1].

Most polymeric surfactants used in food are of natural origin, e.g. proteins obtained from milk and other vegetable sources, gums such as gum arabic (a mixture of glycoproteins and polysaccharides) obtained from various plants, etc. These naturally occurring polymeric surfactants are widely used as emulsifiers for various food products as well as beverages. Before describing the stabilizing mechanism of these polymeric surfactants, it is essential to describe their structure and interfacial properties. This is particularly the case with protein molecules. The stabilizing mechanism

DOI 10.1515/9783110487282-015

of these molecules adsorbed at the interface is described in terms of their interfacial rheological properties as will be discussed below.

14.2 Structure of proteins

A protein is a linear chain of amino acids that assumes a three-dimensional shape dictated by the primary sequence of the amino acids in the chain [2]. The side chains of the amino acids play an important role in directing the way in which the protein folds in solution. The hydrophobic (nonpolar) side chains avoid interaction with water, while the hydrophilic (polar) side chains seek such interaction. This results in a folded globular structure with the hydrophobic side chains inside and the hydrophilic side chains outside [2]. The final shape of the protein (helix, planar or "random coil") is a product of many interactions which form a delicate balance [2]. These interactions and structural organizations are briefly discussed below [2].

Three levels of structural organization have been suggested:
(i) Primary structure, referring to the amino acid sequence.
(ii) Secondary structure, denoting the regular arrangement of the polypeptide backbone.
(iii) Tertiary structure is the three-dimensional organization of globular proteins.

A quaternary structure consisting of the arrangement of aggregates of the globular proteins may also be distinguished. The regular arrangement of the protein polypeptide chain in the secondary structure is determined by the structural restrictions. The C–N bonds in the peptide amide groups have a partial double bond character that restricts the free rotation about the C–N bond. This influences the formation of secondary structures. The polypeptide backbone forms a linear group, if successive peptide units assume identical relative orientations. The secondary structures are stabilized by hydrogen bonds between peptide amide and carbonyl groups. In the α-helix, the C=O bond is parallel to the helix axis and a straight hydrogen bond is formed with the N–H group and this is the most stable geometrical arrangement. The interaction of all constituent atoms of the main chain, which are closely packed together, allows the van der Waals attraction to stabilize the helix. This shows that the α-helix is the most abundant secondary structure in proteins. Several other structures may be identified and these are designated as π-helix, β-sheet, etc.

The classification of proteins is based on the secondary structures: α-proteins with α-helix only, e.g. myoglobin; β-proteins mainly with β-sheets, e.g. immunoglobin; α + β proteins with α-helix and β-sheet region that exist apart in the sequence, e.g. lysozome.

The protein structure is stabilized by covalent disulphide bonds and a complexity of noncovalent forces, e.g. electrostatic interactions, hydrogen bonds, hydrophobic interactions and van der Waals forces. Both the average hydrophobicity and the charge frequency (parameter of hydrophobicity) are important in determining the physical properties such as solubility of the protein. The latter can be expressed as the equilibrium between hydrophilic (protein-solvent) and hydrophobic (protein-protein) interactions.

Protein denaturation can be defined as the change in the native conformation (i.e. in the region of secondary, tertiary and quaternary structure) which takes place without change of the primary structure, i.e. without splitting the peptide bonds. Complete denaturation may correspond to totally unfolded protein.

When the protein is formed, the structure produced adopts the conformation with the least energy. This structure is referred to as the native or naturated form of the protein. Modification of the amino acid side chains or their hydrolysis may lead to different conformations. Similarly, addition of molecules that interact with the amino acids may cause conformational changes (denaturation of the protein). Proteins can be denaturated by adsorption at interfaces, as a result of hydrophobic interaction between the internal hydrophobic core and the nonpolar surfaces.

Many examples of proteins used in interfacial adsorption studies may be quoted: Small and medium size globular proteins, e.g. those present in milk such as β-lactoglobulin, α-lactoalbumin and serum albumin, and egg white, e.g. lysozyme and ovalalbumin. At pH values below the isoelectric point (4.2–4.5), these proteins associate to form dimers, trimers and higher aggregates. α-lactoalbumin is stabilized by Ca^{2+} against thermal unfolding. X-ray analysis of lysozyme showed that all charged and polar groups are located at the surface, whereas the hydrophobic groups are buried in the interior. Bovine serum albumin (which represents about 5% of whey proteins in bovine milk) forms a triple domain structure which includes three very similar structural domains, each consisting of two large double loops and one small double loop. Below pH 4, the molecule becomes fully uncoiled within the limits of its disulphide bonds. Ovalbumin, the major component of egg white, is a monomeric phosphoglycoprotein with a molecular weight of 43 kDa. During storage of eggs, even at low temperatures, ovalbumin is modified by SH/SS exchange into a variant with greater heat stability, called s-ovalbumin.

These protein-forming micelles, namely casein, is the major protein fraction in bovine milk (about 80% of the total milk protein). Several components may be identified, namely $\alpha_{s,1}$- and $\alpha_{s,2}$-caseins, β-casein and κ-casein. A protolytic breakdown product of β-casein is γ-casein. Similar to ovalbumin, caseins are phosphoproteins. Large spherical casein micelles are formed by association of α_s-, β- and κ-casein in the presence of free phosphate and calcium ions. The molecules are held together by electrostatic and hydrophobic interactions. The α_s- and β-caseins are surrounded by the flexible hydrophilic κ-casein, which forms the surface layer of the micelle. The high negative charge of the κ-casein prevents collapse of the micelle by electrostatic repulsion. The micelle diameter varies between 50 and 300 nm.

Several oligomeric plant storage proteins can be identified. They are classified according to their sedimentation behaviour in the analytical ultracentrifuge, namely

11 S, 7 S and 2 S proteins. Both 11 S and 7 S proteins are oligomeric globular proteins. The 11 S globulins are composed of 6 noncovalently linked subunits, each of which contains a disulphide bridged pair of a rather hydrophilic acidic 30–40 kDa α-polypeptide chain and a more hydrophilic basic 20 kDa β-polypeptide chain. The molar mass and size of the protein as well as its shape depend on the nature of the plant from which it is extracted. These plant proteins can be used as emulsifying and foaming agents.

14.3 Interfacial properties of proteins at the liquid/liquid interface

Since proteins are used as emulsifying agents for oil-in-water emulsions, it is important to understand their interfacial properties, in particular the structural change that may occur on adsorption. The properties of protein adsorption layers differ significantly from those of simple surfactant molecules. In the first place, surface denaturation of the protein molecule may take place resulting in unfolding of the molecule, at least at low surface pressures. Secondly, the partial molar surface area of proteins is large and can vary depending on the conditions for adsorption. The number of configurations of the protein molecule at the interface exceeds that in bulk solution, resulting in a significant increase in the nonideality of the surface entropy. Thus, one cannot apply thermodynamic analysis, e.g. Langmuir adsorption isotherm, for protein adsorption. The question of reversibility versus irreversibility of protein adsorption at the liquid interface is still subject to a great deal of controversy. For that reason protein adsorption is usually described using statistical mechanical models. Scaling theories could also be applied.

One of the most important investigations of protein surface layers is to measure their interfacial rheological properties (e.g. their viscoelastic behaviour). Several techniques can be applied to study the rheological properties of protein layers, e.g. using constant stress (creep) or stress relaxation measurements. At very low protein concentrations, the interfacial layer exhibits Newtonian behaviour, independent of pH and ionic strength. At higher protein concentrations, the extent of surface coverage increases and the interfacial layers exhibit viscoelastic behaviour revealing features of solid-like phases. Above a critical protein concentration, protein-protein interactions become significant resulting in a "two-dimensional" structure formation. The dynamics of formation of protein layers at the liquid-liquid interface should be considered in detail when one applies the protein molecules as stabilizers for emulsions. Several kinetic processes must be considered: solubilization of nonpolar molecules resulting in the formation of associates in the aqueous phase; diffusion of solutes from bulk solution to the interface; adsorption of the molecules at the interface; orientation of the molecules at the liquid-liquid interface; formation of aggregation structures, etc.

14.4 Proteins as emulsifiers

When a protein is used as an emulsifier, it may adopt various conformations depending on the interaction forces involved. The protein may adopt a folded or unfolded conformation at the oil/water interface. In addition, the protein molecule may interpentrate into the lipid phase to various degrees. Several layers of proteins may also exist. The protein molecule may bridge one drop interface to another. The actual structure of the protein interfacial layer may be complex, combining any or all of the above possibilities. For these reasons, measuring protein conformations at various interfaces still remains a difficult task, even when using several techniques such as UV, IR and NMR spectroscopy as well as circular dichroism [2].

At an oil/water interface, the assumption is usually made that the protein molecule undergoes some unfolding and this accounts for the lowering of the interfacial tension on protein adsorption. As mentioned above, multilayers of protein molecules may be produced and one should take into account the intermolecular interactions as well as the interaction with the lipid (oil) phase.

Proteins act in a similar way to polymeric stabilizers (steric stabilization). However, the molecules with compact structures may precipitate to form small particles which accumulate at the oil/water interface. These particles stabilize the emulsions (sometimes referred to as Pickering emulsions) by a different mechanism. As a result of the partial wetting of the particles by the water and the oil, they remain at the interface. The equilibrium location at the interface provides the stability since their displacement into the dispersed phase (during coalescence) results in an increase in the wetting energy.

From the above discussion, it is clear that proteins act as stabilizers for emulsions by different mechanisms depending on their state at the interface. If the protein molecules unfold and form loops and tails, they provide stabilization in a similar way to synthetic macromolecules. On the other hand, if the protein molecules form globular structures, they may provide a mechanical barrier that prevents coalescence. Finally, precipitated protein particles that are located at the oil/water interface provide stability as a result of the unfavourable increase in the wetting energy on their displacement. It is clear that in all cases, the rheological behaviour of the film plays an important role in the stability of the emulsions (see below).

14.5 Protein-polysaccharide interactions in food colloids

Proteins and polysaccharides are present in nearly all food colloids [2]. The proteins are used as emulsion and foam stabilizers, whereas the polysaccharide acts as a thickener and also for water-holding. Both proteins and polysaccharides contribute to the structural and textural characteristics of many food colloids through their aggregation and gelation behaviour. Several interactions between proteins and polysaccharides may be distinguished, ranging from repulsive to attractive interactions. The repulsive interactions may arise from excluded volume effects and/or electrostatic interaction. These repulsive interactions tend to be weak except at very low ionic strength (expanded double layers) or with anionic polysaccharides at pH values above the isoelectric point of the protein (negatively charged molecules). The attractive interaction can be weak or strong and either specific or nonspecific. A covalent linkage between protein and polysaccharide represents a specific strong interaction. A nonspecific protein-polysaccharide interaction may occur as a result of ionic, dipolar, hydrophobic or hydrogen bonding interaction between groups on the biopolymers. Strong attractive interaction may occur between a positively charged protein (at a pH below its isoelectric point) and an anionic polysaccharide. In any particular system, the protein-polysaccharide interaction may change from repulsive to attractive as the temperature or solvent conditions (e.g. pH and ionic strength) change.

Aqueous solutions of proteins and polysaccharides may exhibit phase separation at finite concentrations. Two types of behaviour may be recognized, namely coacervation and incompatibility. Complex coacervation involves spontaneous separation into solvent-rich and solvent-depleted phases. The latter phase contains the protein-polysaccharide complex which is caused by nonspecific attractive protein-polysaccharide interaction, e.g. opposite charge interaction. Incompatibility is caused by spontaneous separation into two solvent-rich phases, one composed predominantly of protein and the other predominantly of polysaccharide. Depending on the interactions, a gel formed from a mixture of two biopolymers may contain a coupled network, an interpenetrating network or a phase-separated network. In food colloids the two most important proteinaceous gelling systems are gelatin and casein micelles. An example of a covalent protein-polysaccharide interaction is that produced when gelatin reacts with propylene glycol alginate under mildly alkaline conditions. Noncovalent nonspecific interaction occurs in mixed gels of gelatin with sodium alginate or low-methoxy pectin. In food emulsions containing protein and polysaccharide, any of the mentioned interactions may take place in the aqueous phase of the system. This results in specific structures with desirable rheological characteristics and enhanced stability. The nature of the protein-polysaccharide interaction affects the surface behaviour of the biopolymers and the aggregation properties of the dispersed droplets.

Weak protein-polysaccharide interactions may be exemplified by a mixture of milk protein (sodium casinate) and a hydrocolloid such as xanthan gum. Sodium casinate acts as the emulsifier and xanthan gum (with a molecular weight in the region of 2×10^6 Da) is widely used as a thickening agent and a synergistic gelling agent (with locust bean gum). In solution, xanthan gum exhibits pseudoplastic behaviour that is maintained over a wide range of temperature, pH and ionic strength. Xanthan gum at concentrations exceeding 0.1% inhibits creaming of emulsion droplets by producing a gel-like network with a high residual viscosity. At lower xanthan gum concentrations ($< 0.1\%$), creaming is enhanced as a result of depletion flocculation.

Other hydrocolloids such as carboxymethyl cellulose (with a lower molecular weight than xanthan gum) are less effective in reducing creaming of emulsions.

Covalent protein-polysaccharide conjugates are sometimes used to avoid any flocculation and phase separation that is produced with weak nonspecific protein-polysaccharide interactions. An example of such conjugates is that produced with globulin-dextran or bovine serum albumin-dextran. These conjugates produce emulsions with smaller droplets and narrower size distribution and they stabilize the emulsion against creaming and coalescence.

14.6 Polysaccharide-surfactant interactions

One of the most important aspects of polymer-surfactant systems is their ability to control stability and rheology over a wide range of composition [1, 2]. Surfactant molecules that bind to a polymer chain generally do so in clusters that closely resemble the micelles formed in the absence of polymer [1]. If the polymer is less polar or contains hydrophobic regions or sites, there is an intimate contact between the micelles and the polymer chain. In such a situation, the contact between one surfactant micelle and two polymer segments will be favourable. The two segments can be in the same polymer chain or in two different chains, depending on the polymer concentration. For a dilute solution, the two segments can be in the same polymer chain, whereas in more concentrated solutions the two segments can be in two polymer chains with significant chain overlap. The crosslinking of two or more polymer chains can lead to network formation and dramatic rheological effects.

Surfactant-polymer interaction can be treated in different ways, depending on the nature of the polymer. A useful approach is to consider the binding of surfactant to a polymer chain as a co-operative process. The onset of binding is well defined and can be characterized by a critical association concentration (CAC). This decreases with increasing alkyl chain length of the surfactant, which implies an effect of polymer on surfactant micellization [3]. The polymer is considered to stabilize the micelle by short- or long-range (electrostatic) interaction. The main driving force for surfactant self-assembly in polymer-surfactant mixtures is generally the hydrophobic interaction between the alkyl chains of the surfactant molecules. Ionic surfactants often interact significantly with both nonionic and ionic polymers. This can be attributed to the unfavourable contribution to the energetics of micelle formation from the electrostatic effects and their partial elimination due to charge neutralization or lowering of the charge density. For nonionic surfactants, there is little to gain in forming micelles in the presence of a polymer and hence the interaction between nonionic surfactants and polymers is relatively weak. However, if the polymer chain contains hydrophobic segments or groups, e.g. with block copolymers, the hydrophobic polymer-surfactant interaction will be significant.

For hydrophobically modified polymers (such as hydrophobically modified hydroxyethyl cellulose or polyethylene oxide), the interaction between the surfactant micelles and the hydrophobic chains on the polymer can result in the formation of cross links, i.e. gel formation. This is schematically represented in Fig. 14.1. However, at high surfactant concentrations, there will be more micelles that can interact with the individual polymer chains and the cross links are broken.

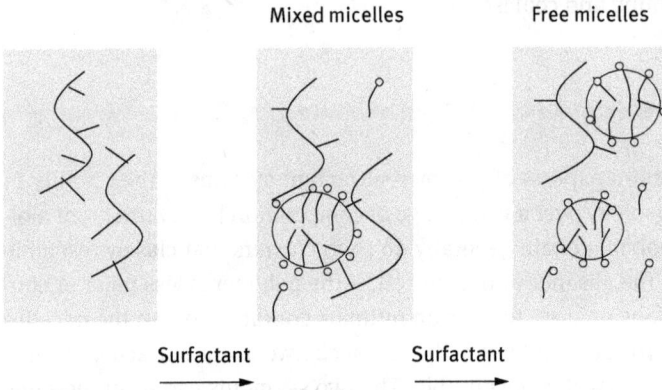

Fig. 14.1: Schematic representation of the interaction of polymers with surfactants.

The interactions illustrated in Fig. 14.1 are manifested in the variation of viscosity with surfactant concentration. Initially, the viscosity shows an increase with increasing surfactant concentration, reaching a maximum and then decreasing with any further increase in surfactant concentration. The maximum is consistent with the formation of cross links and the decrease after that indicates the destruction of these cross links (see Fig. 14.1).

14.7 Emulsion stability using polymeric surfactants

It has long been argued that interfacial rheology, namely interfacial viscosity and elasticity, plays an important role in emulsion stability. This is particularly the case with polymeric surfactants such as hydrocolloids and proteins that are commonly used in food emulsions. The interfacial viscosity is the ratio between shear stress and shear rate in the plane of the interface, i.e. it is a two-dimensional viscosity (the unit for interfacial viscosity is surface Pas or surface poise). A liquid-liquid interface has viscosity if the interface itself contributes to the resistance to shear in the plane of the interface [4]. Most macromolecules adsorbed at the interface are viscous, showing a high-induced interfacial viscosity. Usually the interfacial viscosity is higher than the bulk viscosity. Macromolecular films give high interfacial viscosity due to their orientation at the interface. Usually, the macromolecule adopts a train-loop-tail configuration and the film resists compression as a result of the lateral repulsion between the loops and/or tails. With proteins, more rigid interfacial films are produced, particularly when these molecules adsorb unfolded, and in such cases very high surface viscosities are produced.

Interfacial films show both viscosity and elasticity. Films are elastic if they resist deformation in the plane of the interface and if the surface tends to recover its natural shape where the deforming forces are removed [4]. Similar to bulk materials, interfacial elasticity can be measured by static and dynamic methods. Another important interfacial rheological parameter is the dilational elasticity, ε, that is given by the equation

$$\varepsilon = A\left(\frac{d\gamma}{dA}\right). \tag{14.1}$$

A good example where surface rheology was applied to investigate emulsion stability was the work of Biswas and Haydon [5]. These authors have systematically investigated the rheological characteristics of various proteins (which are relevant to food emulsions), namely albumin, poly(φ-L-Lysine) and arabinic acid at the O/W interface and correlated these measurements with the stability of the oil droplets at a planer O/W interface.

The viscoelastic properties of the adsorbed films were studied using two-dimensional creep and stress relaxation measurements in a specially designed rheometer. In the creep experiments, a constant torque (expressed in mN m^{-1}) was applied and the resulting deformation γ (in radians) was recorded as a function of time. The creep recovery was recorded by following the deformation when the torque was withdrawn. In the stress relaxation experiments, a certain deformation γ was produced in the film by applying an initial strain, and the deformation was kept constant by decreasing the stress. Figure 14.2 shows a typical creep curve for bovine serum albumin films. The curve shows an initial, instantaneous, deformation, characteristic of an elastic body, followed by a nonlinear flow that gradually declined and approached the steady flow behaviour of a viscous body. After 30 minutes, when the external force was withdrawn, the film tended to revert to its original state, with an instantaneous recovery followed by a slow one. The original state, however, was not obtained even after 20 hours and the film seemed to have undergone some flow. This behaviour illustrates the viscoelastic property of the bovine serum albumin film.

From the creep curves one can obtain the instantaneous modulus G_0 (σ/γ_{inst}) and the surface viscosity η_s from the slope of the straight line (which gives the shear rate) and the applied stress. Biswas and Haydon [5] also found a striking effect of the pH on the rigidity of the protein film. This is illustrated in Fig. 14.3, where the shear modulus G_0 and interfacial viscosity η_s are plotted as a function of pH. Both show an increase with increasing pH reaching a maximum at pH ≈ 6 (the isoelectric point of the protein) at which the protein molecules show maximum rigidity at the interface. Biswas and Haydon [5] then measured the rate of coalescence of petroleum ether drops at a planar O/W interface by measuring the lifetime of a droplet beneath the interface. The stabil-

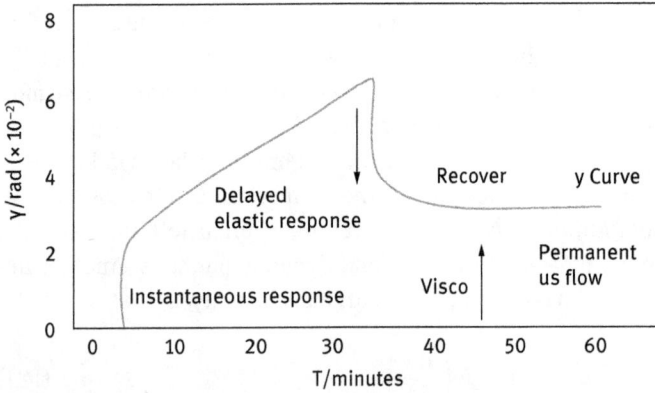

Fig. 14.2: Creep curve for protein film at the O/W interface.

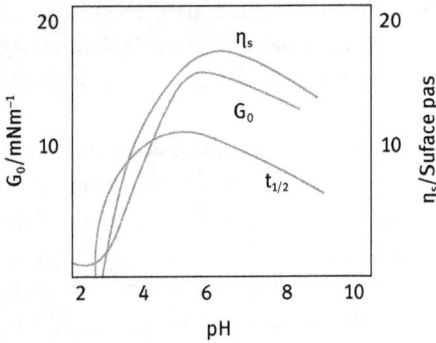

Fig. 14.3: Variation of $t_{1/2}$ and G_0 and η_s with pH.

ity of the emulsion was assessed by measuring the residence time, t, of several oil droplets at a planer O/W interface containing the adsorbed protein. Figure 14.3 shows the variation of $t_{1/2}$ (time taken for half the number of oil droplets to coalesce with the oil at the O/W interface) with pH. Good correlation between $t_{1/2}$ and G_0 and η_s is obtained.

Biswas and Haydon [5] derived a relationship between coalescence time τ and surface viscosity η_s, instantaneous modulus G_0 and adsorbed film thickness h:

$$\tau = \eta_s \left[3C' \frac{h^2}{A} - \frac{1}{G_0} - \phi(t) \right], \tag{14.2}$$

where $3C'$ is a critical deformation factor, A is the Hamaker constant and $\phi(t)$ is the elastic deformation per unit stress.

Equation (14.2) shows that τ increases with increasing η_s but most importantly it is directly proportional to h^2. These results show that viscoelasticity is necessary but not sufficient to ensure stability against coalescence. To ensure stability of an emulsion one must make sure that h is large enough and film drainage is prevented.

References

[1] Tadros, Th. F., "Applied Surfactants", Wiley-VCH, Germany (2005).
[2] Krog, N. J. and Riisom, T. H., in "Encyclopedia of Emulsion Technology", P. Becher (ed.), Marcel Dekker, NY, Vol. 2, pp. 321–365 (1985).
[3] Goddard, E. D. and Ananthapadmanqabhan, K. P. (eds.), "Polymer-Surfactant Interaction", CRC Press, Boca Raton (1992).
[4] Tadros, Th. F., "Rheology of Dispersions", Wiley-VCH, Germany (2010).
[5] Biswas, B. and Haydon, D. A., Proc. Royal Soc., A271, 296 (1963).

Index

www.ingramcontent.com/pod-product-compliance
Lightning Source LLC
Chambersburg PA
CBHW061348210326
41598CB00035B/5917